Integrative Approaches to Biotechnology

Biotechnology is one of the fastest emerging fields that has attracted attention of conventional biologists, biochemists, microbiologists, medical and agricultural scientists. The coming decades are likely to witness a boom in biotechnology, which is expected to surpass information technology as the new engine of the global economy. Biotechnology is experiencing a revolution that will affect every facet of our lives, from crop improvement to commerce, drugs and sustainable development. New approaches and a plethora of information available at a frantic pace demand its dissemination to the scientific community. The current book has been written with the specific objective of providing information on the recent developments in biotechnology to the readers.

The proposed book presents a multidisciplinary approach to the latest information and developments in biotechnology in an easy-to-read, succinct format. The book has been divided into 6 sections and 15 chapters giving an in-depth analysis of the latest research and developments in the biotechnological realm. The topics have been presented in a lucid, easy-to-read methodical way with illustrations and suitable case studies to provide additional help and clarity. The authors have tried to present state-of-the-art and integrative information in a manner that familiarizes the reader with the important concepts and tools of recent biotechnological studies. Apart from biotechnological personnel, the book would also be useful for readers of diverse disciplines such as bioinformatics, agriculture, environmental science, pharmaceutical sciences, biochemistry and general biology.

Features

- A systematic overview of the recent state-of-the-art technologies.
- Novel contents with maximum coherence.
- Extensive use of examples and case studies to illustrate how each technique has been used in practice.
- Incorporation of the latest information on these topics from recent research papers.

This book serves as a reference book and presents information in an accessible way for students, researchers and scientific investigators in biotechnology. It may also be used as a textbook for postgraduate-level courses in biological sciences.

Integrative Approaches to Biotechnology

Edited by

Atul Bhargava
Shilpi Srivastava

CRC Press
Taylor & Francis Group
Boca Raton London New York

CRC Press is an imprint of the
Taylor & Francis Group, an **informa** business

First edition published 2024
by CRC Press
2385 NW Executive Center Drive, Suite 320, Boca Raton FL 33431

and by CRC Press
4 Park Square, Milton Park, Abingdon, Oxon, OX14 4RN

CRC Press is an imprint of Taylor & Francis Group, LLC

Library of Congress Cataloging-in-Publication Data
Names: Bhargava, Atul, 1975- editor. | Srivastava, Shilpi, editor.
Title: Integrative approaches to biotechnology / edited by Atul Bhargava, Shilpi Srivastava.
Description: First edition. | Boca Raton : CRC Press, 2024. | Includes bibliographical references and index.
Identifiers: LCCN 2023014714 (print) | LCCN 2023014715 (ebook) | ISBN 9781032349589 (hardback) | ISBN 9781032349664 (paperback) | ISBN 9781003324706 (ebook)
Subjects: LCSH: Biotechnology.
Classification: LCC TP248.15 .I58 2024 (print) | LCC TP248.15 (ebook) | DDC 660.6--dc23/eng/20230629
LC record available at https://lccn.loc.gov/2023014714
LC ebook record available at https://lccn.loc.gov/2023014715

ISBN: 978-1-032-34958-9 (hbk)
ISBN: 978-1-032-34966-4 (pbk)
ISBN: 978-1-003-32470-6 (ebk)

DOI: 10.1201/9781003324706

Typeset in Times
by MPS Limited, Dehradun

Contents

Preface

"An investment in knowledge always pays the best interest"

The above quote by the American scientist Benjamin Franklin fittingly describes the power of knowledge. It is in the pursuit of knowledge that this work was conceptualized.

Since the start of our academic journeys, we have been fascinated by the secrets of science and the power of knowledge. Biotechnology, the technological application that uses biological systems, living organisms, or derivatives thereof, to make or modify products or processes for specific use, has advanced rapidly in recent decades with enormous information available with each passing day. The biotechnology realm encompasses microbiology, biochemistry, genetics, molecular biology, tissue engineering, immunology, agriculture, environment, cell and tissue culture physiology, and, in some instances, is also dependent on knowledge and methods from several sciences outside the biological sphere such as chemical engineering, bioprocess engineering and information technology. Biotechnology has been the engine of growth in the 21st century and has been an integral component of human life since thousands of years. This field has been knowingly or unknowingly used by mankind in everyday life. It has been in use ever since the dawn of civilization when man evolved methods to ease his life and living. Ancient civilizations used biotechnology through the use of yeasts to prepare leavened bread, the fermentation techniques for brewing and cheese making in China, and the use of *Spirulina* to make cakes. Later microorganisms and enzymes were used to make bio-based products in chemical, food, detergent, paper, textiles and biofuels industry. Biotechnology has fascinated us to a great extent and propelled us to delve more and more into the unknown. **The global biotechnology market was estimated at USD 850 billion in 2022 and is expected to be worth around USD 1,683.52 billion by the year 2030.**

Since biotechnology is a rapidly emerging field with new approaches and discoveries taking place every decade, these technological advances need to be introduced to the researchers for their benefit. Fewer books on the subject are currently available that incorporate recent developments in its different sub-branches. Since biotechnology is a multidisciplinary field that is witnessing newer applications and technologies day by day, it is prudent enough to bring out a book on these newer technological breakthroughs that can guide the readers to these technologies in a simple way. Thus, the present book has been conceptualized to provide a comprehensive discussion of recent developments in this branch of science so that the reader may get a detailed view of the recent advances in this rapidly emerging field.

The proposed book introduces these recent studies to academicians and aspiring biotechnologists in a simple and lucid way that is easily understandable to the reader. The authors have tried to present complex and updated information in a manner that familiarizes the reader with the important concepts and tools of recent biotechnological studies. The book establishes novel topics from fundamental concepts to novel methodologies along with throwing light on the applied aspects. The proposed book covers all major aspects of this field, from agriculture to environment, biochemistry, bioinformatics, nanobiotechnology, medical and microbial biotechnology. Using relevant case studies, the proposed book aims to explain both the applications and implications of biotechnology so that even a lay reader can have an idea of how the upcoming studies can add to human welfare. The chapters are comprehensible, straightforward, with illustrations to make the topics simple, and provide additional help and clarity. Apart from core biotechnology students, the book would be useful for readers of diverse disciplines such as nanotechnology, pharmaceutical sciences, environmental biotechnology and general biology.

We hope that this book will help emerging biotechnologists to understand the various facets of this fascinating science which is bound to rule the world in the 21st century.

Atul Bhargava
Shilpi Srivastava

Acknowledgments

Atul Bhargava

I am highly grateful to Prof. (Dr.) Anand Prakash, Vice Chancellor, Mahatma Gandhi Central University, Motihari, for his inspiring leadership and facilities provided in the University which helped me in accomplishing this task.

My sincere thanks are due to Dr. Art Pal, Campus Director (Chanakya Parisar), Mahatma Gandhi Central University for his never-ending support and affection showered on me. I am blessed and fortunate enough to have my support system in the form of my wonderful faculty colleagues Dr. Brijesh Pandey, Dr. Shahana Majumder, Dr. Ram Prasad, Dr. Akhilesh Kumar Singh and Dr. Saurabh Singh Rathore for creating a lively atmosphere and relentless support during the writing of this text.

On a very personal note, I am grateful to my brother Mr. Akhilesh Bhargava, sister-in-law Dr. Meenakshi Bhargava, my niece Ms. Anushka Sharma and my son Mr. Akshay Bhargava for their patience and perseverance, as well as their constant support during the writing of this book. They wholeheartedly supported me during my overburdened schedule and were with me during thick and thin.

Shilpi Srivastava

I am deeply indebted to my parent organization Amity University Uttar Pradesh and Dr. Aseem Chauhan, Chairman, Amity University Uttar Pradesh (Lucknow Campus) for providing me a suitable forum to fulfill my ambitions. I owe my deepest sense of veneration and gratitude to Dr. Balwinder Shukla, VC, and Dr. Sunil Dhaneshwar, Pro VC, AUUP, Lucknow Campus, India, for giving me an opportunity to be a part of their prestigious and renowned institution without which this would not have been possible.

I also wish to thank Dr. J.K. Srivastava, Director, Amity Institute of Biotechnology, Lucknow, for his constant support and encouragement. I thank my colleagues and friends at Amity University, namely, Dr. Prachi Srivastava, Dr. Rachna Chaturvedi, Dr. Jyoti Prakash, Dr. Ruchi Yadav and Dr. Garima Awasthi for their support and psycho-stimulant company.

Both of us would also like to extend gratitude toward all our contributing authors who have worked tirelessly to produce chapters of high excellence. We express our appreciation for the editorial guidance of Dr. Renu Upadhyay (Commissioning Editor, CRC Press) and Ms. Jyotsna Jangra (Editorial Assistant, CRC Press) whose efforts have helped us in shaping the text to bring out the best to the readers.

Shilpi Srivastava
Atul Bhargava

About the Editors

Dr. Atul Bhargava is Head of the Department of Botany, School of Life Sciences at Mahatma Gandhi Central University, Bihar, India. He completed his PhD in 2005 from National Botanical Research Institute, Lucknow, and has over 16 years of teaching and research experience. His research interests include genetic improvement of crop plants, nanobiotechnology and phytoremediation. Dr. Bhargava has published 55 research papers on these topics in international, peer-reviewed journals, as well as 5 books and numerous book chapters. He also has more than 4000 citations of his work. Dr. Atul Bhargava is also serving as an editorial board member of several international journals of repute.

Dr. Shilpi Srivastava is Assistant Professor at Amity Institute of Biotechnology, Amity University Uttar Pradesh Lucknow Campus, Lucknow. Her research areas include nanobiotechnology, biological synthesis of nanoparticles and their antimicrobial activity. She has 16 years of research and teaching experience, and has published 5 books, several book chapters and numerous research papers in peer-reviewed journals. She has attended and presented her research work in various national and international conferences. Dr. Srivastava is a referee for a number of international journals and is also a member of the editorial boards of several international journals.

Contributors

Vineet Awasthi
Amity Institute of Biotechnology
Amity University Uttar Pradesh
Lucknow Campus, Lucknow, India

Prarabdh C. Badgujar
Department of Food Science and Technology, National Institute of Food Technology Entrepreneurship and Management
Haryana, India

R. Bhardwaj
Socorro, Bardez
North Goa, India

U. Bhardwaj
School of Sciences
Noida International University
Yamuna Expressway, Gautam Budh Nagar
Uttar Pradesh, India

Atul Bhargava
Department of Botany
Mahatma Gandhi Central University
Motihari, Bihar, India

Rachna Chaturvedi
Amity Institute of Biotechnology
Amity University Uttar Pradesh
Lucknow Campus, Lucknow, India

N. Dhingra
Department of Agriculture
Medi-Caps University
Indore, Madhya Pradesh, India

Francisco Fuentes
Facultad de Agronomía e Ingeniería Forestal, Facultad de Ingeniería
Facultad de Medicina. Pontificia Universidad Católica de Chile
Código, Santiago, Chile

Prekshi Garg
Amity Institute of Biotechnology
Amity University Uttar Pradesh
Lucknow Campus, Lucknow, India

Upagya Gyaneshwari
Departments of Biotechnology, School of Life Sciences Mahatma
Gandhi Central University
Motihari, Bihar, India

Supriya Karpathak
Department of Respiratory Medicine
King George's Medical University
Lucknow, Uttar Pradesh, India

Gurjeet Kaur
Amity Institute of Biotechnology
Amity University Uttar Pradesh
Lucknow Campus, Lucknow, India

Mahadeo Kumar
Biochemistry Laboratory, Animal Facility, Regulatory Toxicology Group, CSIR-Indian Institute of Toxicology Research (CSIR-IITR)
Lucknow, Uttar Pradesh, India

Babli Kumari
Departments of Biotechnology, School of Life Sciences
Mahatma Gandhi Central University
Motihari, Bihar, India

Deborah Lanterbecq
Laboratoire de Biotechnologie et Biologie Appliquée, CARAH ASBL
Rue Paul Pastur
Belgium

Ramanuj Maurya
Department of Botany
University of Lucknow
Lucknow, Uttar Pradesh, India

Arushi Mishra
Bioinformatics Laboratory, Department of Biological Sciences
Birla Institute of Technology and Science (BITS)
Pilani Rajasthan, India

Benoît Moreau
Laboratoire de "Chimie verte et Produits Biobasés", Haute Ecole Provinciale de Hainaut-Condorcet,
Département AgroBioscience et Chimie, 11, Rue de la Sucrerie
Belgium

Pawan Kumar Pal
ICAR-Indian Institute of Pulses Research
Kanpur, Uttar Pradesh, India

Priti Pal
Shri Ramswaroop Memorial College of Engineering and Management, Tiwariganj
Lucknow, Uttar Pradesh, India

Brijesh Pandey
Departments of Biotechnology, School of Life Sciences
Mahatma Gandhi Central University
Motihari, Bihar, India

Anand Prakash
Departments of Biotechnology, School of Life Sciences
Mahatma Gandhi Central University
Motihari, Bihar, India

Jyoti Prakash
Amity Institute of Biotechnology
Amity University Uttar Pradesh
Lucknow Campus, Lucknow, India

Vinay B. Raghvendra
Department of Biotechnology (PG)
Teresian College, Siddarthanagar
Mysore, India

T.P. Sari
Department of Food Science and Technology
National Institute of Food Technology Entrepreneurship and Management
Haryana, India

Minaxi Sharma
Laboratoire de "Chimie verte et Produits Biobasés", Haute Ecole Provinciale de Hainaut-Condorcet
Département AgroBioscience et Chimie, 11,
Rue de la Sucrerie
Belgium
and
Laboratoire de Biotechnologie et Biologie Appliquée
CARAH ASBL
Rue Paul Pastur
Belgium

Monika Sharma
Department of Botany, Sri Avadh Raj Singh Smarak Degree College
Vishunpur Bairya, Gonda, UP, India

Nevadita Sharma
B.S. Anangpuria Institute of Technology and Management
Faridabad, Haryana, India
and
The Public Health Research Institute at the International
Centre for Public Health (ICPH)
Newark, NJ, USA

Nishant Sharma
Translational Health Science and Technology Institute
Faridabad, Haryana, India
and
The Public Health Research Institute at the International
Centre for Public Health (ICPH)
Newark, NJ, USA

S. Sharma
Department of Biology, College of Arts and Sciences
Georgia State University
Atlanta, GA, USA

Aryan Shukla
Amity Institute of Biotechnology
Amity University Uttar Pradesh
Lucknow Campus, Lucknow, India

Srishti Shukla
Amity Institute of Biotechnology
Amity University Uttar Pradesh
Lucknow Campus, Lucknow, India

Ajay Kumar Singh
Department of Bioinformatics, School of Life Sciences
Central University of South Bihar
Bihar, India

Akhilesh Kumar Singh
Departments of Biotechnology, School of Life Sciences
Mahatma Gandhi Central University
Motihari, Bihar, India

Suchitra Singh
Department of Bioinformatics, School of Life Sciences
Central University of South Bihar
Bihar, India

Anurag Kumar Srivastav
Department of Clinical Immunology and Rheumatology
Sanjay Gandhi Postgraduate Institute of Medical Sciences, Lucknow
Uttar Pradesh, India

Prachi Srivastava
Amity Institute of Biotechnology
Amity University Uttar Pradesh
Lucknow Campus, Lucknow, India

Shilpi Srivastava
Amity Institute of Biotechnology
Amity University Uttar Pradesh
Lucknow Campus, Lucknow, India

Kumari Swati
Departments of Biotechnology, School of Life Sciences
Mahatma Gandhi Central University
Motihari, Bihar, India

Kiran Verma
Department of Food Science and Technology
National Institute of Food Technology Entrepreneurship and Management
Sonipat, Haryana, India

Shalini Singh Visen
Amity Food and Agriculture Foundation
Lucknow, Uttar Pradesh, India

Piyush Kumar Yadav
Department of Bioinformatics, School of Life Sciences
Central University of South Bihar
Bihar, India

Ruchi Yadav
Amity Institute of Biotechnology
Amity University Uttar Pradesh
Lucknow Campus, Lucknow, India

Section I

Plant Biotechnology

1 Insights in Plant Epigenetics

Shilpi Srivastava
Amity Institute of Biotechnology, Amity University Uttar Pradesh, Lucknow Campus, Lucknow, India

Francisco Fuentes
Facultad de Agronomía e Ingeniería Forestal, Facultad de Ingeniería, Facultad de Medicina. Pontificia Universidad Católica de Chile, Código, Santiago, Chile

Atul Bhargava
Department of Botany, Mahatma Gandhi Central University, Motihari, Bihar, India

CONTENTS

1.1 INTRODUCTION

Genetic variability, the heritable variation of genetic information of individuals and populations, has been considered as a prerequisite for plant improvement. Genetic selection based on natural variation has been the hot cake of breeders. Plant breeders exhaustively utilize this variability to develop plants having the desired characteristics. Till some years ago genetic variation was considered as the sole factor responsible for revealing phenotypic traits (Goulet et al. 2017). However, recent studies have clearly pointed out that heritable phenotypic variation is not wholly based on DNA sequence polymorphism and has often no direct correlation with DNA polymorphisms (Springer and Schmitz 2017).

Epigenetics, also sometimes known as epigenomics, refers to the study of heritable phenotypic alterations that do not involve changes in the DNA sequence but often involve variations affecting gene activity and expression. These heritable events comprise an extensive range of protein complexes and regulatory mechanisms not involving alterations in either the coding sequence of a gene or the upstream promoter region (Rapp and Wendel 2005). Epigenetic variations normally

DOI: 10.1201/9781003324706-2

involve a variety of chromatin marks such as cytosine methylation, modification of histones, chromatin remodeling and non-coding RNAs (Rajewsky et al. 2017). Although several studies have amply demonstrated that environmental conditions can take precedence over epigenetic regulation, there is increasing evidence to suggest that novel adaptations can arise in nature since natural selection acts on epigenetic variation in the same way as on genetic components (Verhoeven et al. 2010; Dowen et al. 2012; Furrow 2014; Kronholm and Collins 2016; Balao et al. 2018). Thus, epigenetics offers an additional source of natural variation, which is of extreme importance in small populations that lacks variation and/or occupies a fragmented landscape (Verhoeven and Preite 2014; Herrera et al. 2016). Epigenetics has important implications in evolutionary perspectives since epigenetically acquired traits exhibit transgenerational heritability and have the potential to be inherited trans-generationally without changes in the DNA sequence (Jones 2012).

1.2 HISTORICAL PERSPECTIVES

The term "epigenetics" was coined in the early 1940s by Conrad Waddington who took a cue from the Aristotelian theory of epigenesis (Waddington 1942, 1968). The term was originally used by Waddington to refer to alterations occurring in the cells during developmental stages in response to environmental stimuli but are not encoded in the DNA. Since at that time, the role of gene in organismic development was largely unknown, Waddington was of the opinion that embryological development should encompass networks of gene interactions (Jablonka and Lamb 2006). In succeeding years, the term was used to relate certain inheritance patterns such as imprinting, X-chromosome inactivation, paramutation and transgene silencing that were not governed by either Mendelian or quantitative genetics. Apart from sporadic usage after its introduction (Løvtrup 1974), the first three decades rarely witnessed the use of the term "epigenetics". The concept started to take a new shape in the early 1990s when B.K. Hall redefined epigenetics as "the sum of the genetic and non-genetic factors acting upon cells to selectively control the gene expression that produces increasing phenotypic complexity during development". The work of V.E.A. Russo et al. (1996) further narrowed the scope of epigenetics as "the study of mitotically and/or meiotically heritable changes in gene function that cannot be explained by changes in DNA sequence". However, research contributions of Holliday (1990, 1994) brought the scope quite close to Waddington's original concept. Holliday in his 1994 work entitled "Epigenetics: an overview" defined epigenetics as "the study of the changes in gene expression, which occur in organisms with differentiated cells, and the mitotic inheritance of given patterns of gene expression". Holliday's concept included DNA–protein interactions, DNA rearrangements and even mitochondrial inheritance. By the beginning of the 21st century, research in epigenetics had gained momentum but several researchers equated the term with "epigenetic inheritance" which presented considerable technical problems. Such was the confusion regarding the usage of the term "epigenetics" that some workers were in favor of its abandonment and suggested use of other terms such as nucleic, epinucleic and extranucleic information, rather than epigenetic information (Lederberg 2001).

1.3 SOURCE OF EPIGENETICS

Epialleles are known to arise from either non-genetic or genetic sources (Figure 1.1) (Taudt et al. 2016; Springer and Schmitz 2017). Non-genetic sources of epialleles comprise of natural epialleles due to improper maintenance of methylation states or through the off-target effects of small RNAs. Apart from these, the non-genetic sources may also include developmental or environmental factors that initiate chromatin changes or may impact the stability of epigenetic states. Genetic sources of epialleles comprise of transposable elements (TEs), first discovered by Barbara McClintock in maize in the late 1940s, whose insertions modify regional chromatin and structural

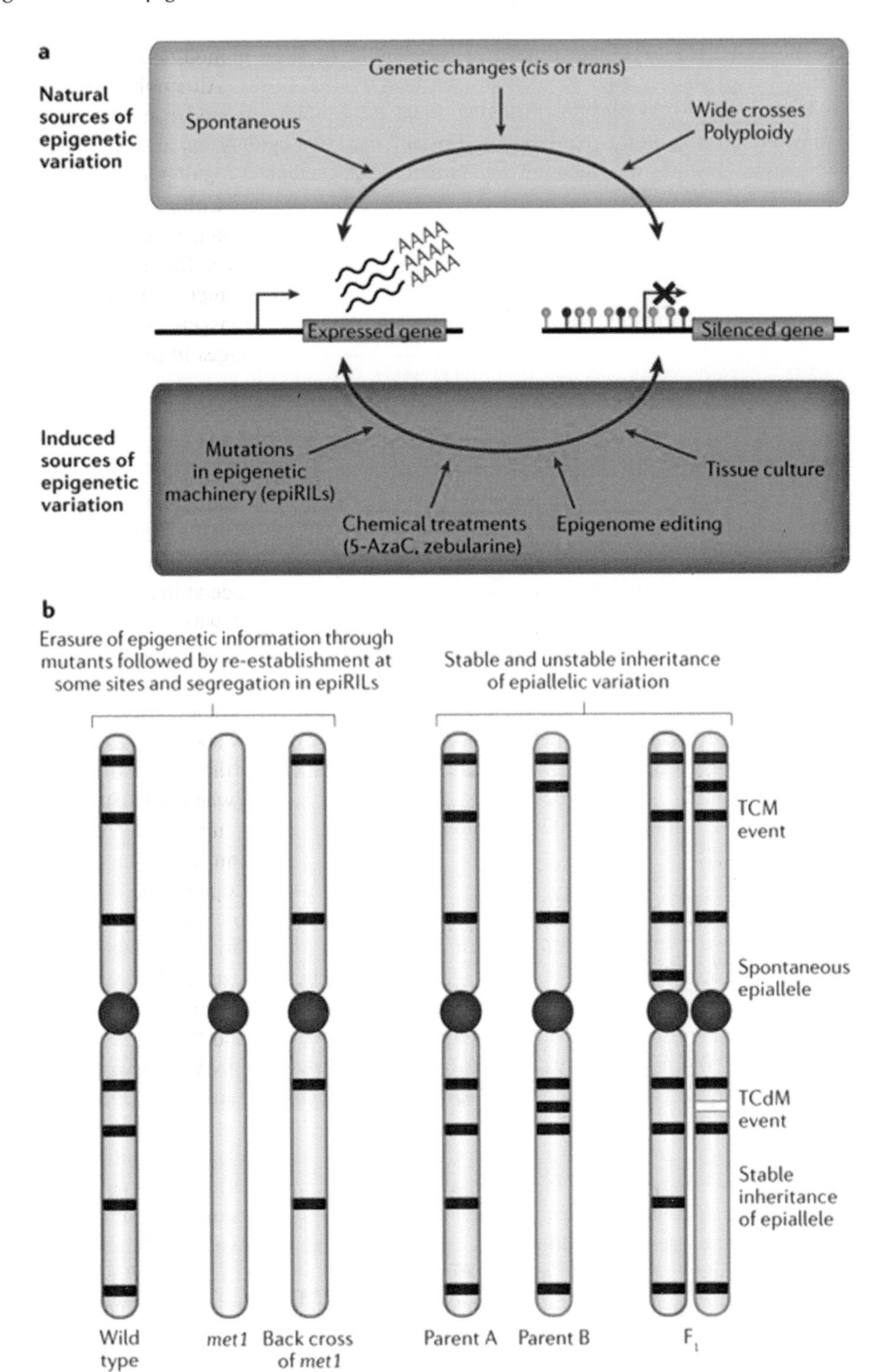

FIGURE 1.1 Sources of epigenetic variation (Reprinted with permission from Springer and Schmitz 2017).

rearrangements. The TEs are major targets of epigenetic silencing mechanisms and the shifting pattern of DNA methylation leads to silencing and reactivation of the TEs resulting in epigenetic changes (Muyle et al. 2021). Besides this, functional RNA-mediated mechanisms are also speculated to play a major role in the regulation of TEs activity (Hsu et al. 2021). The crosstalk among DNA methylation, histone modification and small RNA-mediated regulation protects against invasion by TEs and controls the orchestration of gene expression (Hsu et al. 2021).

1.3.1 DNA Methylation

DNA methylation is a chemical change involving the addition of methyl groups at the 5th position of carbon of the existing cytosine residue in the DNA helix through a covalent bond to form 5-methylcytosine (5mC) (Villicaña and Bell 2021). DNA methylation is considered to be a critical process that has an enormous bearing on gene expression. This phenomenon is catalyzed by the methyl-binding proteins (MBPs) like DNA methyltransferases (MET) and chromomethylases (CMT). In plants, DNA methylation occurs all cytosine sequence contexts, including CG, CHH and CHG (where H stands for nucleotides C, T or A). Exhaustive studies in *Arabidopsis* point to the fact that methylation activation in plants is mediated through de novo RNA-directed DNA methylation pathway (RdDM) and its maintenance primarily requires MET1 at the CG context, CMT3 at the CHG context, and domains rearranged methyltransferase2 (DRM2) or CMT2 at the CHH context (Duan et al. 2018; Zhang et al. 2018). Except for paramutations that generate phenotypic diversity having non-Mendelian inheritance, DNA methylation is normally passed on in Mendelian pattern. DNA methylation is further classified into de novo (previously unmethylated cytosine residues are methylated) and maintenance methylation (existing methylation patterns are retained after DNA replication) (Gupta and Salgotra 2022). Several agronomic and phenotypic traits such as flowering, secondary growth, seed germination, seed development and seed yield are known to be influenced by DNA methylation (Finnegan et al. 1998; Kawakatsu et al. 2017; Rajkumar et al. 2020; Miryeganeh 2022; Inácio et al. 2022).

1.3.2 Post-translational Histone Modifications and Chromatin Remodeling

Nucleosome is the basic, structural unit of chromatin in eukaryotes wherein the highly condensed genomic DNA is wrapped around a core of eight histone proteins. The octamer consists of two molecules each of the polymer histones viz. H2A, H2B, H3 and H4 (Liu et al. 2016). High amounts of lysine and arginine in histones confer extreme basicity to their amino (N-terminal) tails which protrude from the nucleosome core and facilitate protein–protein and protein–DNA interactions, along with several forms of post-translational modifications (Grabsztunowicz et al. 2017; Demetriadou et al. 2020).

Histone modifications are known to play an important role in epigenetic regulation. A number of covalent modifications to histone tails such as acetylation, methylation, phosphorylation, biotinylation, sumoylation ubiquitination, glycosylation and ribosylation of adenosine diphosphate (ADP) are significant epigenetic modifications that control gene expression (Alfalahi et al. 2022). Among the above-mentioned modifications acetylation, phosphorylation and ubiquitination enhance the process of transcription, while biotinylation and sumoylation lead to repression of gene transcription. Histone modification is carried out by methylation of the 9th lysine residue that induces condensation of chromosomes and leads to transcription blockage. The diverse types of histone modifications are aided by particular histone modification enzymes which are of prime importance in specifying chromatin function. Histone acetylation and deacetylation are catalyzed by the use of enzymes such as histone acetyltransferases (HATs) and histone deacetylases (HDACs) (Sun et al. 2012a; Haery et al. 2015). Hyperacetylation of histones relaxes chromatin structure and is associated with transcriptional activation, whereas hypoacetylation of histones induces chromatin compaction and resulting gene repression.

Chromatin structure is dynamic and is involved in active interactions with the transcription factors. The dynamic nature of chromatin is facilitated through active histone modifications especially due to the post-translational modifications of amino-terminal ends of histone proteins. This sets the stage for intense dynamism between histone and DNA modifications, leading to combinatorial possibilities for gene regulation (Kouzarides 2007; Kim and Kaang 2017). Chromatin structure can be modified either when interactions between nucleosomes are broken or when different protein factors are engaged in the unwinded nucleosomes. It has been often observed that chromatin remodeling facilitates the passage of transcriptional factors to the nucleosome octamer core ultimately leading to altered gene expression (Secco et al. 2017). This is best exemplified during stress when the stress signals modify the chromatin structure to facilitate the accessibility of DNA to various chromatin remodeling factors (CRFs). The CRFs either remove the histones or induce modifications in the chromatin structure by destabilizing the nucleosome in an adenosine triphosphate (ATP)-dependent energy-derived process (Verma et al. 2022). The disruption of nucleosome structure determines the extent of modulation of chromatin (Goldstein et al. 2013). Studies have also indicated that the epigenetic changes induced through chromatin remodeling are even transmissible to succeeding generations.

1.3.3 Non-coding RNAs

Eukaryotic genomes are known to transcribe about 90% of the genomic DNA. However, only 1–2% of these transcripts encode for proteins, whereas majority of them are not translated and are transcribed as non-coding RNAs (ncRNAs). These ncRNAs are classified into housekeeping non-coding RNAs and regulatory non-coding RNAs (Figure 1.2). The ncRNAs are a cluster of RNAs that do not encode functional proteins but play an important role in the regulation of gene expression and epigenetic modification (Wei et al. 2017) (Table 1.1). The regulatory RNAs are further subdivided into two categories, namely, short-chain non-coding RNAs (siRNAs, miRNA and piRNAs) and long non-coding RNA (lncRNAs).

The ncRNAs are known to be effective tools for facilitating cell division and differentiation and play a pivotal role in regulating plant response to environmental stresses both at transcriptional and posttranscriptional levels (Cavalli and Heard 2019). In Arabidopsis (Family: Brassicaceae), a subset of miRNAs has been identified that targets the transcripts encoding resistance proteins, immune receptors that identify pathogen effectors and initiate suitable defense responses (Wang and Qi 2018). The ncRNAs induce epigenetic changes through mechanisms such as DNA methylation, histone modification and chromatin remodeling. In case of LncRNAs, the epigenetic

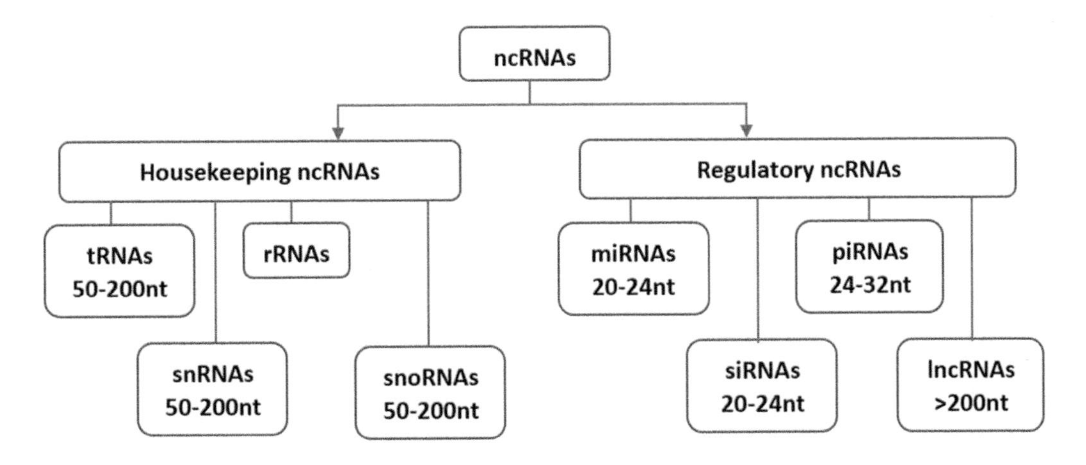

FIGURE 1.2 Classification of non-coding RNAs (ncRNAs) (Reprinted from Waititu et al. 2020).

TABLE 1.1
Regulatory ncRNAs Produced from Eukaryotic Genomes and Their Characteristics and Functions

Type	Long Name	Length (nt)	Characteristics	Function
miRNA	MicroRNA	20–24	Pri-miRNA produced in the nucleus as capped and polyadenylated ssRNA with an imperfectly paired stem-loop structure	Perfect complementarity: Ago2-mediated cleavage of mRNA
			Processing by Drosha and Dicer lead to a production of mature dsRNA with exact ends	Non-perfect complementarity: Suppression of translation or mRNA degradation (deadenylation, decapping and exonucleocytic degradation)
			Effector phase occurs primarily in the cytoplasm mediated by Ago proteins	Minor functions in transcriptional silencing and translational activation
piRNA	PIWI-interacting RNA	24–31	Precursor ssRNA, which is modified to contain 3′-terminal 2′-O-methyl	Silencing of transposable elements in the germline
			Strong preference for uridine at the 5′ end	
siRNA	Small interfering RNA	20–24	Canonical form long, linear, perfectly base-paired dsRNA	Perfect match: endonucleocytic cleavage
			Processed by Dicer into mature siRNA with heterogenous end composition	Non-perfect match or endonuclease-inactive RISC: translational repression or exonucleocytic degradation
			Effector functions occur primarily in the cytoplasm supported by Ago proteins	Induction of heterochromatin formation
				Silencing of the same locus from which they are derived
PAR ([a]PASR, TSSa-RNA, tiRNA, PROMPT)	Promoter-associated RNA	16–200	Weakly expressed ssRNAs	Partly unknown but indications of transcriptional regulation (example interaction with Polycomb group of proteins)
			Short half-life	
			Bidirectional expression reflecting PolII distribution	
eRNA	Enhancer RNA	100–9000	ssRNA produced bidirectionally from enhancer regions enriched for H3K4me1, PolII and coactivators such as p300	Mostly unknown but plays a role in transcriptional gene activation
			Short half-life	
			Evolutionarily conserved sequences	
			Dynamically regulated upon signaling	
			Expression correlates positively with nearby mRNA expression	
lncRNA		>200	Precursor ssRNA	Chromatin remodeling

Long non-coding RNA	Many lncRNAs are subject to splicing, polyadenylation and other posttranscriptional modifications	Transcriptional regulation
	Mostly nuclear RNAs but a subset also located in the cytoplasm	Posttranscriptional regulation (splicing, TF localization)
	Not evolutionary conserved with the exception of large intergenic ncRNAs, lincRNAs (H3K4me3-H3K36me3 signature)	Precursors for siRNAs
		Component of nuclear organelles (paraspeckles, nuclear speckles)

Source: Reprinted with permission from Kaikkonen et al. (2011).

[a] PASR, promoter-associated small RNA; TSSa-RNA, transcription start site-associated RNA; tiRNA, transcription initiation RNA; PROMTs, promoter upstream transcript.

changes are mediated through controlling transcription by engaging chromatin remodeling complexes (Ponting et al. 2009). The repressive polycomb group (PcG) is an exhaustively researched transcriptional complex that initiates and maintains epigenetic changes. The sncRNAs are also known to have a major role in the expressed hybrid vigor by managing DNA methylation via RNA-directed DNA methylation pathway.

1.4 EPIGENETICS IN PLANT DEVELOPMENT AND CROP IMPROVEMENT

Variability forms the raw material for genetic improvement of plants. Conventional plant breeders focus on accumulating the desirable alleles for improvement of traits of interest. However, rapid strides in epigenetic research have opened new avenues for crop improvement utilizing a unique type of variation having no direct correlation with DNA polymorphisms. Epigenetics could provide increased opportunity for breeders to raise plants having high vigor in the desired traits.

1.4.1 Plant Response to Abiotic Stress

Abiotic stress is harmful to plant growth and severely limits distribution of plants as well as crop productivity across environments (Bhargava and Srivastava 2013; Zhang et al. 2022). The environmental factors causing abiotic stress in plants include temperature, drought, soil salinity and acidity, UV, light intensity, CO_2 and mineral availability, all of which are important determinants of plant growth (Robert-Seilaniantz et al. 2010; Driedonks et al. 2015; Ding et al. 2019). Abiotic stresses trigger a number of morphological, physiological, biochemical and molecular changes in plants leading to negative effects on growth, development, reproduction and productivity (Bhargava et al. 2003; Bhargava and Srivastava 2020; Gong 2021).

Plants cope up with environmental stresses during different developmental stages by considerably modifying their physiology and developmental parameters through the cumulative effect of countless genes. Plants have the unique ability and inherent mechanism to perceive diverse environmental challenges, transmit the stress signals to appropriate cells and tissues and make suitable adjustments in their growth and developmental processes for ensuring survival and reproduction. The modifications in gene expression lead to the formation of stress memory and become stored in the cell memories. This helps the organism in predicting the future conditions based on their past experiences and enables better understanding of the phenomenon from past experiences (Latzel et al. 2016). These memorized conditions, although sometimes transient and being reset between generations, are religiously inherited from the parent generation to the progeny of the stress-treated plants (Hagmann et al. 2015; Friedrich et al. 2019).

In recent years, credible evidence points toward the significant role of epigenetic modifications in adaptation during stressful conditions (Chang et al. 2020). Rapid advances in epigenetic research have greatly helped us in understanding the complex mechanisms involved in adaptation to environmental stresses (Rehman and Tanti 2020). Epigenetics facilitates understanding of the complex mechanisms involved in adaptation to environmental stresses and is known to play a major role in preparing the plant for stressful conditions (Figure 1.3). A plethora of data has indicated toward involvement of epigenetic mechanisms in the response of plants to abiotic stresses by chromatin modifications leading to alteration in the regulation of stress-responsive genes at the transcriptional and posttranscriptional levels (Sahu et al. 2013; Lamke and Baurle 2017; Luo and He 2020) (Table 1.2). Akhter et al. (2021) were of the opinion that epigenetic factors like DNA methylation aid in crop improvement by neutralizing the effect of cold, drought, salt and heat-induced stresses. The miRNA regulatory networks in plants are also known to assist them in adapting to stressful conditions (Shriram et al. 2016). The sRNAs recognize complementary sequences in nucleic acids and lead to posttranscriptional gene silencing by mRNA or transcriptional gene silencing (TGS) by DNA methylation via the RNA-directed DNA methylation pathway (Singroha and Sharma 2019).

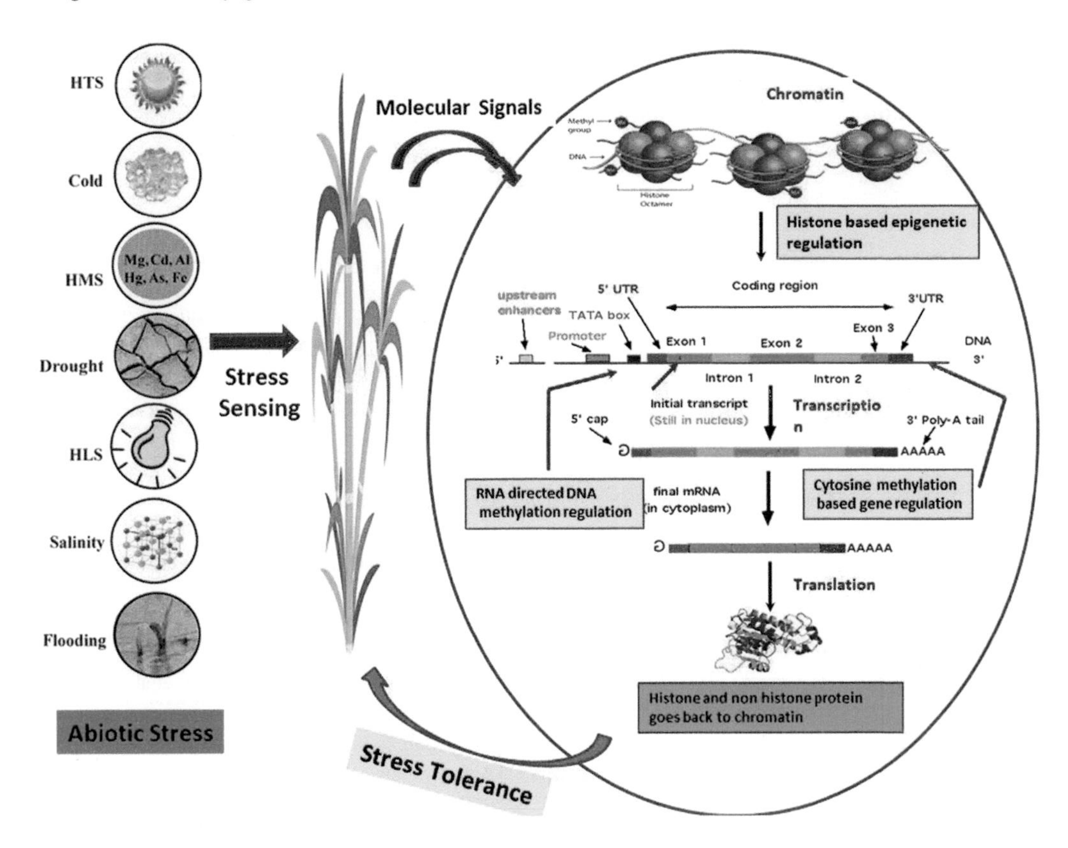

FIGURE 1.3 Epigenetic regulatory mechanism of abiotic stress in plants (Reprinted with permission from Verma et al. 2022).

1.4.2 Epigenetics and Biotic Stress and Interactions

Biotic stress refers to the stress induced in plants by living organisms viz. viruses, bacteria, fungi, nematodes, insects and weeds (Singla and Krattinger 2016). The agents inducing biotic stress deprive the host (plant) of its nutrients leading to reduced development, vigor and death of the host plant. Despite lacking an adaptive immune system, plants have evolved multitude strategies for counteracting biotic stresses. The genetic basis of these defense mechanisms is firmly entrenched in the plant's genetic code that encodes hundreds of biotic stress resistance genes.

Plant defense against living organisms is attributed to diverse defense mechanisms which confer protection against potential pathogens (Iqbal et al. 2021). A unique feature in plants is their ability to remember the initial stress that induced defense responses among them, a phenomenon known as primability (Hilker et al. 2016). Primability is correlated with reprogramming of the gene expression and prepares or primes the plant to future stresses. This leads to a more rapid and potent response against further such attacks (Martinez-Medina et al. 2016). Epigenetics plays a key role in the management of gene expression and memory to strengthen primability in plants (Roberts and López Sánchez 2019). Table 1.3 demonstrates the role of epigenetics in biotic stress management in plant systems.

An interesting example is with respect to role of epigenetics with respect to herbivory. Plants seem to be quite "intelligent" when it comes to attract pollinators and deter herbivores. This entails rapid recognition of the insect community to prevent potential threats. DNA methylation comes into full play and mediates such responses. Epigenetics comes into play whenever herbivory induces stress on the plant which triggers methylation changes in the defense-related genes as

TABLE 1.2
Epigenetic Modification in Crop Plants for Abiotic Stresses

Sr. No.	Crop	Trait	Changes Induced	References
1	Alfalfa	Drought stress	Increase in expression of miRNA156 assisted in drought tolerance	Arshad et al. (2018)
2	Alfalfa	Salinity stress	Cytosine methylation alters under the salt stress condition	Al-Lawati et al. (2016)
3	Arabidopsis	ABA sensitivity	Histone acetylation and deacetylation regulate the expression of genes controlling production of ABA	Zhou et al. (2005)
4	Arabidopsis	Cold stress	Histone acetylation of cold-responsive genes increases the transcription level to combat the cold stress	Pavangadkar et al. (2010)
5	Arabidopsis	Cold stress	GCN5 and ADA2 genes undergo histone acetylation for providing freezing tolerance	Vlachonasios et al. (2003)
6	Arabidopsis	Cold stress	This study confirmed that H3K27me3 does not inhibit the expression of cold-responsive genes such as COR15A and ATGOLS3 during the cold stress	Kwon et al. (2009)
7	Arabidopsis	Heat stress	Several gene activate with heat stress due to changes in histone modifications and heterochromatin condensation	Scheid et al. (2010)
8	Arabidopsis	Heat stress	RNA-dependent DNA methylation enhances the transcriptional response to high temperature	Popova et al. (2013)
9	Arabidopsis	Heat stress	H3K4 methylation has been shown to play a vital role in detecting recurring heat stress in plants and further activates the required heat shock proteins	Lämke et al. (2016)
10	Arabidopsis	Heat stress	Modification of histone and mRNA methylation occurs in responses to heat stress	Migicovsky et al. (2014)
11	Arabidopsis	Drought stress	Reduction in methylation of ANACO19 and ANACO55 genes leads to transcriptional upregulation under drought	Ramirez-Prado et al. (2019)
12	Arabidopsis	Drought stress	H3J4me3 and Ser5P have shown to regulate the response of water stress-responsive genes under drought and normal conditions in the plant	Ding et al. (2012 a,b)
13	Arabidopsis	Drought stress	Histone deacetylase HDA9 plays an important role in regulating sensitivity of plant to drought and salt stress with regulation of histone acetylation of number of genes	Zheng et al. (2016)
14	Arabidopsis	Drought stress	Histone deacetylase gene HD2 increases its expression under drought and cold stress causing various morphological changes in the plant for coping up with them	Han et al. (2016)
15	Arabidopsis	Drought stress	SWI2/SNF2 assist in chromatin remodeling to enhance drought tolerance by reducing increasing the sensitivity to abscisic acid	Han et al. (2013)
16	Arabidopsis	Drought stress	ABO1 gene undergoes tRNA modification to enhance the drought tolerance in plant	Chen et al. (2006)

TABLE 1.2 (Continued)
Epigenetic Modification in Crop Plants for Abiotic Stresses

Sr. No.	Crop	Trait	Changes Induced	References
17	Arabidopsis	Drought stress	MST1 undergoes chromatin modification to increase the positive regulation of proteins conferring tolerance to drought stress	Alexandre et al. (2009)
18	Arabidopsis	Drought stress	AtCHR12 undergoes chromatin remodeling under the response of drought	Mlynárová et al. (2007)
19	Arabidopsis	Drought stress	Increase in acetylation and methylation of RD29B, RD20 and RAP2.4 gene regions increases the drought tolerance	Kim et al. (2008)
20	Arabidopsis	Salinity stress	Light-induced expression of H3K4me3 maintain the activity of salt-induced transcriptional factor P5CS1	Feng et al. (2016)
21	Arabidopsis	Salinity stress	Primed plants undergo histone modifications which reduces the uptake of salt	Sani et al. (2013)
22	Arabidopsis	Salinity stress	SKB1 helps in increasing the salt tolerance in plant by altering the methylation status in LSM4 gene families	Zhang et al. (2011)
23	Arabidopsis	Salinity stress	HDA6 and HDA19 displays deacetylation under the salt stress	Chen and Wu (2010)
24	Arabidopsis	Salinity stress	Cytosine methylation in the promoter region of gene AtHKT1 increases the salt stress tolerance	Baek et al. (2011)
25	Barley	Drought stress	Increase in concentration of H3K4me3 results in increase of HSP17 for regulating the drought	Temel et al. (2017)
26	Barley	Salinity stress	HD2 gene families in barley respond to the exogenous application of hormones, ultimately resulting in tolerance to various abiotic stresses	Demetriou et al. (2009)
27	Faba bean	Drought stress	DNA methylation decreases under the drought-tolerant responses	Abid et al. (2017)
28	Hemp	Nickel stress	Decrease in methylation was observed under stress using methylation-sensitive amplification polymorphism	Aina et al. (2004)
29	Maize	Aluminum stress	Increase in DNA methylation marks as assessed with restriction enzyme digestion random amplification	Taspinar et al. (2018)
30	Maize	Cold stress	DNA methylation of cold-responsive genes increases the plants tolerance to cold stress	Shan et al. (2013)
31	Maize	Cold stress	DNA demethylation changes the expression of several genes with regulates the plant under cold stress especially through modifications in the roots	Steward et al. (2002)
32	Maize	Drought stress	Histone modifications with H3K36me3 and H3K9ac assists in drought tolerance mechanism	Xu et al. (2017)
33	Maize	Drought stress	Chromatin state is changed with action of histone deacetylases resulting in removal of acetyl groups for assisting in drought tolerance in plants	Forestan et al. (2018)
34	Maize	Heat stress		Hou et al. (2019)

(Continued)

TABLE 1.2 (Continued)
Epigenetic Modification in Crop Plants for Abiotic Stresses

Sr. No.	Crop	Trait	Changes Induced	References
			Increase in amount of H3K4me2 and H3K9ac leads to upregulation of heat stress factor genes	
35	Maize	Salinity stress	Histone acetylation causes activation of ZmHATB and ZMGCN5 genes which increases tolerance to salt stress	Li et al. (2014a, b)
36	Maize	Zinc stress	DNA methylation was across the genome was observed under zinc stress using restriction enzyme digestion random amplification	Erturk et al. (2015)
37	Mangrove	Salinity stress	Cytosine methylation throughout the genome assist in salt stress tolerance	Lira-Medeiros et al. (2010)
38	Onion	Aluminum stress	Aluminum stress results in DNA damage and alteration in methylation pattern	Murali Achary and Panda (2010)
39	Pea	Drought stress	Methylation occurs at second cytosine level under water deficit	Labra et al. (2002)
40	Popular	Copper stress	Methylation-sensitive amplification polymorphism was used to detect cytosine DNA methylation	Cicatelli et al. (2014)
41	Popular	Drought stress	Genome-wide acetylation of lysine 9 residue aid in combating the drought stress	Li et al. (2019a, b)
42	Rapeseed	Heat stress	Difference in DNA methylation patterns is the primary reasons for heat stress-tolerant varieties	Gao et al. (2014)
43	Rapeseed	Salinity stress	Increase in cytosine methylation in the salt tolerant genotypes	Marconi et al. (2013)
44	Rice	Cadmium stress	Hypomethylation of cadmium-resistant gene *OsCTF* aid for detoxification and reduction accumulation of cadmium	Feng et al. (2020)
45	Rice	Cadmium stress	Increase in DNA methylation in CG, CHH and CHG context	Feng et al. (2016)
46	Rice	Cold stress	Demethylation of cold-responsive genes increases their expression levels in the plants and provide the cold tolerance in some races of rice	Dai et al. (2015)
47	Rice	Cold stress	Histone acetyltransferase and histone deacetylases regulates the acetylation of several genes which regulates the cold, and salt tolerance ability of rice plant	Fu et al. (2007)
48	Rice	Cold stress	Several genes families in rice have acetylation mechanism to confer resistance to abiotic stresses such as cold	Liu et al. (2012)
49	Rice	Copper stress	CHG hypomethylation occurs under copper stress and induces heritable changes	Ou et al. (2012)
50	Rice	Copper and cadmium stress	DNA methylation occurs across several gene regions causing heritable changes and showed transgenerational inheritance	Cong et al. (2019)
51	Rice	Drought stress	Histone acetylation maintains the of genes with histone acetyltransferases and histone	Fang et al. (2014)

TABLE 1.2 (Continued)
Epigenetic Modification in Crop Plants for Abiotic Stresses

Sr. No.	Crop	Trait	Changes Induced	References
			deacetylases increases the drought tolerance ability of rice plant	
52	Rice	Drought stress	Modification in the expression of several stress-responsive genes with H3K4me3 modification for increasing the response to drought stress	Zong et al. (2013)
53	Rice	Drought stress	Histone acetyltransferase plays an important role in activating several genes localizing in nucleus and cytosol for coping with drought stress	Liu et al. (2012)
54	Rice	Drought stress	DNA methylation in the DK151 and IR64 rice lines occurs under drought stress	Wang et al. (2011)
55	Rice	Oxidative stress	OsSRT1 increases the acetylation and decreases the demethylation to enhance the plants tolerance to oxidative stress	Huang et al. (2007)
56	Rice	Salinity stress	Methylation of different salt stress regulating genes increases and provides the capacity to cope with salt stress	Karan et al. (2012)
57	Soybean	Heat stress	Roots hairs get hypomethylated under the heat stress	Hossain et al. (2017)
58	Soybean	Salinity stress	Increase rate of acetylation and reduced demethylation of 49 transcription factors assist in salinity stress tolerance	Song et al. (2012)
59	Tobacco	Aluminum stress	Demethylation occurs over the CG sites under the aluminum toxicity	Choi and Sano (2007)
60	Tomato	Cold stress	Methylation changes occur in the promoter regions of cold-responsive genes	Zhang et al. (2016a, b)
61	Tomato	Drought stress	Asr2 gene acquires stable epialleles by methylation at specific sites under the water-deprived conditions for increasing the stress-tolerant ability	González et al. (2013)
62	Tomato	Drought stress	Cytosine methylation of several genes alleviate the plant water supply under the drought conditions	González et al. (2013)
63	Tomato	Drought stress	DNA methylation with RNA-dependent pathways assists in drought tolerance	Benoit et al. (2019)
64	Triticale	Aluminum stress	DNA methylation increases in root cells as identified with methylation-sensitive amplification polymorphism	Agnieszka (2018)
65	Triticale	Aluminum stress	Demethylation of CG islands	Bednarek et al. (2017)
66	Wheat	Salinity stress	Cytosine methylation of potassium transporters reduces the uptake of salts	Kumar et al. (2017)
67	Wheat	Heat stress	Lysine induces heritable histone demethylation (LSD1) under heat stress	Wang et al. (2016a, b)

Source: Reprinted with permission from Samantara et al. (2021).

TABLE 1.3
Epigenetic Modification in Crop Plants during Biotic Stresses

Sr. No.	Crop	Biotic Factor	Changes Induced	References
1	Arabidopsis	*Hyaloperonospora arabidopsidis* (*Hpa*)	DNA methylation increase resistance toward Hpa	López Sánchez et al. (2016)
2	Arabidopsis	*Alternaria brassicicola* and *Botrytis cinerea*	Histone ubiquitination, histone H3K36 methyltransferase mediates induction of the JA/ethylene pathway genes expands affectability to necrotic fungi	Berr et al. (2010)
3	Arabidopsis	*Pseudomonas syringae* (*Pst*)	DNA methylation increases sensitivity in somatic recombination frequency to bacterial pathogen	Dowen et al. (2012)
4	Arabidopsis	*Botrytis cinerea*	Activating histone marks H3K4me3, H3K9ac and repressive H3K27me3 and Chromatin remodeling factors increases sensitivity to pathogen	Walley et al. (2008)
5	Arabidopsis	*Heterodera schachtii*	DNA methylation and small RNA increased susceptibility to beet cyst nematode	Hewezi et al. (2017)
6	Arabidopsis	*Fusarium oxysporum*	DNA demethylases increase resistance to fungal	Lee and Hwang (2005)
7	Arabidopsis	*Pseudomonas strains*	small RNA including MicroRNA regulates resistance to bacterial pathogens	Li et al. (2010)
8	Arabidopsis	*Alternaria brassicicola*	Histone deacetylase HDA19 increases sensitivity to pathogen	Zhou et al. (2005)
9	Arabidopsis	*Helicoverpa zea* and *Pieris rapae*	transgenerational priming responses (TPR) results in a reduction in caterpillar weight gain	Rasmann et al. (2012)
10	*Arabidopsis* mutants	*Heterodera schachtii, Meloidogyne javanica* and *Meloidogyne incognita*	small RNAs to decrease susceptibility to the nematodes.miRNA genes can reduce the size of giant-cells and number of galls per plant.	Díaz-Manzano et al. (2018)
11	*Brassica rapa*	*Plasmodiophora brassicae* and *Bombus terrestris*	DNA methylation changes were associated with flower morphology and aroma variation	Kellenberger et al. (2016)
12	Chickpea	*Fusarium oxysporum*	DNA methylation may be implicated in the resistance to pathogen	Mohammadi et al. (2015)
13	Cucumber	*Hop stunt viroid* (HSVd)	Epigenetic control of rRNA genes and TEs in gametic cells of cucumber is impaired by virus	Martinez-Medina et al. (2016)
14	Maize	*Fusarium graminearum*	Histone modifications the chromatin defense against to pathogens	Wang et al. (2017)
15	Medicago	Cell differentiation and nodule development	miRNA, mtr-miR169a targets the expression of NF-YA1/HAP2	Combier et al. (2006)
16	Musa	*Mycosphaerella fijiensis*	DNA methylation related with resistance to sigatoka negra toxins.	Giménez et al. (2006)
17	Pepper	Xanthomonas camestris pv. vesicatoria	Effected systemic acquired resistance (SAR) by avirulent strain	Lee and Hwang (2005)

TABLE 1.3 (Continued)
Epigenetic Modification in Crop Plants during Biotic Stresses

Sr. No.	Crop	Biotic Factor	Changes Induced	References
18	Rice	*Xanthomonas oryza*	Epigenetic regulation of PigmS fine-tunes disease resistance	Deng et al. (2017)
19	Rice	*Xanthomonas oryza*	Overexpression of a histone lysine demethylase *JMJ705* increases sensitivity to blight	Li et al. (2013)
20	Rice (transgenic)	*Magnaporthe oryzae* and *Xanthomonas oryzae pathovar oryzae*	Histone *HDT701* deacetylase gene enhanced resistance to pathogens	Ding et al. (2012a, b)
21	Soybean	*Mungbean yellow mosaic India virus* (MYMIV)	Higher level of Intergenic Region (IR)-specific DNA methylation	Yadav and Chattopadhyay (2011)
22	Soybean	auxin factors, GmARF8a and GmARF8b	miRNA167 regulates nodule and lateral root formation	Wang et al. (2015)
23	Tobacco	*Tobacco mosaic virus*	Pathogen resistance homologous recombination frequency	Kathiria et al. (2010)
24	Tobacco (transgenic)	*Tomato leaf curl virus*	Methylation of cytosine in the Intergenic Region (IR) of virus	Bian et al. (2006)
25	Tomato	*Botrytis cinerea*	Histone methylation increases sensitivity to pathogen-activating histone marks H3K4me3, H3K9ac and repressive H3K27me3	Crespo-Salvador et al. (2018)
26	Tomato	*Tomato yellow leaf curl Sardinia virus*	Alterations in DNA methylation patterns implies in defense and stress response	Mason et al. (2008)
27	Wheat	*Puccinia striiformis* f. sp. *tritici* (*Pst*)	lncRNAs differential expression and genes related to the defense of wheat against rust	Zhang et al. (2013)
28	Wheat	*Blumeria graminis* f. sp. tritici (*Bgt*)	TaHDA6 inhibits histone acetylation in protection-related gene promoters and thus, it negatively regulates its expression and the plant's defense response to powdery mildew	Jiao et al. (2019)
29	Watermelon	*Cucumber Green Mottle Mosaic Virus* (CGMMV)	DNA methylation variety in response to viral contamination in watermelon	Sun et al. (2019)

Source: Reprinted with permission from Samantara et al. (2021).

exemplified in *Brassica rapa* where we encounter herbivore-induced DNA methylation against leaf damage by the caterpillar *Pieris brassicae* (Kellenberger et al. 2016). Likewise increased number of trichomes that secrete a sticky and potentially noxious fluid has been observed in yellow monkey flower (*Mimulus guttatus*) fed by insects in the previous generation (Holeski 2007). Another interesting observation seems to be the presence of abnormal floral phenotypes having distinctive methylation profiles in *Solanum ruiz-lealii*, a trait having evolutionary implications since bumblebees rarely visit plants bearing aberrant flowers (Marfil et al. 2009).

1.4.3 Epigenetics in Improvement of Agronomic Traits

Improvement in the desired agronomic traits has been one of the specific objectives of plant breeders. Epigenetics has an important role to play in this respect and is either directly or indirectly responsible in the improvement of both quantitative and qualitative traits. Table 1.4 illustrates the importance of epigenetics in crop improvement. Both epigenetics and epigenomics help in elucidation of diverse mechanisms through which environment has a controlling effect on the phenotype. The term epibreeding refers to the use of epigenetic variation for crop improvement (Kapazoglou et al. 2018; Dalakouras and Vlachostergios 2021). The heritable epigenetic changes concomitant with alterations in gene expression result in phenotypic variations which are the contributing factors in crop improvement (Kakoulidou et al. 2021; Gupta and Salgotra 2022). However, the extent of genetic diversity varies between crops. Some of the economically important crops have low genetic diversity and experience rapid genetic erosion. In these species, phenotypic variability can be induced within some generations through epibreeding and some trait alterations may pass on to the succeeding generations. This would be of immense utility in overcoming the limitations and constraints due to narrow genetic base in certain crops and enhance the effectiveness genetic improvement.

TABLE 1.4
Role of Epigenetics in the Improvement of Agronomic Traits in Different Crop Species

Plant	Trait	References
Apple	Production of anthocyanin	Telias et al. (2011)
Arabidopsis	Breaking seed dormancy	Nonogaki (2014)
	Rapid germination	van Zanten et al. (2014)
Banana	Ripening	Fu et al. (2018)
Cotton	Altered photoperiodic response	Wang et al. (2017)
	Fiber initiation	Kumar et al. (2018)
	Fiber differentiation	Wang et al. (2016a, b)
Maize	Pigment synthesis	Cocciolone et al. (2001)
	Regulation of plant height, leaf development and fertility	Forestan et al. (2018)
Pigeon pea	High heterosis	Sinha et al. (2020)
Pine apple	Increased somatic embryogenesis	Luan et al. (2020)
Potato	Tuber yield	Bhogale et al. (2014)
Rapeseed	Bolting and flowering time	Si et al. (2021)
	Enhanced seed protein as well linoleic and/or palmitic acid content	Amoah et al. (2012)
Rice	Flowering period	Sun et al. (2012b)
	Short and dense panicles	Luan et al. (2019)
	Thickened aleurone layer and improved nutritional value	Liu et al. (2018a, b)
Soybean	Enhanced yield and stability	Raju et al. (2018)
Sweet orange	Fruit development and ripening	Huang et al. (2019)
Tomato	Delayed ripening	Manning et al. (2006)
	Enhanced vigor	Kundariya et al. (2020)
	Vitamin E biosynthesis and accumulation	Miura et al. (2009); Quadrana et al. (2014)
Watermelon	Sex determination	Martin et al. (2009)

1.5 EPIGENETICS IN PLANT EVOLUTION

It has been generally observed that epigenetic processes allow speedier adaption in comparison to DNA-based changes. Due to direct effect of DNA methylation and indirect effect of chromatin modifications on the mutation rate in organisms thereby increasing the probability of acquiring beneficial mutations, epigenetic mechanisms are known to directly affect genome evolution (Ashe et al. 2021). There have also been some reports of the role of epigenetic mechanisms in the evolution of sex chromosomes as well as in the regulation of sexual phenotypes. Epigenetics can modify the adaptation dynamics and substantially vary the evolutionary outcomes as has been the case of asexually reproducing organisms. However, dearth of research warrants similar studies in sexually reproducing organisms including plants (Stajic and Jansen 2021).

Plants differ fundamentally from animals in the nature of meristematic growth and the fact that future plant generations arise from those very cells that produced the somatic shoot tissues. Somatic events seem to pass on to the next generation since they can dedifferentiate and form new shoot meristems during normal development as well as in response to injury. This enables plants to accumulate epigenetic mutations which can revert to a previous expression state (Catoni et al. 2017). Thus, plants are uniquely unstable in possessing both expressed and silenced epialleles which allows them to express an advantageous phenotype, without entirely altering the ancestral genetic state for possible expression in its progeny. The relative state of the epigenetic changes facilitates selection that favors a precise expression state. After reproduction, epigenetic drift creates variation in the progenies by producing a range of expression states distributed around the parental level (Das and Messing 1994). Plants are most suitably placed for exploitation of the epigenetic variation and adaptations since they play a major role in augmenting plant genetic evolution (Figure 1.4).

However, it is important to note that not all epigenetic modifications are functionally relevant. Studies in *Arabidopsis* have shown that of the different variations in the methylation patterns in the evolved offspring only a few had significant phenotypic effects and were associated with changes

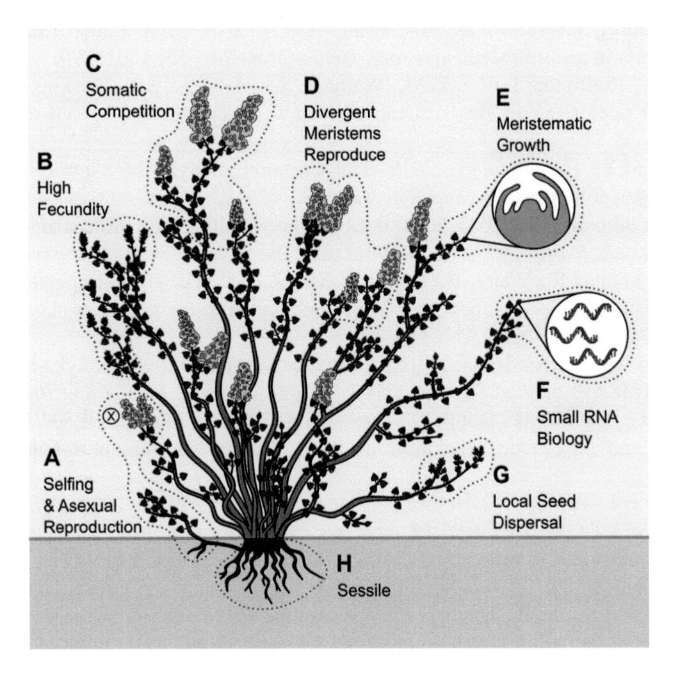

FIGURE 1.4 Advantages of epigenetic changes in plant genome (Reprinted with permission from Minow and Colasanti 2020).

in gene expression (Schmid et al. 2018). Therefore, extensive research is required for ascertaining the mechanisms of epigenetic control of gene expression along with investigation of other chromatin states.

1.6 CONCLUSION

Epigenetics is the study of chemical modifications and resultant heritable changes in chromatin function that do not involve changes in DNA sequence but influence gene expression. It offers innumerable opportunities for manipulating plant genomes for crop improvement. If suitably harnessed epigenetic variation can be utilized for stable improvement of agronomic traits. Future research in this emerging field will require more comprehensive research agendas, refining in technological and bioinformatic approaches coupled with development of new model organisms in addition to the few model species that have dominated the research arena.

REFERENCES

Abid, G., Mingeot, D., Muhovski, Y., Mergeai, G., Aouida, M., Abdelkarim, S., et al., 2017. Analysis of DNA methylation patterns associated with drought stress response in faba bean (*Vicia faba* L.) using methylation-sensitive amplification polymorphism (MSAP). *Environ. Exp. Bot.* 142, 34–44.

Agnieszka, N., 2018. The influence of Al3+ on DNA methylation and sequence changes in the triticale (× Triticosecale wittmack) genome. *J. Appl. Genet.* 59, 405–417.

Aina, R., Sgorbati, S., Santagostino, A., Labra, M., Ghiani, A., Citterio, S. 2004. Specific hypomethylation of DNA is induced by heavy metals in white clover and industrial hemp. *Physiol. Plant.* 121, 472–480.

Akhter, Z., Bi, Z., Ali, K., Sun, C., Fiaz, S., Haider, F.U., Bai, J., 2021. In response to abiotic stress, DNA methylation confers epigenetic changes in plants. *Plants* 10, 1096.

Alexandre, C., Möller-Steinbach, Y., Schönrock, N., Gruissem, W., Hennig, L., 2009. *Arabidopsis* MSI1 is required for negative regulation of the response to drought stress. *Mol. Plant* 2, 675–687.

Alfalahi, A.O., Hussein, Z.T., Khalofah, A., Sadder, M.T., Qasem, J.R., Al-Khayri, J.M., Jain, S.M., Almehemdi, A.F., 2022. Epigenetic variation as a new plant breeding tool: A review. *J. King Saud Univ. Sci.* 34, 102302.

Al-Lawati, A., Al-Bahry, S., Victor, R., Al-Lawati, A.H., Yaish, M.W., 2016. Salt stress alters DNA methylation levels in alfalfa (Medicago spp). *Genet. Mol. Res.* 15, 15018299.

Amoah, S., Kurup, S., Rodriguez Lopez, C.M., Welham, S.J., Powers, S.J., Hopkins, C.J., et al., 2012. A hypomethylated population of Brassica rapa for forward and reverse epigenetics. *BMC Plant Biol.* 12, 193.

Arshad, M., Gruber, M.Y., Hannoufa, A., 2018. Transcriptome analysis of microRNA156 overexpression alfalfa roots under drought stress. *Sci. Rep.* 8, 9363.

Ashe, A., Colot, V., Oldroyd, B.P., 2021. How does epigenetics influence the course of evolution? *Philos. Trans. R. Soc. Lond. B Biol. Sci.* 376, 20200111.

Baek, D., Jiang, J., Chung, J.S., Wang, B., Chen, J., Xin, Z., Shi, H., 2011. Regulated AtHKT1 gene expression by a distal enhancer element and DNA methylation in the promoter plays an important role in salt tolerance. *Plant Cell Physiol.* 52, 149–161.

Bednarek, P.T., Orłowska, R., Niedziela, A., 2017. A relative quantitative methylation-sensitive amplified polymorphism (MSAP) method for the analysis of abiotic stress. *BMC Plant Biol.* 17, 79.

Benoit, M., Drost, H.G., Catoni, M., Gouil, Q., Lopez-Gomollon, S., Baulcombe, D., Paszkowski, J., 2019. Environmental and epigenetic regulation of Rider retrotransposons in tomato. *PLoS Genet.* 15, e1008370.

Berr, A., McCallum, E.J., Alioua, A., Heintz, D., Heitz, T., Shen, W.H., 2010. *Arabidopsis* histone methyltransferase SET DOMAIN GROUP8 mediates induction of the jasmonate/ethylene pathway genes in plant defense response to necrotrophic fungi. *Plant Physiol.* 154, 1403–1414.

Bhargava, A., Srivastava, S., 2013. *Quinoa: Botany, Production and Uses*. CABI, Oxforshire, UK.

Bhargava, A., Srivastava, S., 2020. Response of Amaranthus sp. to salinity stress. In: Hirich, A., Choukr-Allah, R., Ragab, R. (Eds.) *Emerging Research in Alternative Crops*. Springer, Switzerland. pp. 245–264.

Bhargava, A., Shukla, S., Katiyar, R.S., Ohri, D., 2003. Selection parameters for genetic improvement in Chenopodium grain in sodic soil. *J. Appl. Horticul.* 5, 45–48.

Bhogale, S., Mahajan, A.S., Natarajan, B., Rajabhoj, M., Thulasiram, H.V., Banerjee, A.K., 2014. MicroRNA156: A potential graft-transmissible microRNA that modulates plant architecture and tuberization in Solanum tuberosum ssp. andigena. *Plant Physiol.* 164, 1011–1027.

Bian, X.Y., Rasheed, M.S., Seemanpillai, M.J., Rezaian, M.A., 2006. Analysis of silencing escape of Tomato leaf curl virus: An evaluation of the role of DNA methylation. *Mol. Plant-Microbe Interact.* 19, 614–624.

Catoni, M., Griffiths, J., Becker, C., Zabet, N.R., Bayon, C., Dapp, M., Lieberman-Lazarovich, M., Weigel, D., Paszkowski, J., 2017. DNA sequence properties that predict susceptibility to epiallelic switching. *The EMBO J.* 36(5), 617–628.

Cavalli, G., Heard, E., 2019. Advances in epigenetics link genetics to the environment and disease. *Nature* 571(7766), 489–499.

Chang, Y.N., Zhu, C., Jiang, J., Zhang, H., Zhu, J.K., Duan, C.G., 2020. Epigenetic regulation in plant abiotic stress responses. *J. Integr. Plant Biol.* 62, 563–580.

Chen, L.-T., Wu, K., 2010. Role of histone deacetylases HDA6 and HDA19 in ABA and abiotic stress response. *Plant Signal. Behav.* 5(10), 1318–1320.

Chen, Z., Zhang, H., Jablonowski, D., Zhou, X., Ren, X., Hong, X., et al., 2006. Mutations in ABO1/ELO2, a subunit of holo-elongator, increase abscisic acid sensitivity and drought tolerance in *Arabidopsis* thaliana. *Mol. Cell. Biol.* 26, 6902–6912.

Choi, C.S., Sano, H., 2007. Abiotic-stress induces demethylation and transcriptional activation of a gene encoding a glycerophosphodiesterase-like protein in tobacco plants. *Mol. Genet. Genom.* 277, 589–600.

Cicatelli, A., Todeschini, V., Lingua, G., Biondi, S., Torrigiani, P., Castiglione, S., 2014. Epigenetic control of heavy metal stress response in mycorrhizal versus non-mycorrhizal poplar plants. *Environ. Sci. Poll. Res.* 21, 1723–1737.

Cocciolone, S.M., Chopra, S., Flint-Garcia, S.A., McMullen, M.D., Peterson, T., 2001. Tissue-specific patterns of a maize Myb transcription factor are epigenetically regulated. *Plant J.* 27, 467–478.

Combier, J.P., Frugier, F., De Billy, F., Boualem, A., El-Yahyaoui, F., Moreau, S., et al., 2006. MtHAP2-1 is a key transcriptional regulator of symbiotic nodule development regulated by microRNA169 in *Medicago truncatula. Genes Dev.* 20, 3084–3088.

Cong, W., Miao, Y., Xu, L., Zhang, Y., Yuan, C., Wang, J., et al., 2019. Transgenerational memory of gene expression changes induced by heavy metal stress in rice (*Oryza sativa* L.). *BMC Plant Biol.* 19, 282.

Crespo-Salvador, O., Escamilla-Aguilar, M., López-Cruz, J., López-Rodas, G., González- Bosch, C., 2018. Determination of histone epigenetic marks in *Arabidopsis* and tomato genes in the early response to *Botrytis cinerea. Plant Cell Rep.* 37, 153–166.

Dai, L.F., Chen, Y.L., Luo, X.D., Wen, X.F., Cui, F.L., Zhang, F.T., et al., 2015. Level and pattern of DNA methylation changes in rice cold tolerance introgression lines derived from Oryza rufipogon Griff. *Euphytica* 205, 73–83.

Dalakouras, A., Vlachostergios, D., 2021. Epigenetic approaches to crop breeding: Current status and perspectives. *J. Exp. Bot.* 72, 5356–5371.

Das, O.P., Messing, J., 1994. Variegated phenotype and developmental methylation changes of a maize allele originating from epimutation. *Genetics*, 136, 1121–1141.

Demetriadou, C., Koufaris, C., Kirmizis, A., 2020. Histone N-alpha terminal modifications: genome regulation at the tip of the tail. *Epigenet. Chromat.* 13, 29.

Demetriou, K., Kapazoglou, A., Tondelli, A., Francia, E., Stanca, M.A., Bladenopoulos, K., Tsaftaris, A.S., 2009. Epigenetic chromatin modifiers in barley: I. Cloning, mapping and expression analysis of the plant specific HD2 family of histone deacetylases from barley, during seed development and after hormonal treatment. *Physiol. Plant.* 136, 358–368.

Deng, Y., Zhai, K., Xie, Z., Yang, D., Zhu, X., Liu, J., et al., 2017. Epigenetic regulation of antagonistic receptors confers rice blast resistance with yield balance. *Science* 355, 962–965.

Díaz-Manzano, F.E., Cabrera, J., Ripoll, J.-J., del Olmo, I., Andr´es, M.F., Silva, A.C., et al., 2018. A role for the gene regulatory module microRNA172/TARGET OF EARLY ACTIVATION TAGGED 1/FLOWERING LOCUS T (miRNA172/TOE1/FT) in the feeding sites induced by *Meloidogyne javanica* in Arabidopsis *thaliana. New Phytol.* 217, 813–827.

Ding, Y., Shi, Y., Yang, S., 2019. Advances and challenges in uncovering cold tolerance regulatory mechanisms in plants. *New Phytol.* 222, 1690–1704.

Ding, B., Bellizzi, M., del, R., Ning, Y., Meyers, B.C., Wang, G.L., 2012a. HDT701, a histone H4 deacetylase, negatively regulates plant innate immunity by modulating histone H4 acetylation of defense-related genes in rice. *Plant Cell* 24, 3783–3794.

Ding, Y., Fromm, M., Avramova, Z., 2012b. Multiple exposures to drought "train" transcriptional responses in *Arabidopsis*. *Nat. Commun.* 3, 1–9.

Dowen, R.H., Pelizzola, M., Schmitz, R.J., Lister, R., Dowen, J.M., Nery, J.R., Dixon, J.E., Ecker, J.R., 2012. Widespread dynamic DNA methylation in response to biotic stress. *Proc. Natl. Acad. Sci. USA* 109, E2183–E2191.

Driedonks, N., Xu, J., Peters, J.L., Park, S., Rieu, I., 2015. Multi-level interactions between heat shock factors, heat shock proteins, and the redox system regulate acclimation to heat. *Front. Plant Sci.* 6, 999.

Duan, C.G., Zhu, J.K., Cao, X., 2018. Retrospective and perspective of plant epigenetics in China. *J. Genet. Genom.* 45, 621–638.

Erturk, F.A., Agar, G., Arslan, E., Nardemir, G., 2015. Analysis of genetic and epigenetic effects of maize seeds in response to heavy metal (Zn) stress. *Environ. Sci. Pollut. Res.* 22, 10291–10297.

Fang, H., Liu, X., Thorn, G., Duan, J., Tian, L., 2014. Expression analysis of histone acetyltransferases in rice under drought stress. *Biochem. Biophys. Res. Commun.* 443, 400–405.

Feng, S.J., Liu, X.S., Ma, L.Y., khan, Iullah, Rono, J.K., Yang, Z.M., 2020. Identification of epigenetic mechanisms in paddy crop associated with lowering environmentally related cadmium risks to food safety. *Environ. Pollut.* 256, 113464.

Feng, X.J., Li, J.R., Qi, S.L., Lin, Q.F., Jin, J.B., Hua, X.J., 2016. Light affects salt stress-induced transcriptional memory of P5CS1 in *Arabidopsis*. *Proc. Natl. Acad. Sci. USA* 113, E8335–E8343.

Finnegan, E.J., Genger, R.K., Kovac, K., Peacock, W.J., Dennis, E.S., 1998. DNA methylation and the promotion of flowering by vernalization. *Proc. Natl. Acad. Sci. USA* 95, 5824–5829.

Forestan, C., Farinati, S., Rouster, J., Lassagne, H., Lauria, M., Dal Ferro, N., Varotto, S., 2018. Control of maize vegetative and reproductive development, fertility, and rRNAs silencing by histone deacetylase 108. *Genetics* 208, 1443–1466.

Friedrich, T., Faivre, L., Bäurle, I., Schubert, D., 2019. Chromatin-based mechanisms of temperature memory in plants. *Plant Cell Environ.* 42, 762–770.

Fu, C.C., Han, Y.C., Guo, Y.F., Kuang, J.F., Chen, J.Y., Shan, W., Lu, W.J., 2018. Differential expression of histone deacetylases during banana ripening and identification of MaHDA6 in regulating ripening-associated genes. *Postharvest Biol. Technol.* 141, 24–32.

Fu, W., Wu, K., Duan, J., 2007. Sequence and expression analysis of histone deacetylases in rice. *Biochem. Biophys. Res. Commun.* 356, 843–850.

Furrow, R.E., 2014. Epigenetic inheritance, epimutation, and the response to selection. *PLoS One* 9, 7–10.

Gao, G., Li, J., Li, H., Li, F., Xu, K., Yan, G., et al., 2014. Comparison of the heat stress induced variations in DNA methylation between heat-tolerant and heat-sensitive rapeseed seedlings. *Breed. Sci.* 64, 125–133.

Giménez, C., Palacios, G., Colmenares, M., 2006. Musa methylated DNA sequences associated with tolerance to Mycosphaerella fijiensis toxins. *Plant Mol. Biol. Rep.* 24, 33–43.

Goldstein, M., Derheimer, F.A., Tait-Mulder, J., Kastan, M.B., 2013. Nucleolin mediates nucleosome disruption critical for DNA double-strand break repair. *Proc. Natl. Acad. Sci. USA* 110, 16874–16879.

Gong, Z., 2021. Plant abiotic stress: new insights into the factors that activate and modulate plant responses. *J. Interg. Plant Biol.* 63, 429–430.

González, R.M., Ricardi, M.M., Iusem, N.D., 2013. Epigenetic marks in an adaptive water stress-responsive gene in tomato roots under normal and drought conditions. *Epigenetics* 8, 864–872.

Goulet, B.E., Roda, F., Hopkins, R., 2017. Hybridization in plants: old ideas, new techniques. *Plant Physiol.* 173, 65–78.

Grabsztunowicz, M., Koskela, M.M., Mulo, P., 2017. Post-translational modifications in regulation of chloroplast function: recent advances. *Front. Plant Sci.* 8, 240.

Gupta, C., Salgotra, R.K., 2022. Epigenetics and its role in effecting agronomical traits. *Front. Plant Sci.* 13, 925688.

Haery, L., Thompson, R.C., Gilmore, T.D., 2015. Histone acetyltransferases and histone deacetylases in B- and T-cell development, physiology and malignancy. *Genes Cancer*, 6, 184–213.

Hagmann, J., Becker, C., Muller, J., Stegle, O., Meyer, R.C., Wang, G., Schneeberger, K., Fitz, J., Altmann, T., Bergelson, J., Borgwardt, K., Weigel, D., 2015. Century-scale methylome stability in a recently diverged *Arabidopsis* thaliana lineage. *PLoS Genet.* 11, e1004920.

Han, S.K., Sang, Y., Rodrigues, A., Wu, M.F., Rodriguez, P.L., Wagner, D., 2013. The SWI2/SNF2 chromatin remodeling ATPase BRAHMA represses abscisic acid responses in the absence of the stress stimulus in *Arabidopsis*. *Plant Cell* 24, 4892–4906.

Han, Z., Yu, H., Zhao, Z., Hunter, D., Luo, X., Duan, J., Tian, L., 2016. AtHD2D gene plays a role in plant growth, development, and response to abiotic stresses in *Arabidopsis* thaliana. *Front. Plant Sci.* 7, 310. 10.3389/fpls.2016.00310.

Herrera, C.M., Medrano, M., Bazaga, P., 2016. Comparative spatial genetics and epigenetics of plant populations: heuristic value and a proof of concept. *Mol. Ecol.* 25, 1653–1664.

Hewezi, T., Lane, T., Piya, S., Rambani, A., Rice, J.H., Staton, M., 2017. Cyst nematode parasitism induces dynamic changes in the root epigenome. *Plant Physiol.* 174, 405–420.

Hilker, M., Schwachtje, J., Baier, M., Balazadeh, S., Bäurle, I., Geiselhardt, S., Hincha, D.K., Kunze, R., Mueller-Roeber, B., Rillig, M.C., Rolff, J., Romeis, T., Schmülling, T., Steppuhn, A., van Dongen, J., Whitcomb, S.J., Wurst, S., Zuther, E., Kopka, J., 2016. Priming and memory of stress responses in organisms lacking a nervous system. *Biol. Rev.* 91, 1118–1133.

Holeski, L.M., 2007. Within and between generation phenotypic plasticity in trichome density of *Mimulus guttatus*. *J. Evol. Biol.* 20, 2092–2100.

Holliday, R., 1990. Mechanisms for the control of gene activity during development. *Biol. Rev.* 65, 431–471.

Holliday, R., 1994. Epigenetics: an overview. *Dev. Genet.* 15, 453–457.

Hossain, M.S., Kawakatsu, T., Kim, K., Do, Zhang, N., Nguyen, C.T., Khan, S.M., et al., 2017. Divergent cytosine DNA methylation patterns in single-cell, soybean root hairs. *New Phytol.* 214, 808–819.

Hou, H., Zhao, L., Zheng, X., Gautam, M., Yue, M., Hou, J., et al., 2019. Dynamic changes in histone modification are associated with upregulation of Hsf and rRNA genes during heat stress in maize seedlings. *Protoplasma* 256, 1245–1256.

Hsu, P.S., Yu, S.H., Tsai, Y.T., Chang, J.Y., Tsai, L.K., Ye, C.H., Song, N.Y., Yau, L.C., Lin, S.P., 2021. More than causing (epi) genomic instability: emerging physiological implications of transposable element modulation. *J. Biomed. Sci.* 28, 58.

Huang, H., Liu, R., Niu, Q., Tang, K., Zhang, B., Zhang, H., et al., 2019. Global increase in DNA methylation during orange fruit development and ripening. *Proc. Natl. Acad. Sci. USA* 116, 1430–1436.

Huang, L., Sun, Q., Qin, F., Li, C., Zhao, Y., Zhou, D.X., 2007. Down-regulation of a Silent Information Regulator2-related histone deacetylase gene, OsSRT1, induces DNA fragmentation and cell death in rice. *Plant Physiol.* 144, 1508–1519.

Inácio, V., Santos, R., Prazeres, R., Graça, J., Miguel, C.M., Morais-Cecílio, L., 2022. Epigenetics at the crossroads of secondary growth regulation. *Front. Plant Sci.* 13, 970342.

Iqbal, Z., Iqbal, M.S., Hashem, A., Abd Allah, E.F., Ansari, M.I., 2021. Plant defense responses to biotic stress and its interplay with fluctuating dark/light conditions. *Front. Plant Sci.* 12, 631810.

Jablonka, E., Lamb, M.J., 2006. The changing concept of epigenetics. *Ann. NY Acad. Sci.* 981, 82–96.

Jiao, L., Zhi, P., Wang, X., Fan, Q., Chang, C., 2019. Wheat WD40-repeat protein TaHOS15 functions in a histone deacetylase complex to fine-tune defense responses to Blumeria graminis f.sp. tritici. *J. Exp. Bot.* 70, 255–268.

Jones, P.A., 2012. Functions of DNA methylation: Islands, start sites, gene bodies and beyond. *Nature Rev. Genet.* 13, 484–492.

Kaikkonen, M.U., Lam, M.T., Glass, C.K., 2011. Non-coding RNAs as regulators of gene expression and epigenetics. *Cardiovasc. Res.* 90, 430–440.

Kakoulidou, I., Avramidou, E.V., Baranek, M., Brunel-Muguet, S., Farrona, S., Johannes, F., et al., 2021. Epigenetics for crop improvement in times of global change. *Biology* 10, 766.

Kapazoglou, A., Ganopoulos, I., Tani, E., Tsaftaris, A., 2018. Epigenetics epigenomics and crop improvement. In: Kuntz, M. (Ed.) *Advances in Botanical Research*. Academic Press, San Diego, CA, pp. 287–324.

Karan, R., DeLeon, T., Biradar, H., Subudhi, P.K., 2012. Salt stress induced variation in DNA methylation pattern and its influence on gene expression in contrasting rice genotypes. *PLoS One* 7, e40203.

Kathiria, P., Sidler, C., Golubov, A., Kalischuk, M., Kawchuk, L.M., Kovalchuk, I., 2010. Tobacco mosaic virus infection results in an increase in recombination frequency and resistance to viral, bacterial, and fungal pathogens in the progeny of infected tobacco plants. *Plant Physiol.* 153, 1859–1870.

Kawakatsu, T., Nery, J.R., Castanon, R., Ecker, J.R., 2017. Dynamic DNA methylation reconfiguration during seed development and germination. *Genome Biol.* 18, 171.

Kellenberger, R.T., Schlüter, P.M., Schiestl, F.P., 2016. Herbivore-induced DNA demethylation changes floral signaling and attractiveness to pollinators in Brassica rapa. *PLoS ONE* 11, e0166646.

Kim, J.M., To, T.K., Ishida, J., Morosawa, T., Kawashima, M., Matsui, A., et al., 2008. Alterations of lysine modifications on the histone H3 N-tail under drought stress conditions in Arabidopsis *thaliana*. *Plant Cell Physiol.* 49, 1580–1588.

Kim, S., Kaang, B.K., 2017. Epigenetic regulation and chromatin remodeling in learning and memory. *Exp. Mol. Med.* 49, e281.

Kouzarides, T., 2007. Chromatin modifications and their function. *Cell* 128, 693–705.

Kronholm, I., Collins, S., 2016. Epigenetic mutations can both help and hinder adaptive evolution. *Mol. Ecol.* 25, 1856–1868.

Kumar, S., Beena, A.S., Awana, M., Singh, A., 2017. Salt-induced tissue-specific cytosine methylation downregulates expression of HKT genes in contrasting wheat (Triticum aestivum L.) genotypes. *DNA Cell Biol.* 36(4), 283–294.

Kumar, V., Singh, B., Singh, S.K., Rai, K.M., Singh, S.P., Sable, A., Pant, P., Saxena, G., and Sawant, S.V. (2018). Role of GhHDA5 in H3K9 deacetylation and fiber initiation in Gossypium hirsutum. *Plant J.* 95, 1069–1083.

Kundariya, H., Yang, X., Morton, K., Sanchez, R., Axtell, M.J., Hutton, S.F., Fromm, M., Mackenzie, S.A., 2020. MSH1-induced heritable enhanced growth vigor through grafting is associated with the RdDM pathway in plants. *Nat. Commun.* 11(1), 5343.

Kwon, C.S., Lee, D., Choi, G., Chung, W.-I., 2009. Histone occupancy-dependent and independent removal of H3K27 trimethylation at cold-responsive genes in *Arabidopsis*. *Plant J.* 60(1), 112–121.

Labra, M., Ghiani, A., Citterio, S., Sgorbati, S., Sala, F., Vannini, C., et al., 2002. Analysis of cytosine methylation pattern in response to water deficit in pea root tips. *Plant Biol.* 4(6), 694–699.

Lamke, J., Baurle, I., 2017. Epigenetic and chromatin-based mechanisms in environmental stress adaptation and stress memory in plants. *Genome Biol.* 18, 124.

Lämkc, J., Brzezinka, K., Altmann, S., B¨aurle, I., 2016. A hit-and-run heat shock factor governs sustained histone methylation and transcriptional stress memory. *EMBO J.* 35(2), 162–175.

Latzel, V., Rendina González, A.P., Rosenthal, J., 2016. Epigenetic memory as a basis for intelligent behavior in clonal plants. *Front. Plant Sci.* 7, 1354.

Lederberg, J., 2001. The meaning of epigenetics. *Scientist* 17, 6.

Lee, S.C., Hwang, B.K., 2005. Induction of some defense-related genes and oxidative burst is required for the establishment of systemic acquired resistance in *Capsicum annuum*. *Planta* 221(6), 790–800.

Li, Y., Zhang, Q.Q., Zhang, J., Wu, L., Qi, Y., Zhou, J.M., 2010. Identification of microRNAs involved in pathogen-associated molecular pattern-triggered plant innate immunity. *Plant Physiol.* 152(4), 2222–2231.

Li, T., Chen, X., Zhong, X., Zhao, Y., Liu, X., Zhou, S., et al., 2013. Jumonji C domain protein JMJ705-mediated removal of histone H3 lysine 27 trimethylation is involved in defense-related gene activation in rice. *Plant Cell* 25(11), 4725–4736.

Li, H., Yan, S., Zhao, L., Tan, J., Zhang, Q., Gao, F., et al., 2014a. Histone acetylation associated up-regulation of the cell wall related genes is involved in salt stress induced maize root swelling. *BMC Plant Biol.* 14(1), 1–14.

Li, R.Q., Huang, J.Z., Zhao, H.J., Fu, H.W., Li, Y.F., Liu, G.Z., Shu, Q.Y., 2014b. A down-regulated epi-allele of the genomes uncoupled 4 gene generates a xantha marker trait in rice. *Theor. Appl. Genet.* 127(11), 2491–2501.

Li, J., Wang, M., Li, Y., Zhang, Q., Lindsey, K., Daniell, H., et al., 2019a. Multi-omics analyses reveal epigenomics basis for cotton somatic embryogenesis through successive regeneration acclimation process. *Plant Biotechnol. J.* 17, 435–450.

Li, S., Lin, Y.C.J., Wang, P., Zhang, B., Li, M., Chen, S., et al., 2019b. The AREB1 transcription factor influences histone acetylation to regulate drought responses and tolerance in Populus trichocarpa. *Plant Cell* 31, 663–686.

Lira-Medeiros, C.F., Parisod, C., Fernandes, R.A., Mata, C.S., Cardoso, M.A., Ferreira, P.C.G., 2010. Epigenetic variation in mangrove plants occurring in contrasting natural environment. *PLoS One* 5, e10326.

Liu, G., Xing, Y., Zhao, H., Wang, J., Shang, Y., Cai, L., 2016. A deformation energy-based model for predicting nucleosome dyads and occupancy. *Sci. Rep.* 6, 24133.

Liu, J., Wu, X., Yao, X., Yu, R., Larkin, P.J., Liu, C.M., 2018a. Mutations in the DNA demethylase OsROS1 result in a thickened aleurone and improved nutritional value in rice grains. *Proc. Natl. Acad. Sci. USA* 115, 11327–11332.

Liu, W., Duttke, S.H., Hetzel, J., Groth, M., Feng, S., Gallego-Bartolome, J., et al., 2018b. RNA-directed DNA methylation involves co-transcriptional small-RNA-guided slicing of polymerase v transcripts in *Arabidopsis*. *Nat. Plants* 4(3), 181–188.

Liu, X., Luo, M., Zhang, W., Zhao, J., Zhang, J., Wu, K., et al., 2012. Histone acetyltransferases in rice (Oryza sativa L.): phylogenetic analysis, subcellular localization and expression. *BMC Plant Biol.* 12 (1), 1–17.

López Sánchez, A., Stassen, J.H.M., Furci, L., Smith, L.M., Ton, J., 2016. The role of DNA (de)methylation in immune responsiveness of *Arabidopsis*. *Plant J.* 88, 361–374.

Løvtrup, S. 1974. *Epigenetics: A Treatise on Theoretical Biology*. John Wiley and Sonsm, London.

Luan, A., Chen, C., Xie, T., He, J., He, Y., 2020. Methylation analysis of CpG islands in pineapple SERK1 promoter. *Genes* 11(4), 425.

Luan, X., Liu, S., Ke, S., Dai, H., Xie, X.M., Hsieh, T.F., Zhang, X.Q. , 2019. Epigenetic modification of ESP, encoding a putative long noncoding RNA, affects panicle architecture in rice. *Rice (New York).* 12,20.

Luo, X., He, Y., 2020. Experiencing winter for spring flowering: A molecular epigenetic perspective on vernalization. *J. Integr. Plant Biol.* 62, 104–117.

Manning, K., Tör, M., Poole, M., Hong, Y., Thompson, A.J., King, G.J., Giovannoni, J.J., Seymour, G.B., 2006. A naturally occurring epigenetic mutation in a gene encoding an SBP-box transcription factor inhibits tomato fruit ripening. *Nature Genet.* 38, 948–952.

Marconi, G., Pace, R., Traini, A., Raggi, L., Lutts, S., Chiusano, M., et al., 2013. Use of MSAP markers to analyse the effects of salt stress on DNA methylation in rapeseed (Brassica napus var. oleifera). *PLoS One* 8(9), e75597.

Marfil, C.F., Camadro, E.L., Masuelli, R.W., 2009. Phenotypic instability and epigenetic variability in a diploid potato of hybrid origin, *Solanum ruiz-lealii. BMC Plant Biol.* 9, 21.

Martin, A., Troadec, C., Boualem, A., Rajab, M., Fernandez, R., Morin, H., et al., 2009. A transposon-induced epigenetic change leads to sex determination in melon. *Nature* 461, 1135–1138.

Martinez-Medina, A., Flors, V., Heil, M., Mauch-Mani, B., Pieterse, C., Pozo, M.J., Ton, J., van Dam, N.M., Conrath, U., 2016. Recognizing plant defense priming. *Trends Plant Sci.* 21, 818–822.

Mason, G., Noris, E., Lanteri, S., Acquadro, A., Accotto, G.P., Portis, E., 2008. Potentiality of methylation-sensitive amplification polymorphism (MSAP) in identifying genes involved in tomato response to tomato yellow leaf curl Sardinia virus. *Plant Mol. Biol. Rep.* 26, 156–173.

Migicovsky, Z., Yao, Y., Kovalchuk, I., 2014. Transgenerational phenotypic and epigenetic changes in response to heat stress in *Arabidopsis* thaliana. *Plant Signal. Behav.* 9, e27971.

Minow, M., Colasanti, J., 2020. Does variable epigenetic inheritance fuel plant evolution? *Genome* 63, 253–262.

Miryeganeh M., 2022. Epigenetic mechanisms of senescence in plants. *Cells* 11, 251.

Miura, K., Agetsuma, M., Kitano, H., Yoshimura, A., Matsuoka, M., Jacobsen, S.E., Ashikari, M., 2009. A metastable DWARF1 epigenetic mutant affecting plant stature in rice. *Proc. Natl. Acad. Sci. USA* 106(27), 11218–11223.

Mlynárová, L., Nap, J.P., Bisseling, T., 2007. The SWI/SNF chromatin-remodeling gene AtCHR12 mediates temporary growth arrest in *Arabidopsis* thaliana upon perceiving environmental stress. *Plant J.* 51, 874–885.

Mohammadi, P., Bahramnejad, B., Badakhshan, H., Kanouni, H., 2015. DNA methylation changes in fusarium wilt resistant and sensitive chickpea genotypes (Cicer arietinum L.). *Physiol. Mol. Plant Pathol.* 91, 72–80.

Murali Achary, V.M., Panda, B.B., 2010. Aluminium-induced DNA damage and adaptive response to genotoxic stress in plant cells are mediated through reactive oxygen intermediates. *Mutagenesis* 25, 201–209.

Muyle, A., Seymour, D., Darzentas, N., Primetis, E., Gaut, B.S., Bousios, A., 2021. Gene capture by transposable elements leads to epigenetic conflict in maize. *Mol. Plant* 14, 237–252.

Nonogaki, H., 2014. Seed dormancy and germination-emerging mechanisms and new hypotheses. *Front. Plant Sci.* 5, 233.

Ou, X., Zhang, Y., Xu, C., Lin, X., Zang, Q., Zhuang, T., et al., 2012. Transgenerational inheritance of modified DNA methylation patterns and enhanced tolerance induced by heavy metal stress in rice (Oryza sativa L.). *PLoS One* 7, e41143.

Pavangadkar, K., Thomashow, M.F., Triezenberg, S.J., 2010. Histone dynamics and roles of histone acetyltransferases during cold-induced gene regulation in *Arabidopsis*. *Plant Mol. Biol.* 74, 183–200.

Ponting, C.P., Oliver, P.L., Reik, W., 2009. Evolution and functions of long noncoding RNAs, *Cell* 136, 629–641.

Popova, O.V., Dinh, H.Q., Aufsatz, W., Jonak, C., 2013. The RdDM pathway is required for basal heat tolerance in *Arabidopsis*. *Mol. Plant* 6, 396–410.

Quadrana, L., Almeida, J., Asís, R., et al., 2014. Natural occurring epialleles determine vitamin E accumulation in tomato fruits. *Nat. Commun.* 5, 4027.

Rajewsky, N., Jurga, S., Barciszewski, J., 2017. *Plant Epigenetics*. Springer International Publishing, AG.

Rajkumar, M.S., Gupta, K., Khemka, N.K., Garg, R., Jain, M., 2020. DNA methylation reprogramming during seed development and its functional relevance in seed size/weight determination in chickpea. *Commun. Biol.* 3, 340.

Raju, S.K.K., Shao, M.-R., Sanchez, R., Xu, Y.-Z., Sandhu, A., Graef, G., Mackenzie, S., 2018. An epigenetic breeding system in soybean for increased yield and stability. *Plant Biotech. J.* 16, 1836–1847.

Ramirez-Prado, J.S., Latrasse, D., Rodriguez-Granados, N.Y., Huang, Y., Manza-Mianza, D., Brik-Chaouche, R., Jaouannet, M., Citerne, S., Bendahmane, A., Hirt, H., Raynaud, C., and Benhamed, M., 2019. The Polycomb protein LHP1 regulates *Arabidopsis* thaliana stress responses through the repression of the MYC2-dependent branch of immunity. *Plant J.* 100(6), 1118–1131.

Rapp, R.A., Wendel, J.F., 2005. Epigenetics and plant evolution. *New Phytol.* 168, 81–91.

Rasmann, S., De Vos, M., Casteel, C.L., Tian, D., Halitschke, R., Sun, J.Y., Agrawal, A.A., Felton, G.W., and Jander, G., 2012. Herbivory in the previous generation primes plants for enhanced insect resistance. *Plant Physiol.* 158(2), 854–863.

Rehman, M., Tanti, B., 2020. Understanding epigenetic modifications in response to abiotic stresses in plants. *Biocatalysis Agric. Biotechnol.* 27, 101673.

Roberts, M.R., López Sánchez, A., 2019. Plant epigenetic mechanisms in response to biotic stress. In: Alvarez-Venegas, R., De-la-Peña, C., Casas-Mollano, J. (Eds.) *Epigenetics in Plants of Agronomic Importance: Fundamentals and Applications*. Springer, Cham. pp. 65–113.

Robert-Seilaniantz, A., Bari, R., Jones, J.D.G., 2010. Abiotic or biotic stresses. In: Pareek, A., Sopory, S.K., Bohnert, H.J., Govindjee (Eds.) *Abiotic Stress Adaptation in Plants: Physiological, Molecular and Genomic Foundation*. Springer, New York, USA. pp. 103–122.

Russo, V.E.A., Martienssen, R.A., Riggs, A.D., 1996. *Epigenetic Mechanisms of Gene Regulation*. Cold Spring Harbor Laboratory Press, NY.

Sahu, P.P., Pandey, G., Sharma, N., Puranik, S., Muthamilarasan, M., Prasad, M., 2013. Epigenetic mechanisms of plant stress responses and adaptation. *Plant Cell Rep.* 32(8), 1151–1159.

Samantara, K., Shiv, A., de Sousa, L.L., Sandhu, K.S., Priyadarshini, P., Mohapatra, S.R., 2021. A comprehensive review on epigenetic mechanisms and application of epigenetic modifications for crop improvement. *Environ. Exp. Bot.* 188, 104479.

Sani, E., Herzyk, P., Perrella, G., Colot, V., Amtmann, A., 2013. Hyperosmotic priming of *Arabidopsis* seedlings establishes a long-term somatic memory accompanied by specific changes of the epigenome. *Genome Biol.* 14(6), 1–24.

Scheid, O.M., Pecinka, A., Dinh, H.Q., Baubec, T., Rosa, M., Lettner, N., 2010. Epigenetic regulation of repetitive elements is attenuated by prolonged heat stress in *Arabidopsis*. *Plant Cell* 22 (9), 3118–3129.

Schmid, M.W., Heichinger, C., Coman Schmid, D., Guthörl, D., Gagliardini, V., Bruggmann, R., Aluri, S., Aquino, C., Schmid, B., Turnbull, L.A., Grossniklaus, U., 2018. Contribution of epigenetic variation to adaptation in *Arabidopsis*. *Nature Commun.* 9(1), 4446.

Secco, D., Whelan, J., Rouached, H., Lister, R., 2017. Nutrient stress-induced chromatin changes in plants. *Curr. Opin. Plant Biol.* 39, 1–7.

Shan, X., Wang, X., Yang, G., Wu, Y., Su, S., Li, S., et al., 2013. Analysis of the DNA methylation of maize (Zea mays L.) in response to cold stress based on methylation-sensitive amplified polymorphisms. *J. Plant Biol.* 56(1), 32–38.

Shriram, V., Kumar, V., Devarumath, R.M., Khare, T.S., Wani, S.H. 2016. MicroRNAs as potential targets for abiotic stress tolerance in plants. *Front. Plant Sci.* 7, 817.

Si, S., Zhang, M., Hu, Y., Wu, C., Yang, Y., Luo, S., Xiao, X., 2021. BrcuHAC1 is a histone acetyltransferase that affects bolting development in Chinese flowering cabbage. *J. Genet.* 100, 56.

Singla J., Krattinger S.G., 2016. Biotic stress resistance genes in wheat. In: Wrigley, C., Corke, H., Seetharaman, K., Faubion J. (Eds.). *Encyclopedia of Food Grains*. Vol. 4. Elsevier. pp. 388–392.

Singroha, G., Sharma, P. 2019. Epigenetic modifications in plants under abiotic stress. In: Meccariello, R. (Ed.) *Epigenetics*. IntechOpen. pp. 1–14.

Sinha, P., Singh, V.K., Saxena, R.K., Kale, S.M., Li, Y., Garg, V., Meifang, T., Khan, A.W., Kim, K.D., Chitikineni, A., Saxena, K.B., Sameer Kumar, C.V., Liu, X., Xu, X., Jackson, S., Powell, W., Nevo, E., Searle, I.R., Lodha, M., Varshney, R.K., 2020. Genome-wide analysis of epigenetic and transcriptional changes associated with heterosis in pigeonpea. *Plant Biotechnol. J.* 18, 1697–1710.

Song, Y., Ji, D., Li, S., Wang, P., Li, Q., Xiang, F., 2012. The dynamic changes of DNA methylation and histone modifications of salt responsive transcription factor genes in soybean. *PLoS One* 7(7), e41274.

Springer, N.M., Schmitz, R.J., 2017. Exploiting induced and natural epigenetic variation for crop improvement. *Nat. Rev. Genet.* 18(9), 563–575.
Springer, N., Schmitz, R., 2017. Exploiting induced and natural epigenetic variation for crop improvement. *Nat. Rev. Genet.* 18, 563–575.
Stajic, D., Jansen, L., 2021. Empirical evidence for epigenetic inheritance driving evolutionary adaptation. *Philos. Trans. R. Soc. Lond. B Biol. Sci.* 376(1826), 20200121.
Steward, N., Ito, M., Yamaguchi, Y., Koizumi, N., Sano, H., 2002. Periodic DNA methylation in maize nucleosomes and demethylation by environmental stress. *J. Biol. Chem.* 277, 37741–37746.
Sun, C., Fang, J., Zhao, T., Xu, B., Zhang, F., Liu, L., Tang, J., Zhang, G., Deng, X., Chen, F., Qian, Q., Cao, X., Chu, C. 2012b. The histone methyltransferase SDG724 mediates H3K36me2/3 deposition at MADS50 and RFT1 and promotes flowering in rice. *Plant Cell* 24(8), 3235–3247.
Sun, Y., Fan, M., He, Y., 2019. DNA methylation analysis of the Citrullus lanatus response to cucumber green mottle mosaic virus infection by whole-genome bisulfite sequencing. *Genes* 10(5), 344.
Sun, W.J., Zhou, X., Zheng, J.H., Lu, M.D., Nie, J.Y., Yang, X.J., Zheng, Z.Q., 2012a. Histone acetyltransferases and deacetylases: molecular and clinical implications to gastrointestinal carcinogenesis. *Acta Biochim. Biophys. Sin.* 44(1), 80–91.
Taspinar, M.S., Aydin, M., Sigmaz, B., Yagci, S., Arslan, E., Agar, G., 2018. Aluminum-induced changes on DNA damage, DNA methylation and LTR retrotransposon polymorphism in maize. *Arab. J. Sci. Eng.* 43, 123–131.
Taudt, A., Colome-Tatche, M., Johannes, F., 2016. Genetic sources of population epigenomic variation. *Nat. Rev. Genet.* 17, 319–332.
Telias, A., Lin-Wang, K., Stevenson, D.E., Cooney, J.M., Hellens, R.P., Allan, A.C., Hoover, E.E., Bradeen, J.M. 2011. Apple skin patterning is associated with differential expression of MYB10. *BMC Plant Biol.* 11, 93.
Temel, A., Janack, B., Humbeck, K., 2017. Drought stress-related physiological changes and histone modifications in barley primary leaves at HSP17 gene. *Agronomy* 7(2), 43.
van Zanten, M., Zöll, C., Wang, Z., Philipp, C., Carles, A., Li, Y., et al., 2014. HISTONE DEACETYLASE 9 represses seedling traits in *Arabidopsis* thaliana dry seeds. *Plant J.* 80(3), 475–488.
Verhoeven, K.J.F., Jansen, J.J., van Dijk, P.J., Biere, A., 2010. Stress-induced DNA methylation changes and their heritability in asexual dandelions. *New Phytol.* 185, 1108–1118.
Verhoeven, K.J.F., Preite, V., 2014. Epigenetic variation in asexually reproducing organisms. *Evolution* 68, 644–655.
Verma, N., Giri, S.K., Singh, G., Gill, R., Kumar, A., 2022. Epigenetic regulation of heat and cold stress responses in crop plants. *Plant Gene.* 29, 100351.
Villicaña, S., Bell, J.T., 2021. Genetic impacts on DNA methylation: research findings and future perspectives. *Genome Biol.* 22, 127.
Vlachonasios, K.E., Thomashow, M.F., Triezenberg, S.J., 2003. Disruption mutations of ADA2b and GCN5 transcriptional adaptor genes dramatically affect *Arabidopsis* growth, development, and gene expression. *Plant Cell*, 15, 626–638.
Waddington, C.H., 1942. The epigenotype. *Endeavour* 1, 18–20.
Waddington, C.H. 1968. The basic ideas of biology. In: Waddington, C.H. (Ed.) *Towards a Theoretical Biology, Vol. 1: Prolegomena.* Edinburgh University Press, Edinburgh. pp. 1–32.
Waititu, J.K., Zhang, C., Liu, J., Wang, H., 2020. Plant non-coding RNAs: origin, biogenesis, mode of action and their roles in abiotic stress. *Intern. J. Mol. Sci.* 21, 8401.
Walley, J.W., Rowe, H.C., Xiao, Y., Chehab, E.W., Kliebenstein, D.J., Wagner, D., Dehesh, K., 2008. The chromatin remodeler SPLAYED regulates specific stress signaling pathways. *PLoS Pathog.* 4, e1000237.
Wang, J.-W., Qi, Y., 2018. Plant non-coding RNAs and epigenetics. *Sci. China Life Sci.* 61, 135–137.
Wang, W.S., Pan, Y.J., Zhao, X.Q., Dwivedi, D., Zhu, L.H., Ali, J., Fu, B.Y., Li, Z.K. 2011. Drought-induced site-specific DNA methylation and its association with drought tolerance in rice (*Oryza sativa* L.). *J. Exp. Bot.* 62, 1951–1960.
Wang, Y., Li, K., Chen, L., Zou, Y., Liu, H., Tian, Y., Li, D., Wang, R., Zhao, F., Ferguson, B.J., Gresshoff, P.M., Li, X. 2015. Microrna167-directed regulation of the auxin response factors GmARF8a and GmARF8b is required for soybean nodulation and lateral root development. *Plant Physiol.* 168, 984–999.
Wang, M., Wang, P., Tu, L., Zhu, S., Zhang, L., Li, Z., Zhang, Q., Yuan, D., Zhang, X., 2016a. Multi-omics maps of cotton fibre reveal epigenetic basis for staged single-cell differentiation. *Nucleic Acids Res.* 44, 4067–4079.

Wang, X., Xin, C., Cai, J., Zhou, Q., Dai, T., Cao, W., Jiang, D., 2016b. Heat priming induces transgenerational tolerance to high temperature stress in wheat. *Front. Plant Sci.* 7, 501.

Wang, C., Yang, Q., Wang, W., Li, Y., Guo, Y., Zhang, D., Ma, X., Song, W., Zhao, J., Xu, M., 2017. A transposon-directed epigenetic change in ZmCCT underlies quantitative resistance to Gibberella stalk rot in maize. *New Phytol.* 215, 1503–1515.

Wei, J.W., Huang, K., Yang, C., Kang, C.S., 2017. Non-coding RNAs as regulators in epigenetics (Review). *Oncol. Rep.* 37, 3–9.

Xu, J., Wang, Q., Freeling, M., Zhang, X., Xu, Y., Mao, Y., Tang, X., Wu, F., Lan, H., Cao, M., Rong, T., Lisch, D., Lu, Y., 2017. Natural antisense transcripts are significantly involved in regulation of drought stress in maize. *Nucleic Acids Res.* 45, 5126–5141.

Yadav, R.K., Chattopadhyay, D., 2011. Enhanced viral intergenic region-specific short interfering RNA accumulation and DNA methylation correlates with resistance against a geminivirus. *Mol. Plant-Microbe Interact.* 24, 1189–1197.

Zhang, Z., Zhang, S., Zhang, Y., Wang, X., Li, D., Li, Q., Yue, M., Li, Q., Zhang, Y.E., Xu, Y., Xue, Y., Chong, K., Bao, S., 2011. *Arabidopsis* floral initiator SKB1 confers high salt tolerance by regulating transcription and pre-mRNA splicing through altering histone H4R3 and small nuclear ribonucleoprotein LSM4 methylation. *Plant Cell* 23, 396–411.

Zhang, H., Chen, X., Wang, C., Xu, Z., Wang, Y., Liu, X., Kang, Z., Ji, W., 2013. Long non-coding genes implicated in response to stripe rust pathogen stress in wheat (Triticum aestivum L.). *Mol. Biol. Rep.* 40, 6245–6253.

Zhang, B., Ticman, D.M., Jiao, C., Xu, Y., Chen, K., Fei, Z., Giovannoni, J.J., Klee, H.J., 2016a. Chilling-induced tomato flavor loss is associated with altered volatile synthesis and transient changes in DNA methylation. *Proc. Natl. Acad. Sci. USA* 113(44), 12580–12585.

Zhang, Z., Liu, X., Guo, X., Wang, X.J., Zhang, X., 2016b. *Arabidopsis* AGO3 predominantly recruits 24-nt small RNAs to regulate epigenetic silencing. *Nat. Plants* 2, 16049.

Zhang, H., Lang, Z., Zhu, J.K., 2018. Dynamics and function of DNA methylation in plants. *Nat. Rev. Mol. Cell Biol.* 19, 489–506.

Zhang, H., Zhu, J., Gong, Z., Zhu, J.K., 2022. Abiotic stress responses in plants. *Nat. Rev. Genet.* 23, 104–119.

Zheng, Y., Ding, Y., Sun, X., Xie, S., Wang, D., Liu, X., Su, L., Wei, W., Pan, L., Zhou, D.X., 2016. Histone deacetylase HDA9 negatively regulates salt and drought stress responsiveness in *Arabidopsis*. *J. Exp. Bot.* 67, 1703–1713.

Zhou, C., Zhang, L., Duan, J., Miki, B., Wu, K., 2005. Histone Deacetylase19 is involved in jasmonic acid and ethylene signaling of pathogen response in *Arabidopsis*. *Plant Cell* 17, 1196–1204.

Zong, W., Zhong, X., You, J., Xiong, L., 2013. Genome-wide profiling of histone H3K4-tri-methylation and gene expression in rice under drought stress. *Plant Mol. Biol.* 81, 175–188.

2 Recent Advances in Biotechnological Interventions for Nutritional Enhancement of Linseed (*Linum usitatissimum* L.)

Ramanuj Maurya
Department of Botany, University of Lucknow, Lucknow, Uttar Pradesh, India

Pawan Kumar Pal
ICAR-Indian Institute of Pulses Research, Kanpur, Uttar Pradesh, India

CONTENTS

DOI: 10.1201/9781003324706-3

2.1 INTRODUCTION

Linseed or flax (*Linum usitatissimum* L.) is a diploid (2n = 30) having small genome size of ~370 Mb and self-pollinated annual plant (Chandrawati et al. 2014). It has been cultivated mostly for the seed oil, stem fibers and medically related compounds since ancient times (Chandrawati et al. 2014). The plant has been utilized by humans for some 30,000 years since the Paleolithic era, was later brought into domestication around 7,000 years ago in the Near East and then spread to the Fertile Crescent where it was grown for its seed oil and stem fibers. The center of origin of cultivated flax is believed to be the Middle East, although secondary diversity centers were identified in the Mediterranean basin, Ethiopia, Central Asia and India (Vavilov 1926; Zohary and Hopf 2000). Globally, the important flaxseed–linseed cultivating and producing countries include Canada, India, China, the United Kingdom, Ethiopia and the United States of America. In India, it is mainly grown as an industrial oil seed crop in marginal soils and rainfed conditions. Linseed is an important industrial crop plant as its oil with high linolenic acid content (45–65%) is used for manufacturing rapidly drying paints, stains, inks, varnishes and the polymer linoleum (Rowland 1998). Besides its oil, the linseed fibers (phloem fiber) are used by industries for producing high-quality linen fabrics, pulp, biofuels (Diederichsen and Ulrich 2009), raw materials of thermal insulations (Kymalainen and Sjoberg 2008) and bioplastics (Kwiatkowska et al. 2009). Dietary consumption of flaxseed prevents coronary heart disease due to high mucilage and lignans content (Diederichsen 2008; Bassett et al. 2009). Linseed oil is valuable for chemical and food industries due to high linolenic acid (Kumar et. al. 2018). According to Westcott and Muir (2003), linseed contains mainly five types of fatty acids, namely, palmitic, stearic, oleic, linoleic and linolenic acid. High percentage of linolenic acid causes oxidative instability and is used as drying oil in industries such as soaps, varnishes, paints and inks (Juita et al. 2012). However, the high level of linolenic acid in the oil causes it to be unsuitable for use in edible products because of undesirable odors and flavor that result from the auto-oxidation of this unsaturated fatty acid (Chandrawati et al. 2014, Green 1986a, 1986b). Efforts have been made to develop edible purpose linseed by reducing the level of linolenic acid content to be 5% and several varieties with low linolenic acid have been developed and released for commercial cultivation in Canada and Australia (Dribnenki et al. 1999, 2004; Green and Marshall 1981; Rowland 1991). Recently, efforts are being made to breed and develop dual-purpose linseed varieties which could serve both high-quality fiber and oil. The availability and knowledge about the extent of genetic diversity of genetic resource material play an important role in identifying parental lines and developing new varieties with desirable traits. Similarly, the first high LIN flax, NuLin™ 50 variety was developed and registered in Canada (www.viterra.ca). Solin variety developed for high palmitic acid and variegated seed color has a loss of function mutation in fatty acid desaturase 3 and is low in omega-3 (2–3%) and high in omega-6 (70%). Whole linseed plant part is utilized commercially, either directly or after bio-processing.

2.2 CULTIVATION

In 2019, linseed occupied an area of 32.63 lakh per ha yielding 31.82 lakh tons with an average productivity of 975 kg per ha in the world (FAO STAT 2018). In 2017, of the 2.65 million tonnes of linseed produced, Canada (34%) was the highest producer, followed by the Russian Federation (15%) and China (13%) (FAO STAT 2017). India is the sixth largest producer in the world with contribution to global linseed area and production of 13% and 5.5%, respectively (Singh et al. 2019; Kaur et al. 2018). Recently, as per FAO STAT (2019), Kazakhstan has the largest area under linseed closely followed by Russian Federation and Canada (Table 2.1). Other countries with sizable linseed areas are China, India, the United States, Ethiopia, Afghanistan, Belarus, France, Ukraine, etc. The total world linseed production during 2019 was 3.41 million tons with an average productivity of 1156 kg/ha. In terms of production, Kazakhstan stands first followed by Russian

TABLE 2.1
Area, Production and Productivity of Linseed in the World

Country	Area (000 ha)	Production (000 t)	Productivity (kg/ha)
Afghanistan	38.3	55.0	1435
Argentina	9.0	10.0	1100
Armenia	0.2	0.1	805
Australia	7.0	6.0	857
Austria	1.4	1.7	1204
Bangladesh	5.3	3.6	689
Belarus	24.4	9.4	383
Bolivia	0.3	0.1	249
Brazil	4.0	3.8	953
Canada	339.3	486.1	1433
Chile	4.2	1.4	330
China	260.0	340.0	1308
China, mainland	260.0	340.0	1308
Czech Rep.	1.1	1.4	1250
Ecuador	0.2	0.1	328
Egypt	7.0	10.0	1429
Eritrea	0.5	0.2	326
Estonia	0.1	0.1	1000
Ethiopia	69.2	79.7	1153
Finland	0.5	0.4	800
France	21.8	45.5	2083
Greece	0.0	0.1	3500
Hungary	1.1	1.5	1383
India	172.7	99.1	574
Iraq	0.3	0.1	380
Kazakhstan	1245.0	1007.2	809
Kenya	1.0	1.0	1005
Kyrgyzstan	0.0	0.1	2407
Latvia	0.1	0.1	1000
Lithuania	0.4	0.3	750
Mexico	0.0	0.0	889
Nepal	12.2	12.5	1026
New Zealand	1.9	2.5	1319
Pakistan	2.6	1.9	726
Peru	1.2	1.1	932
Poland	3.7	4.8	1300
Romania	3.1	6.2	2013
Russian Federation	810.9	658.6	812
Slovakia	0.8	0.8	976
Spain	0.0	0.0	500
Sweden	2.2	4.3	1964
Switzerland	0.2	0.3	2134
Tunisia	1.0	3.1	3135
Ukraine	16.9	15.4	911
United Kingdom of Great Britain and Northern Ireland	15.0	27.0	1800
United States	129.1	162.4	1258

(Continued)

TABLE 2.1 (Continued)
Area, Production and Productivity of Linseed in the World

Country	Area (000 ha)	Production (000 t)	Productivity (kg/ha)
Uruguay	2.0	2.5	1250
Uzbekistan	3.0	1.0	333
World	**3480.2**	**3408.5**	**1156**

Source: FAO STAT 2019.

Federation and Canada. Highest productivity of linseed was in Greece (3500 kg/ha) followed by Tunisia (3135 kg/ha), Kyrgyzstan (2407 kg/ha), Switzerland (2134 kg/ha), France (2083 kg/ha) and Romania (2013 kg/ha).

In India, linseed is cultivated on 2–3 lakh hectares with Madhya Pradesh, Chhattisgarh, Uttar Pradesh, Maharashtra and Bihar being the major linseed-growing states. Madhya Pradesh is the leading producer followed by Jharkhand and Uttar Pradesh (Table 2.2). The total area, production and productivity during 2018–19 were 172.7 thousand hectares, 99.1 thousand tons and 574 kg/ha, respectively. There has been almost 49% reduction in linseed area in the country during the last 8 years with some reduction in the total production. However, the productivity of linseed has increased gradually from 403 kg/ha during 2010–11 to 574 kg/ha during 2018–19. Globally, the important flaxseed–linseed cultivating and producing countries include Canada, India, China, the United Kingdom, Ethiopia and the United States of America. World's largest producer of linseed is Canada followed by India.

TABLE 2.2
State-Wise Area, Production and Productivity of Linseed in India

State	Area (000 ha)	Production (000 t)	Productivity (kg/ha)
Assam	4.8	2.9	608
Bihar	9.7	8.2	849
Chhattisgarh	17.8	4.6	260
Himachal Pradesh	0.8	0.2	300
Jammu & Kashmir	0.2	0.2	1111
Jharkhand	39.8	19.5	491
Karnataka	1.9	0.4	217
Madhya Pradesh	38.0	27.1	712
Maharashtra	7.0	2.7	380
Meghalaya	0.1	0.1	596
Nagaland	5.9	4.8	814
Odisha	11.6	5.6	483
Rajasthan	3.4	3.4	1009
Tripura	0.1	0.0	774
Uttar Pradesh	28.0	17.7	631
West Bengal	3.9	1.8	446
All India	**172.7**	**99.1**	**574**

Source: FAO STAT 2018, 2019.

2.3 GENOMIC RESOURCES IN LINSEED

The genome resources for linseed or flax include an impressive array of molecular markers that are more reliable than morphological and biochemical markers that provide an opportunity to identify the genetic variation present in available genetic resources used to locate quantitative trait locus (QTL) and molecular-assisted selection (MAS) (Nadeem et al. 2018; Hoque et al. 2020).

Diversity analysis is very useful component for management and utilization of genetic repertoire. Flax has a small genome (comprising of about 30% of highly repeated tandemly arrayed sequences) packed into 15 pairs of similar-sized small chromosomes (Schweitzer 1979). The genes for the large ribosomal RNAs are localized and reside on a single chromosome while the 5S ribosomal RNA genes seem to be distributed among most members of the chromosome complement (Creissen and Cullis 1987; Schneeberger et al. 1989). Correct identification of genotypes is an essential and important step for the final utilization of cultivars in crop production schemes (UPOV 1991). Following are the major applications of genomics especially in the form of molecular markers toward genetic improvement of linseed.

2.3.1 Restriction Fragment Length Polymorphisms

Restriction fragment length polymorphisms (RFLP) marker is based on hybridization method and searches DNA sequence polymorphism. Initially, this method was used for human genome mapping (Botstein et al. 1980) but later RFLP technique adapt for plant diversity analysis (Kim and Ward 2000; Nkhoma et al. 2020). RFLP is a codominant marker with several advantages including being robust, reliable, detection of unlimited number of loci and transferable results across populations include (Govindaraj et al. 2015). However, the technique is costly, time-consuming, laborious, requiring large amounts of DNA and displaying inadequate polymorphism especially in closely related lines (Collard et al. 2005). Oh et al. (2000) were the first to construct linkage map having genetic length of around 1000 cM using RFLPs and random amplified polymorphic DNA (RAPDs) comprising of 15 linkage groups having 94 (13 RFLP, 80 RAPD and 1 sequence-tagged site) markers.

2.3.2 Random Amplified Polymorphic DNA

RAPD is an efficient and cost-effective molecular tool for estimating genetic diversity. The DNA fragments are amplified by Polymerase Chain Reaction (PCR) using short primers of random sequence. This technique is less time-consuming, easy to work, requires less quantity of template DNA and does not require sequence data for primer construction due to commercial availability of random primers. A series of genotrophs have been studied by PCR through random oligonucleotide primer in flax (Cullis et al. 1999). Fu et al. (2003) evaluated the efficacy of diverse bulking strategies in detection of RAPD variations and determining genetic relationships of five flax landraces. Bulking entities before and after isolation of DNA led to consistent RAPD variations and heterogeneity at around 5.6% of the loci scored among the bulked samples of the same taxa. Despite several limitations, bulking generated compatible genetic associations among the five accessions from its plant-by-plant (PBP) sampling. Ijaz et al. (2013) used 120 RAPD markers for molecular characterization in 40 linseed varieties/genotypes. The RAPDs revealed a high degree of polymorphism of 13%. Iqbal et al. (2014) used RAPD markers to evaluate the genetic variability in 20 linseed genotypes. Of the different primers used, 51 were polymorphic, which showed 64% of overall polymorphism that differed substantially between accessions.

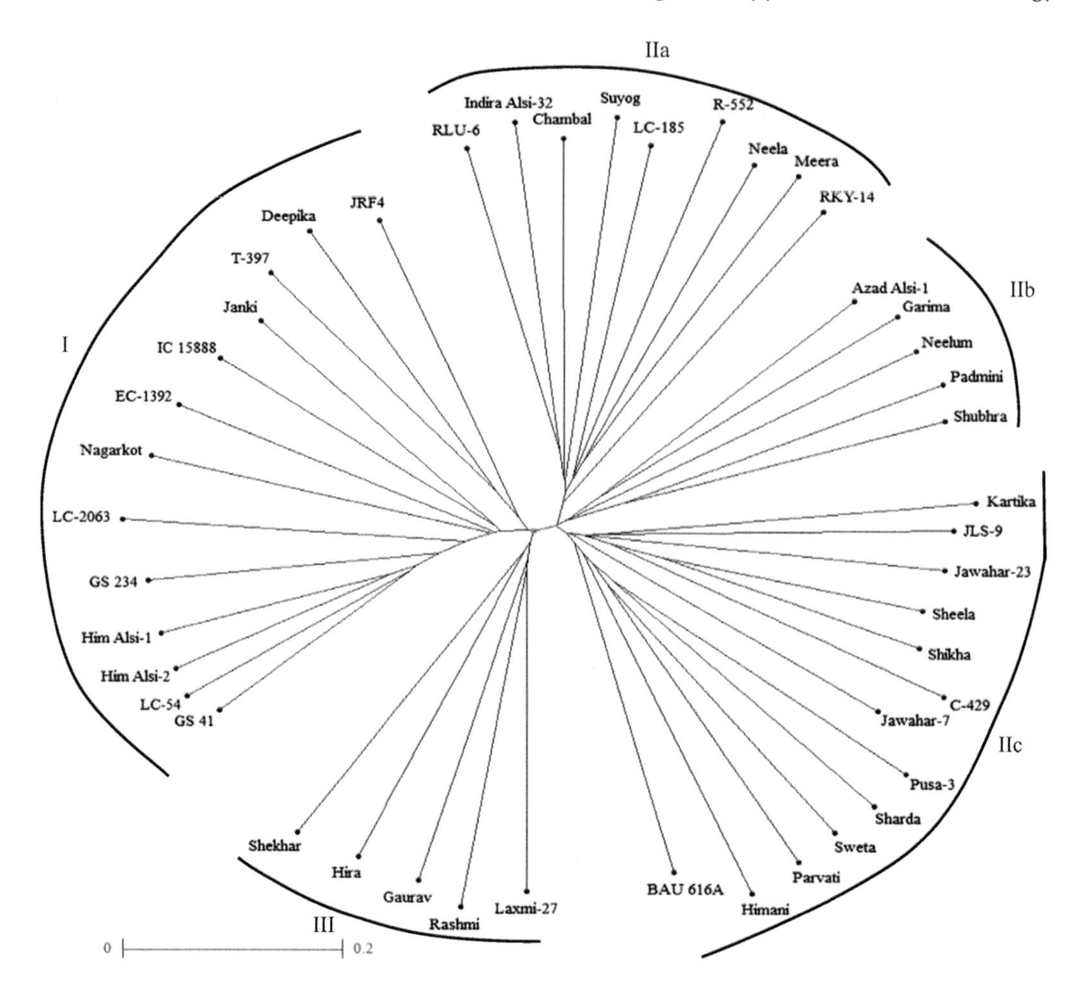

FIGURE 2.1 Genetic relatedness among 45 genotypes of linseed based on neighbor-joining clustering. The scale at the bottom is for NJ distances (Reprinted with permission from Chandrawati et al. 2014).

2.3.3 Amplified Fragment Length Polymorphism

Amplified fragment length polymorphism (AFLP) fingerprinting is a combination of RFLP and PCR-based methods that involves digestion of PCR amplified fragments using specific restriction enzymes and use of ligating primer recognition sequence viz. adapters (Vos et al. 1995). The PCR primers have a core sequence that is part of adapter and restriction enzyme-specific sequence with 1–5 selective nucleotides consequently resulting in variation in restriction sites. Chandrawati et al. (2014) used 45 Indian genotypes of linseed with 16 primers that produced 1129 polymorphic fragments. The neighbor-joining tree constructed using a calculated dissimilarity matrix classified all the genotypes into three clusters (Figure 2.1) with cluster II having the largest number of genotypes (27), followed by cluster I having 13 genotypes and cluster III with 5 genotypes having high oil content and low 1000 seed weight.

2.3.4 Simple Sequence Repeat

Simple sequence repeats (SSRs) or short tandem repeats (STR) are a ubiquitous class of tandemly repeated short motifs of 2–6 nucleotides that are widely distributed in the genome and display high variability even in organisms having little genetic variation (Bhargava and Fuentes 2010). These

TABLE 2.3
List of Linseed SSR Markers Available in Public Domain

SSR Source	Number of Polymorphic SSRs	Reference	Research Institute/Universities
Genomic	10	Wiesner et al. (2001)	Institute of Plant Molecular Biology, Czech Republic
Genomic	23	Roose-Amsaleg et al. (2006)	Institut Technique du Lin, France
EST	248	Cloutier et al. (2009)	Cereal Research Centre, Agriculture and Agri-Food Canada
EST	23	Soto-Cerda et al. (2011b)	Agri-aquaculture Nutritional Genomic Center, Chile
Genomics	60	Soto-Cerda et al. (2011a)	Agri-aquaculture Nutritional Genomic Center, Chile
Genomic	35	Deng et al. (2010)	Institute of Bast Fiber Crops, China
Genomics	38	Deng et al. (2011)	Institute of Bast Fiber Crops, China
Genomic	9	Kale et al. (2012b)	Plant molecular biology group, NCL Pune
EST	145	Cloutier et al. (2012)	Cereal Research Centre, Agriculture and Agri-Food Canada
BES	673	Cloutier et al. (2012)	Cereal Research Centre, Agriculture and Agri-Food Canada
Genomic	10	Singh et al. (2019)	CSIR-National Botanical Research Institute, Lucknow, India

are highly polymorphic, multi-allelic, frequently codominant and reproducible (Fuentes and Bhargava 2011). SSR markers are highly suitable for genetic diversity and linkage map analysis due to its reproducibility abundance and distribution nature characteristics (Cloutier et al. 2009; Soto-Cerda et al. 2011a, 2011b). As shown in Table 2.3, a number of SSR markers are available in linseed. Wu et al. (2017) screened 1574 microsatellites from flax attained through reduced representation genome sequencing (RRGS) for systematically identifying SSR markers. About 1720 SSR loci were identified that comprised nearly 20% of the unigenes in the flax genome. The trinucleotides were the most prominent (965), followed by dinucleotide (606) repeats and only 149 had other motifs. Among the dinucleotide motifs AT/TA motifs were the most represented dinucleotide while CTT/GAA motifs were the most abundant trinucleotide microsatellites. CG/GC motifs were completely absent (Figure 2.2). Cluster analysis carried out according to unweighted pair-group method using unweighted pair-group method with arithmetic mean (UPGMA) clustering procedure separated flax varieties into 2 groups viz. 25 fiber cultivars and 23 linseed cultivars, all corresponding to the fiber and oil categories (Figure 2.3).

2.3.5 Inter Simple Sequence Repeat

This technique is PCR based which involves the amplification of DNA segment present at an amplifiable distance between two identical microsatellite repeat regions oriented in opposite directions. The technique uses microsatellites, usually 16–25 bp long, as primers in a single primer PCR reaction targeting multiple genomic loci to amplify mainly the inter-SSR sequences of different sizes. The microsatellite repeats used as primers can be dinucleotide, trinucleotide, tetranucleotide or penta-nucleotide. The primers used can be either unanchored or more usually anchored at 3' or 5' end with 1–4 degenerate bases extended into the flanking sequences

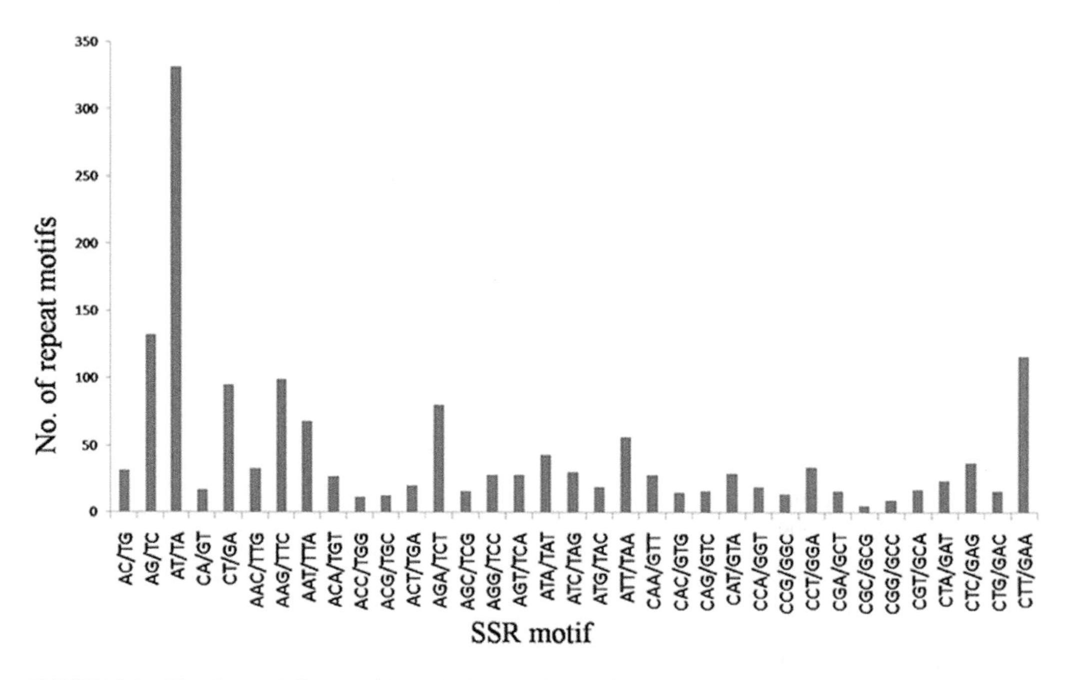

FIGURE 2.2 Numbers of dinucleotide and trinucleotide SSRs classified based on their motifs (Reprinted from Wu et al. 2017, open-access article distributed under the terms of the Creative Commons Attribution License (CC BY).

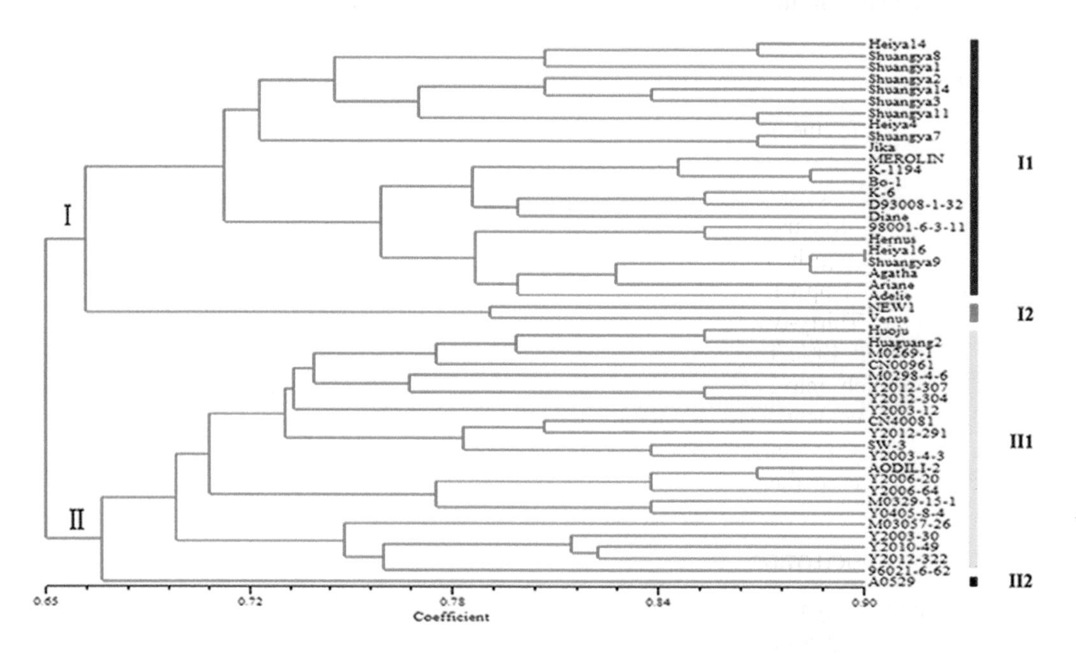

FIGURE 2.3 Genetic diversity of flax accessions based on SSR markers (Reprinted from Wu et al. 2017, open-access article distributed under the terms of the Creative Commons Attribution License (CC BY).

(Zietkiewicz et al. 1994). This technique ensures greater reproducibility and high percentage of polymorphism in the genotypes studied (Reddy et al. 2002).

Wiesnerová and Wiesner (2004) used 9 anchored inter simple sequence repeat (ISSR) primers for fingerprinting of 53 flax cultivars and obtained 62 scorable bands of which 45 were polymorphic. The UPGMA algorithm efficiently separated the germplasm into four groups and eight subgroups. In linseed, an important pathogen *Alternaria alternate,* which cause blight disease, results in 90% yield losses. The use of ISSR markers has revealed high genetic diversity among *A. alternata* isolates of Indian origin (Kale 2012a).

2.3.6 Cleaved Amplified Polymorphic Sequence

Cleaved amplified polymorphic sequences (CAPS) are DNA fragments amplified by PCR using specific 20–25 bp primers, followed by digestion of the PCR products with a restriction enzyme. Subsequently, length polymorphisms resulting from variation in the occurrence of restriction sites are identified by gel electrophoresis of the digested products (Konieczny and Ausubel 1993; Wang et al. 2022). Hausner et al. (1999) developed a CAPS marker for flax rust resistance gene M3 in linseed.

2.3.7 Sequence Characterized Amplified Region

Sequence characterized amplified region (SCAR), a PCR-based monolocus codominant marker, uses two specific primers designed from nucleotide sequences established in cloned RAPD fragments linked to a trait of interest (Kiran et al. 2010). In this technique, a genomic DNA fragment is identified by PCR amplification using a pair of specific oligonucleotide primer. SCAR markers are known for their ease, reliability and reproducibility. Bo et al. (2002) developed a SCAR marker for rust resistance gene M4 in flax.

2.3.8 InDels

InDel, a term used for the insertion or deletion of bases, is structural variations dispersed profusely throughout the genome that arise as a result of polymerase slippage, transposons and unequal crossing-over (Mullaney et al. 2010; Rockah-Shmuel et al. 2013). InDels often lead to the gain/loss of function in the organism. Jiang et al. (2022) identified 17,110 InDel markers using whole-genome re-sequencing data of two accessions with the flax reference genome. InDels ranged in length from 1 to 277 bp with single-nucleotide being the most common, followed by dinucleotide and trinucleotide. Thirty-two pairs of polymorphic primers were observed along with polymorphism rate of about 40%. The principal component analysis (PCA) efficiently segregated the flax accessions separately into fiber and oil seed types. Two InDels were found to be associated with traits related to oil content, while two candidate genes linked to these InDels were thought to play a major role in flax lipid metabolism. This pioneer research effectively pointed out toward the importance of InDel markers in germplasm identification, genetic diversity and MAS.

2.3.9 Single-Nucleotide Polymorphism

Single-nucleotide polymorphisms (SNPs) are highly abundant and distributed throughout the genome in various species including plants. They are an attractive tool for mapping, MAS and map-based cloning. Yi et al. (2017) applied specific length amplified fragment sequencing (SLAF-seq) to develop SNP markers in an F2 population and for constructing a high-density genetic map for *L. usitatissimum.* A total of 196.29 million paired-end reads each of 125 bp × 2 in length were obtained from which 389,288 high-quality polymorphic SLAFs were extracted. A total of 260,380 polymorphic SNPs were developed from the SLAFs. The genetic map constructed by Yi et al. (2017) was

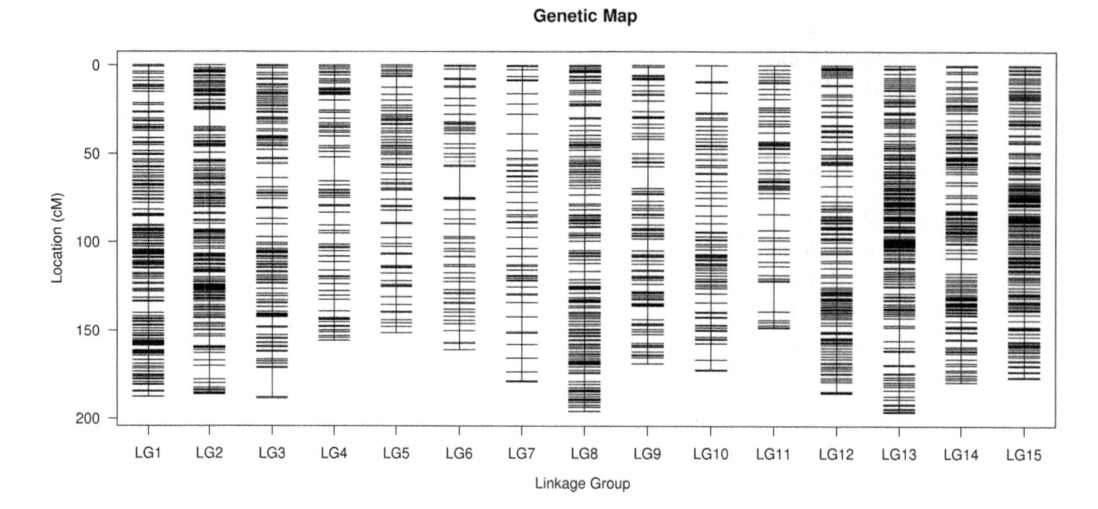

FIGURE 2.4 High-density genetic linkage map of flax (Reprinted from Yi et al. 2017, an open-access article distributed under the terms of the Creative Commons Attribution License (CC BY).

2,632.94 cM in length and included 4,145 SNP markers distributed on different linkage groups (Figure 2.4).

2.4 GENETIC MAPS

Genomics resources such as bacterial artificial chromosome (BAC) libraries, genetic and physical maps, QTL analysis, BAC-end sequences (BESs) and whole-genome sequences are emerging as promising techniques to enhance breeding processes. The Total Utilization Flax GENomics project was conceptualized to develop a comprehensive knowledge of its unique genome with specific goals in applied genomics aiming at the improvement of flax as a total utilization crop. However, a recent estimate of the size of the CDC Bethune flax genome of 0.38 pg/C would translate into only 370 Mb (Michael Deyholos and David Galbraith, personal communication). A physical map represents a genomic region (single locus) or an entire genome, constructed by set(s) of overlapping large-insert clones in which the distances are measured in base pairs. The first genome-wide physical map of flax constructed with BAC clones provides a framework for accessing target loci with economic importance for marker development and positional cloning (Ragupathy et al. 2011). Analysis of the BES has provided insights into the uniqueness of the flax genome. Compared to other plant genomes, the proportion of rDNA (~13.8%) was found to be very high, whereas the proportion of known transposable elements was comparatively less (6.1%). The authors were of the view that the 4,064 genomic SSRs identified from BES would be valuable in saturating existing linkage maps and for anchoring physical and genetic maps (Ragupathy et al. 2011). Genome-wide SNP discovery was successfully accomplished in *L. usitatissimum* using Next Generation Sequencing (NGS) of reduced representation libraries coupled with use of Illumina sequencing platform was carried out by Kumar et al. (2012). About 55,000 SNPs were discovered of which more than 80% were identified in a single genotype out of eight sequenced flax genotypes. Genetic maps are of prime importance since they facilitate identification of genomic regions linked to quantitative traits, allow mapping of genes on chromosomes and provide valuable insights on recombination rates and gene rearrangements. Spielmeyer et al. (1998) constructed a linkage map utilizing a mapping population from doubled-haploid (DH) flax lines. AFLP generated 213 marker loci comprising 18 linkage groups. Cloutier et al. (2011) constructed the first SSR-based linkage map using 114 expressed sequence tag (EST)-derived SSR markers, 5 SNP markers, 5 genes and a single phenotypic trait, utilizing a DH population of 78 individuals.

2.4.1 BAC Libraries and Physical Map

BAC libraries, which are large-insert genomic DNA libraries, provide a platform for physical mapping, map-based cloning of desired gene, genome sequencing and analysis of gene structure. In the case of linseed, the first genome-wide physical map of flax and the generation and analysis of BESs from 43,776 clones provided initial insights into the plant genome (Ragupathy et al. 2011). It provided a framework to access the target locus with economic importance of marker development and positional cloning of traits. The SSR identified from BES helped in anchored physical and genetic map, and the physical map from BAC clones served as scaffold to build and also provide whole-genome shotgun assembly. Cloutier et al. (2012) carried out comparative analyses of 1,506 putative SSRs of which 1,164 were derived from BESs and 342 from ESTs on 16 flax accessions. Trinucleotides were the most abundant nucleotide motifs while dinucleotides were the most polymorphic.

2.4.2 QTL Mapping

An effective approach for studying complex and polygenic forms of trait is known as QTL mapping (Tanksley 1993; Kumar et al. 2017). With QTL mapping, the roles of specific loci in genetically complex traits can be described. In linseed, QTLs have been identified associated with Fusarium wilt resistance and fatty acid composition (Table 2.4).

2.4.3 Transcriptomics

Transcriptome resources are important alternative sources to genome sequence in species where genome sequencing has not been done. Transcriptomic resources like ESTs are extremely useful in understanding genomic dynamics, development of gene-based marker and maps and for transcript profiling to locate the genes involved in the expression of desired traits. Microarray and NGS technologies provide transcriptional profiling, transcript gene, gene discovery, transcript expression and polymorphism detection. NGS technology has been applied for functional genomics analysis in linseed for the development of genomic SSR markers (Kale et al. 2012b). A total of 36,332 reads were obtained having an average size of 2 bp, which represented about 0.17% of the size of the plant genome.

TABLE 2.4
The Use of Different Mapping Populations for the Linkage/QTL Mapping in Linseed

Population	Mapping Population	Marker Loci Mapped	References
CI1303 X Stormont Cirrus	F_2	13 RFLP, 80 RAPD	Cullis et al. (1999)
(CRZY8/RA91) X Glenelg	DH	213 AFLP	Spielmeyer (1998)
SP2047 X UGG5–5	DH	113 SSR	Cloutier et al. (2011)
CDC Bethune x G1186/94	RIL	52 SSR	Sudarshan et al. (2017)
JRF-4 X Chambal	F_2	100 SSR	Singh et al. (2019)

2.4.4 Whole-Genome Sequencing

Genome sequencing for a number of crop species rice (2005), maize (2009), sorghum (2009) and soybean (2010) has been successfully accomplished in the new millennium. The flax variety CDC Bethune was sequenced and the genome size was estimated to be about ~373Mb based on flow cytometry. Wang et al. (2012) assembled the nuclear genome of flax using short shotgun sequence reads and predicted 43,384 protein-coding genes. An alignment of the CDS regions of the predicted genes with flax ESTs showed that about 93% of known ESTs aligned with predicted CDS sequences with high stringency.

2.5 APPLICATION OF GENOMIC RESOURCES TOWARD GENETIC IMPROVEMENT OF LINSEED

Genomic resources are useful for basic and applied research for flax improvement. Marker development, preparation of gene-based genetic map and ESTs are useful for transcript profiling to identify the desired gene for desired trait, as well as development of microarray which helps to study differential expression of different genes at different stages.

2.5.1 Desired Trait Mapping

Desired trait mapping is very useful to locate the tightly linked markers to desired trait through linkage/association mapping. Genomic resources are useful for basic and applied research for flax improvement.

2.5.2 Marker-Assisted Selection

In conventional breeding, the introgression of recessive gene and pyramiding of multiple genes is a cumbersome task. However, marker-assisted selection (MAS) overcomes such problems of multiple gene pyramiding as well as faster genome recovery of recurrent parent (Figure 2.5). In the case of linseed, some studies have been initiated to use the molecular breeding programs.

2.5.3 Mutation Breeding

Mutagenesis, a method of creating novel varieties by causing variations at the gene level in elite germplasm, is a powerful tool for crop improvement (Riaz and Gul 2015; Chaudhary et al. 2019). Various reports are available regarding the application of mutagenesis in linseed for improvement of oil quality (Rowland 1991). Recently, genome sequence of linseed has been published which allowed functional genomic studies using suitable reverse genetics tools. Chemical mutagens, namely, ethyl methane sulfonate (EMS) have been frequently used to create point mutations throughout the flax genome which are identified in target genes using TILLING. At present, two TILLING platforms of linseed have been established and hundreds of mutations in targeted genes have been identified (Chantreau et al. 2013).

2.6 FUTURE PROSPECTS

One of the major challenging tasks is that still low level of genetic diversity has been explored inspite of developing scores of markers for linseed. However, cost-effective SNP genotyping platform and SSR markers are still considered the choice markers for genetic and breeding applications in linseed. Although some genetic maps have been developed for linseed, genetic mapping position is available for only few hundred SSR markers. Another challenge to the linseed community is integrating as many markers as possible on genetic maps; even physical map has

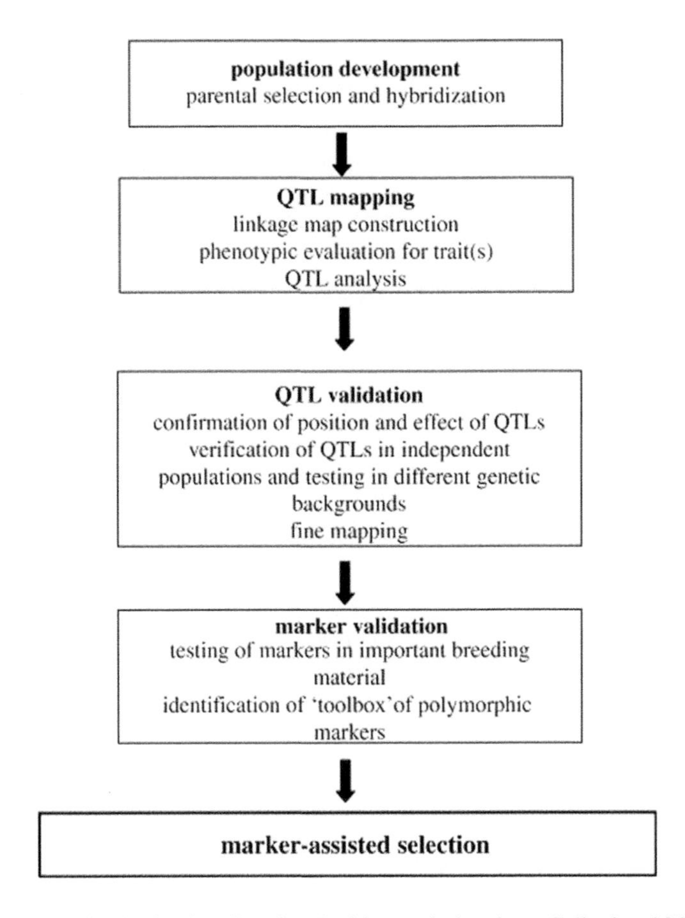

FIGURE 2.5 Marker assisted selection (Reprinted with permission from Collard and Mackill 2008).

been developed for linseed genome as a result, it would be possible to integrate the genetic and physical maps that will facilitate gene cloning as well as molecular breeding in an efficient manner. In terms of linseed breeding, trait mapping and molecular breeding for resistance of disease (rust resistance) have already been initiated.

In the current scenario utilizing molecular breeding to decrease linolenic acid to improve the quality of edible oil needs attention of the linseed community through increase of oleic: linolenic acid ratio which can increase the shelf life of linseed oil and its products along with providing numerous health benefits. Due to rapid advances in sequencing and bioinformatics tools, genome sequence for linseed genome will be soon available. The genome sequences provide opportunities to correlate the phenotype with gene. To summarize, linseed genomics and molecular breeding seem to be promising areas that can lead to improvement in *L. usitatissimum* for yield and quality.

REFERENCES

Bassett, C.M., Rodriguez-Leyva, D., Pierce, G.N., 2009. Experimental and clinical research findings on the cardiovascular benefits of consuming flaxseed. *Appl. Physiol. Nutr. Metab.* 34, 965–974.

Bhargava, A., Fuentes, F.F., 2010. Mutational dynamics of microsatellites. *Mol. Biotechnol.* 44, 250–266.

Bo, T.Y., Ye, H.Z., Wang, S.Q., Yang, J.C., Li, X.B., Zhai, W.X., 2002. Specific molecular markers of the rust resistance gene M4 in flax. *Yi Chuan Xue Bao (Acta Genetica Sinica)*, 29, 922–927.

Botstein, D., White, R.L., Skalnick, M.H., Davies, R.W., 1980. Construction of a genetic linkage map in man using restriction fragment length polymorphism. *Am. J. Hum. Genet.* 32, 314–331.

Chandrawati, M.R., Singh, P.K., Ranade, S.A., Yadav, H.K., 2014. Diversity analysis in Indian genotypes of linseed (*Linum usitatissimum* L.) using AFLP markers. *Gene* 549, 171–178.

Chantreau, M., Grec, S., Gutierrez, L., Dalmais, M., Pineau, C., Demailly, H., Paysant-Leroux, C., Tavernier, R., Trouvé, J.P., Chatterjee, M., Guillot, X., Brunaud, V., Chabbert, B., van Wuytswinkel, O., Bendahmane, A., Thomasset, B., Hawkins, S., 2013. PT-Flax (phenotyping and TILLinG of flax): development of a flax (*Linum usitatissimum* L.) mutant population and TILLinG platform for forward and reverse genetics. *BMC Plant Biol.* 13, 159.

Chaudhary, J., Deshmukh, R., Sonah, H., 2019. Mutagenesis approaches and their role in crop improvement. *Plants* 8, 467.

Cloutier, S., Niu, Z., Datla, R., Duguid, S., 2009. Development and analysis of EST-SSRs for flax (*Linum usitatissimum* L.). *Theor. Appl. Genet.* 119, 53–63.

Cloutier, S., Ragupathy, R., Niu, Z., Duguid, S., 2011. SSR-based linkage map of flax (*Linum usitatissimum* L.) and mapping of QTLs underlying fatty acid composition traits. *Mol. Breed.* 28, 437–451.

Cloutier, S., Miranda, E., Ward, K., Radovanovic, N., Reimer, E., Walichnowski, A., Datla, R., Rowland, G., Duguid, S., Ragupathy, R. 2012. Simple sequence repeat marker development from bacterial artificial chromosome end sequences and expressed sequence tags of flax (*Linum usitatissimum* L.). *Theor. Appl. Genet.* 125, 685–694.

Collard, B.C.Y., Jahufer, M.Z.Z., Brouwer, J.B., Pang, E.C.K., 2005. An introduction to markers, quantitative trait loci (QTL) mapping and marker-assisted selection for crop improvement: the basic concepts. *Euphytica* 142, 169–196.

Collard, B.C.Y., Mackill, D.J., 2008. Marker-assisted selection: an approach for precision plant breeding in the twenty-first century. *Philos. Trans. R. Soc. Lond. B Biol. Sci.* 363, 557–572.

Creissen, G.P., Cullis, C.A., 1987. Genome organization and variation in higher plants. *Ann. Bot.* 60, 103–113.

Cullis, C.A., Swami, S., Song, Y., 1999. RAPD polymorphisms detected among the flax genotrophs. *Plant Mol. Biol.* 41, 795–800.

Deng, X., Long, S., He, D., Li, X., Wang, Y., Liu, J., Chen, X., 2010. Development and characterization of polymorphic microsatellite markers in *Linum usitatissimum*. *J. Plant Res.* 123, 119–123.

Deng, X., Long, S., He, D., Li, X., Wang, Y., Hao, D., Qiu, C., Chen, X., 2011. Isolation and characterization of polymorphic microsatellite markers from flax (*Linum usitatissimum* L.). *Afr. J. Biotechnol.* 10, 734–739.

Diederichsen, A., Fu, F.B., 2008. Flax genetic diversity as the raw material for future success. International Conference on Flax and Other Bast Plants. Saskatoon, SK, Canada, July 21-23, 2008. pp. 270–279.

Diederichsen, A., Ulrich, A., 2009. Variability in stem fibre content and its association with other characteristics in 1177 flax (Linum usitatissimum L.) genebank accessions. *Ind. Crop Prod.* 30, 33–39.

Dribnenki, J.C.P., Green, A.G., 1995. Linola '947' low linolenic acid flax. *Can. J. Plant Sci.* 75, 201–202.

Dribnenki J.C.P., McEachern, S.F., Chen, Y., Green, A.G., Rashid, K.Y., 2004. 2090 low linolenic acid flax. *Can. J. Plant Sci.* 84, 797–799.

Dribnenki, J.C.P., McEachern, S.F., Green, A.G., Kenaschuk, E.O., Rashid, K.Y., 1999. LinolaTM '1084' low linolenic acid flax. *Can. J. Plant Sci.* 79, 607–609.

FAO STAT. 2017. FAO Statistical Data. http://faostat.fao.org

FAO STAT. 2018. FAO Statistical Data. http://faostat.fao.org

FAO STAT. 2019. FAO Statistical Data. http://faostat.fao.org

Fu, Y.B., Guerin, S., Peterson, G.W., Carlson, J.E., Richards, K.W., 2003. Assessment of bulking strategies for RAPD analyses of flax germplasm. *Genet. Res. Crop Evol.* 50, 743–746.

Fuentes, F.F., Bhargava, A., 2011. Morphological analysis of quinoa germplasm grown under lowland desert conditions. *J. Agron. Crop Sci.* 197, 124–134.

Govindaraj, M., Vetriventhan, M., Srinivasan, M., 2015. Importance of genetic diversity assessment in crop plants and its recent advances: an overview of its analytical perspectives. *Genet. Res. Intern.* 2015, 431487.

Green, A.G., 1986a. Genetic control of polyunsaturated fatty acid biosynthesis in flax (*Linum usitatissimum* L.) seed oil. *Theor. Appl. Genet.* 72, 654–661.

Green, A.G., 1986b. A mutant genotype of flax (*Linum usitatissimum* L.) containing very low levels of linolenic acid in its seed oil. *Can. J. Plant Sci.* 66, 499–503.

Green, A.G., Marshall, D.R., 1981. Variation for oil quantity and quality in linseed (Linum usitatissimum). *Aust. J. Agr. Res.* 32, 599–607.

Hausner, G., Rashid, K.Y., Kenaschuk, E.O., Procunier, J.D., 1999. The identification of a cleaved amplified polymorphic sequence (CAPS) marker for the flax rust resistance gene M3. *Plant Pathol.* 21, 187–192.

Hoque, A., Fiedler, J.D., Rahman, M. 2020. Genetic diversity analysis of a flax (*Linum usitatissimum* L.) global collection. *BMC Genom.* 21, 557.

Ijaz, A., Shahbaz, A., Ullah, I., Ali, S., Shaheen, T., Rehman, M., Ijaz, U., Ullah S., 2013. Molecular characterization of linseed germplasm using RAPD DNA fingerprinting markers. *Am.-Eurasian J. Agric. Environ. Sci.*, 13, 1266–1274.

Iqbal, A., Nawaz, S., Babar, A.D., Bukhari, S.A., Amjad, M.S., Khan, M.A., Khan, S.A., Ahmed, N. 2014. Estimation of genetic variability in linseed (*Linum usitatissimum* L.) using molecular marker: Molecular characterization of linseed genotypes. *Intern. J. Agric. Sci.* 4, 151–156.

Jiang, H., Pan, G., Liu, T., Chang, L., Huang, S., Tang, H., Guo, Y., Wu, Y., Tao, J., Chen, A., 2022. Development and application of novel InDel markers in flax (*Linum usitatissimum* L.) through whole-genome re-sequencing. *Genet. Resour. Crop Evol.* 69, 1471–1483.

Juita Dlugogorski, B.Z., Kennedy, E.M., Mackie, J.C., 2012. Low temperature oxidation of linseed oil: a review. *Fire Sci. Rev.* 1, 3.

Kale, S.M., Pardeshi, V.C., Gurgar, G.S., Gupta, V.S., Gohokar, R.T., Ghorpade, P.B., Kadoo, N.Y., 2012a Inter-simple sequence repeat markers reveal high genetic diversity among *Alternaria alternata* isolates of Indian origin. *J. Mycol. Plant. Pathol.* 42, 194–200.

Kale, S.M., Pardeshi, V.C., Prakash, Y.K., Ghorpade, P.B., Jana, M.M., Gupta, V.S., 2012b. Development of genomic simple sequence repeat markers for linseed using next-generation sequencing. *Mol. Breed.* 30, 597–606.

Kaur, V., Kumar, S., Yadav, R., Wankhede, D.P., Arvind, J., Radhamani, J., Rana, J.C., Kumar, A., 2018. Analysis of genetic diversity in Indian and exotic linseed germplasm and identification of trait specific superior accessions. *J. Environ. Biol.* 39, 702–709.

Kim, H., Ward, R., 2000. Patterns of RFLP-based genetic diversity in germplasm pools of common wheat with different geographical or breeding program origins. *Euphytica* 115, 197–208.

Kiran, U., Khan, S., Mirza, K.J., Ram, M., Abdin, M.Z., 2010. SCAR markers: a potential tool for authentication of herbal drugs. *Fitoterapia* 81(8), 969–976.

Konieczny, A., Ausubel, F.M., 1993. Procedure for mapping *Arabidopsis* mutations using co-dominant ecotype-specific PCR-based markers. *Plant J.* 4, 403–410.

Kumar, S., Hash, C.T., Nepolean, T., Satyavathi, C.T., Singh, G., Mahendrakar, M.D., Yadav, R.S., Srivastava, R.K., 2017. Mapping QTLs controlling flowering time and important agronomic traits in pearl millet. *Front. Plant Sci.* 8, 1731.

Kumar, S., Singh, J.K., Vishwakarma, A., 2018. Importance of linseed crops in agricultural sustainability. *Int. JCurrMicrobiol. App. Sci.* 7(12): 1198–1207.

Kumar, S., You, F.M., Cloutier, S. 2012. Genome wide SNP discovery in flax through next generation sequencing of reduced representation libraries. *BMC Genom.* 13, 684.

Kwiatkowska, M.W., Telichowska, K.S., Dymińska, L., Maczka, M., Hanuza, J., Szopa, J., 2009. Biochemical, mechanical, and spectroscopic analyses of genetically engineered flax fibers producing bioplastic (poly-beta-hydroxybutyrate). *Biotechnol. Prog.* 25, 1489–1498.

Kymalainen, H.R., Sjoberg, A.M., 2008. Flax and hemp fibres as raw materials for thermalinsulations. *Build. Environ.* 43, 1261–1269.

Mullaney, J.M., Mills, R.E., Pittard, W.S., Devine, S.E., 2010. Small insertions and deletions (INDELs) in human genomes. *Human Mol. Genet.* 19, R131–R136.

Nadeem, M.A., Nawaz, M.A., Shahid, M.Q., Doğan, Y., Comertpay, G., Yıldız, M., Hatipoğlu, R., Ahmad, F., Alsaleh, A., Labhane, N., Özkan, H., Chung, G., Baloch, F.S., 2018. DNA molecular markers in plant breeding: current status and recent advancements in genomic selection and genome editing. *Biotechnol. Biotechnological Equip.* 32, 261–285.

Nkhoma, N., Shimelis, H., Laing, M.D., Shayanowako, A., Mathew, I. 2020. Assessing the genetic diversity of cowpea [*Vigna unguiculata* (L.) Walp.] germplasm collections using phenotypic traits and SNP markers. *BMC Genet.* 21, 110.

Oh, T., Gorman, M., Cullis, C., 2000. RFLP and RAPD mapping in flax (*Linum usitatissimum*). *Theor. Appl. Genet.* 101, 590–593.

Ragupathy, R., Rathinavelu, R., Cloutier, S., 2011. Physical mapping and BAC-end sequence analysis provide initial insights into the flax (*Linum usitatissimum* L.) genome. *BMC Genom.* 12, 217.

Reddy, M.P., Sarla, N., Siddiq, E.A., 2002. Inter simple sequence repeat (ISSR) polymorphism and its application in plant breeding. *Euphytica* 128, 9–17.

Riaz, A., Gul, A., 2015. Plant mutagenesis and crop improvement. In: Hakeem, K. (Eds.) *Crop Production and Global Environmental Issues*. Springer, Cham. pp. 181–209.

Rockah-Shmuel, L., Tóth-Petróczy, Á., Sela, A., Wurtzel, O., Sorek, R., Tawfik, D.S., 2013. Correlated occurrence and bypass of frame-shifting insertion-deletions (InDels) to give functional proteins. *PLoS Genet.* 9, e1003882.

Roose-Amsaleg, C., Cariou-Pham, E., Vautrin, D., Tavernier, R., Solignac, M., 2006. Polymorphic microsatellite loci in *Linum usitatissimum*. *Mol. Ecol. Notes* 6, 796–799.

Rowland, G.G., 1991. An EMS-induced low-linolenic-acid mutant in McGregor flax (*Linum usitatissimum L.*). *Can. J. Plant Sci.* 71, 393–396.

Rowland, G.G., 1998. *Growing Flax: Production, Management and Diagnostic Guide*. Flax Council of Canada and Saskatchewan Flax Development Commission.

Schneeberger, R.G., Creissen, G.P., Cullis, C.A., 1989. Chromosomal and molecular analysis of 5S RNA gene organization in flax, *Linum usitatissimum*. *Gene* 83, 75–84.

Schweitzer, H.-J., 1979. Die Zosterophyllaceae des rheinischen Unterdevons. *Bonner paläobotanische Mitteilungen*. 3, 1–32.

Singh, N., Kumar, R., Kumar, S., Singh, P.K., Singh, B., Kumar, U., Khatoon, B., Yadav, H.K., 2019. Bulk segregants analysis identifies SSR markers associated with *Alternaria* blight resistance in linseed (*Linum usitatissimum* L.). *J. Environ. Biol.* 40, 1137–1144.

Soto-Cerda, B.J., Carrasco, R.A., Aravena, G.A., Urbina, H.A., Navarro, C.S., 2011a. Identifying novel polymorphic microsatellites from cultivated flax (*Linum usitatissimum* L.) following data mining. *Plant Mol. Biol. Rep.* 29, 753–759.

Soto-Cerda, B.J., Saavdra, H.U., Navarro, C.N., Mora Ortega, P.M., 2011b. Characterization of novel genic SSR markers in *Linum usitatissimum* (L.) and their transferability across eleven *Linum* species. *Electron. J. Biotechnol.* 14, 6.

Spielmeyer, W., Green, A.G., Bittisnich, D., Mendham, N., Lagudah, E.S., 1998. Identification of quantitative trait loci contributing to Fusarium wilt resistance on an AFLP linkage map of flax. *Theor. Appl. Genet.* 97, 633–641.

Sudarshan, G.P., Kulkarni, M., Akhov, L., Ashe, P., Shaterian, H., Cloutier, S., Rowland, G., Wei, Y., Selvaraj, G., 2017. QTL mapping and molecular characterization of the classical D locus controlling seed and flower color in *Linum usitatissimum* (flax). *Sci. Rep.* 7, 15751.

Tanksley, S.D., 1993. Mapping polygenes. *Annu. Rev. Genet.* 27, 205–233.

UPOV. 1991. International Convention for the Protection of New Varieties of Plants of December 2, 1961, as Revised at Geneva on November 10, 1972, on October 23, 1978, and on March 19, 1991.

Vavilov, N.I., 1926. Studies on the origin of cultivated plants. *Bull. Appl. Bot.* 16, 139–248.

Vos, P., Hogers, R., Bleeker, M., Reijans, M., van de Lee, T., Hornes, M., Frijters, A., Pot, J., Peleman, J., Kuiper, M., 1995. AFLP: a new technique for DNA fingerprinting. *Nucl. Acids Res.* 23, 4407–4414.

Wang, Y., Hu, X., Fu, L., Wu, X., Niu, Z., Liu, M., Ru, Z. 2022. Cleaved amplified polymorphic sequence (CAP) marker development and haplotype geographic distribution of *TaBOR1.2* associated with grain number in common wheat in China. *Cereal Res. Commun.* (In press DOI: 10.1007/s42976-022-00301-1)

Wang, Z., Hobson, N., Galindo, L., Zhu, S., Shi, D., McDill, J., Yang, L., Hawkins, S., Neutelings, G., Datla, R., Lambert, G., Galbraith, D.W., Grassa, C.J., Geraldes, A., Cronk, Q.C., Cullis, C., Dash, P.K., Kumar, P.A., Cloutier, S., Sharpe, A.G., Wong, G.K-S., Wang, J., Deyholos, M.K., 2012. The genome of flax (*Linum usitatissimum*) assembled de novo from short shotgun sequence reads. *Plant J.* 72, 461–473.

Westcott, N.D., Muir, N.D., 2003. Chemical studies on the constituents of *Linum* sp. In: Muir, A.D., Westcott, N.D. (eds.) *Flax, The Genus* Linum. Taylor and Francis, New York, pp. 55–73.

Wiesner, I., Wiesnerova, D., Tejklova, E., 2001. Effect of anchor and core sequence in microsatellite primers on flax fingerprinting patterns. *J. Agric. Sci.* 137, 37–44.

Wiesnerová, D., Wiesner, I., 2004. ISSR-based clustering of cultivated flax germplasm is statistically correlated to thousand seed mass. *Mol. Biotechnol.* 26, 207–214.

Wu, J., Zhao, Q., Wu, G., Zhang, S., Jiang, T., 2017. Development of novel SSR markers for flax (*Linum usitatissimum* L.) using reduced-representation genome sequencing. *Front. Plant Sci.* 7, 2018.

Yi, L., Gao, F., Siqin, B., Zhou, Y., Li, Q., Zhao, X., Jia, X., Zhang, H., 2017. Construction of an SNP-based high-density linkage map for flax (*Linum usitatissimum* L.) using specific length amplified fragment sequencing (SLAF-seq) technology. *PLoS One* 12, e0189785.

Zietkiewicz, E., Rafalski, J.A., Labuda, D., 1994. Genome fingerprinting by simple sequence repeat (SSR)-anchored polymerase chain reaction amplification. *Genomics* 20, 176–183.

Zohary, D., Hopf, M., 2000. *Domestication of Plants in the Old World: The Origin and Spread of Cultivated Plants in West Asia, Europe and the Nile Valley*. Oxford University Press, Oxford.

3 Abiotic Stress and Candid Conduct of Proteins

Srishti Shukla and Vineet Awasthi
Amity Institute of Biotechnology, Amity University Uttar Pradesh, Lucknow Campus, Lucknow, India

Shalini Singh Visen
Amity Food and Agriculture Foundation, Lucknow, Uttar Pradesh, India

Gurjeet Kaur
Amity Institute of Biotechnology, Amity University Uttar Pradesh, Lucknow Campus, Lucknow, India

CONTENTS

3.1 INTRODUCTION

Plants exist and exultantly survive in continually altering situations which can be distressing for development and advancement. Plants are constantly subjected to both abiotic and biotic stress which consequently affect their growth and productivity either independently or synergistically (Bhargava and Srivastava 2013, 2014). The aftermath of stress caused can be regulated technically to a varied extent governing the plant's performance. Biotic stress factors can be grossly herbivore or pathogen related. Abiotic stress due to its demographic distribution is quite complex and many a times difficult to target the identified factors. Biotic pressure, for example, herbivore assault, pathogen contamination and abiotic stress such as drought, salinity, cold, supplement deficiency and an overabundance of sodium chloride or harmful metals in soil is a significant cause of the

DOI: 10.1201/9781003324706-4

misfortune of the agricultural crop worldwide (Srivastava and Bhargava 2015). Abiotic stress such as drought, rain, humidity, sunlight, salinity, temperature, soil, pollution, air, and magnetic fields are a few major causes of misfortune especially to the field crops thereby corroborating a blow to the economy.

3.2 EXTRANEOUS FACTORS CONTRIBUTING TO ABIOTIC STRESS IN PLANTS

3.2.1 Fly Ash

Fly ash is produced as a by-product by coal-fired power stations or thermal power plants that have pozzolanic properties. Chemically, a combination of silicon dioxide, aluminum oxide, ferric oxide and calcium oxide may be present infrequently. The size ranges from 0.5 μm to 300 μm. The applications of fly ash are varied from acting as a catalyst for converting polyethylene into a substance similar, to crude oil through pyrolysis, forming concrete, role in Dams, Flowable fill, Mines, Landfills, Geopolymer concrete, etc. (Ahmad et al. 2021). With respect to agriculture, fly ash is regarded as a dual role player. It facilitated the plant growth by enhancing the soil quality and on the contrary creates menace at higher concentrations due to its size and the macro and micronutrient compositions. The lower concentration of fly ash significantly altered the physical and chemical properties of the soil. A significant rise in plant height, plant fresh and dry biomass, no. of leaves, and average area of the leaf, chlorophyll content and biochemical contents (protein, carbohydrate, mineral and leaf water content) in pumpkin crops was reported. The reduction in plant growth was associated with reduced chlorophyll, carotenoid, biochemical content, proline and yield of the pumpkin crop.

3.2.2 Industrial Efflux

The efflux of essential and non-essential heavy metals is known to produce unfavorable to lethal effects on plants, such as inhibition of growth and photosynthesis, accumulation of low biomass, chlorosis, altered water balance and nutrient assimilation, and senescence, ultimately causing plant death (Yang et al. 2012). The Agency for Toxic Substances and Disease Registry (ATSDR) in its priority list of Hazardous Substances (Report: https://www.atsdr.cdc.gov/spl/index.html) has ranked arsenic first and above other toxic metals such as lead, cadmium and mercury. Though arsenic not being more toxic than other metals on the list, it surfaces above them because of combination of frequency, toxicity and potential for human exposure. Other than the geochemical and anthropogenic sources the semiconductor industrial and industrial sources are also responsible for the arsenic and other heavy metal efflux. Arsenic resistance determinants such as transporters and bio-transformations including redox enzymes, methyltransferases and biosynthetic pathways for arsenosugars, arsenolipids and other nontoxic forms of arsenic (Li et al. 2016, Yang et al. 2016) have been identified in microbes. Aqua glyceroporins have been recognized as universal route of As(III) uptake in plants (Bhattacharjee et al. 2008). Nodulin 26-like intrinsic protein (NIP) subfamily is responsible for facilitating As(III) uptake into the plant root (Chen et al. 2017, Ali et al. 2009). A scientific group has reported a correlation between the concentration and toxicity of heavy metals. We must consider that the presence of heavy metals optimality for land plants is diverse (Arif et al. 2016).

Essential heavy metals such as cobalt (Co), copper (Cu), iron (Fe), manganese (Mn), molybdenum (Mo), nickel (Ni) and zinc (Zn) play a positive role in plant growth and development by regulating through ZRT (Zinc regulated transporters) of ZIP family, IRT1 (Iron regulated transporters), NRAMP (natural resistance-associated macrophage protein), CTR (Copper transporter) and COPT1 (Copper transport protein and upregulating the ROS scavenging system inclusive of enzymatic and non-enzymatic antioxidant mechanisms, DNA repair by Rec A, enhancing the total chlorophyll concentration. These beneficial elements improve the plant's nutritional level and respective mechanisms essential for the normal growth and better yield of the plants. Relatively at higher concentrations expected results were observed including DNA damage, ROS production, oxidative stress, cytotoxic effect and lipid peroxidation.

3.2.3 Sewage and Wastewater Irrigation

The effect of sewage water irrigation showed a significant increase in growth and all biochemical parameters than the control plants as photosynthetic pigments, sugars, proteins, total free amino acids, proline and the activity of antioxidant enzymes of maize plants. The increase was generally higher in shoot than root (Gupta et al. 2010). The levels of heavy metals (Fe > Pb > Mn > Cr > Cd) in the wastewater are low on the contrary the plants and soil show greater concentration due to the accumulated levels of heavy metals in them. The species *Colocasia esculentum, Brassica nigra* and *Raphanus sativus* that are grown showed a decrease in total chlorophyll and total amino acid levels in plants. Increase in amounts of soluble sugars, total protein, ascorbic acid and phenol except *B. nigra* for protein in plants irrigated with wastewater. Similar data were reported in spinach (Letshwenyo et al. 2020) and radish (Atamaleki et al. 2021).

3.2.4 Fertilizers

Plant nutrition is important in attaining quality crop and crop yield (Låsztity et al. 1992). Fertilizers used contribute to the quality of the crop to quite an extent thus pointing toward the protein content (Zhang et al. 2017). In support, a study was conducted with winter rye, triticale and wheat cultivar grown in two subsequent years in the presence of 200 kg/ha for nitrogen and 500–1000 kg/ha for phosphorus and potassium. With all three crops, it was observed that nitrogen fertilization increased the yield by 50% of grains and the raw protein content upon comparing with control samples grown without adding nitrogen fertilizers. Although, significant differences were noticed between the different cereals undertaken in the study. The increase in protein content was reported to relate to a decrease in the essential-to-non-essential amino acid ratio. For most of the amino acids aspartic acid (ASP), isoleucine (ILE) and lysine (LYS), the effect of fertilizing is significant other than threonine (THR), histidine (HIS), leucine (LEU) and valine (VAL) contents, the increase was not significant. Comparing essential amino acid contents in both cereals, higher quantities of threonine (THR), isoleucine (ILE) and lysine (LYS) were found in rye, while other amino acids occur in greater amount in triticale. Thus, the role of fertilizers can be established with improved protein or selective amino acid contents. In another study conducted with protein-rich peanuts, there was interaction between manure and NPK 16 doses to total chlorophyll content of leaves. In another set of experiments, cow manure and NPK 16 significantly affected the plant height, activity of nitrate reductase and total chlorophyll but did not significantly affect on number of leaves. Cow manure and NPK 16 affected on increase of plant height, activity of nitrate reductase and total chlorophyll contents of peanuts (Purbajanti et al. 2019).

3.2.5 Pesticides

The precise use of pesticides is recommended for saving crops from the invasion of pests. The usage of pesticides is always done in approximation, until used under controlled environment. It was reported that its use leads to decreased protein content and regulates the redox enzymes (Sule et al. 2022). The use of fungicides to induce the resistance in wheat to rust infection in turn increases the peroxidase, polyphenol oxidase, and total phenol concentration (Tohamey and Ebrahim 2015). The increasing doses of the waste pesticides cause the decrease in the amount of protein and also, the increase in the enzymes, peroxidase, and catalase activities of *Nicotiana* and *Vigna* plants (Mishra et al. 2015). Different concentrations of the insecticides in *Solanum tuberosum* the catalase, peroxidase and superoxide dismutase amount of enzyme depended on the increasing concentrations of insecticides used for the study (Ferreira et al. 2021). The reports are conclusive of the usage of these chemicals hampering the protein content and regulation of redox enzymes.

3.2.6 Urbanization

Growing urbanization may be one of the measures of developing countries. The abiotic factors which are partners to urbanization affect the plant-related ecology due to multiple factors (Miles et al. 2019, Arjona-García et al. 2021). The abiotic changes include water availability, urban heat island effect, pollution, habitat fragmentation, and more. Empirical evidence has been published to support but no substantial data has been reported with respect to urban environments with higher atmospheric pollution levels influencing the plant proteins. It is assumed that with the growing evidence of the changes in the physiology and the quality of the plant, it is expected to bring changes in the respective proteins.

3.3 PLANT RESPONSE TO ABIOTIC STRESSES

The response generated by the plants due to the abiotic stress incorporates physiological, morphological and biochemical which are well-coordinated and engineered methods including phytohormones, transcription factors (TFs), kinase cascades, reactive oxygen species (ROS), post-translation modifications, etc., which trigger the production of proteins. These result in the management of generated stress. Functional genomics further supports the understanding of array of genes involved in the stress signaling responses which are meaningful and directional tools in the development of transgenic plants with better stress resistance and resilience (Osakabe et al. 2013; Zhu 2016).

The generated abiotic stress is eventually the outcome of cascade of events which undergo crosstalk with each other (Figure 3.1). The regulatory circuits comprehend several factors such as signaling routes consisting of protein–protein interaction, stress sensors, promoters and factors involved in the transcription process and resulting proteins or metabolites. Typically, antioxidative mechanisms are helpful for the removal of ROS; however, stress moderates this eradication, inflicting an upregulated intracellular response thus promoting damage. To restore the damage, plant cells use a sophisticated defence mechanism including antioxidative strain-associated defence genes that, in turn, promote alteration in the regular biochemical mechanisms through a defined set of proteins. ROS possibly requires different molecules to transduce and make more extensive defence alerts. Thus, ROS production and antioxidant strategies act synergistically as an additive or antagonist, associated with the regulated oxidative strain.

Response generated to different kinds of pressure are not owing to linear pathways but are complex circuits including various cell organelles, tissues, and the interaction of extra cofactors and signaling molecules to coordinate and disseminate a designated response to a given stimulus. The preliminary sensors tend to activate cytoplasmic Ca^{2+} and protein signaling pathways, leading to pressure-responsive gene expression and physiological modifications as a response to pressure impact. In accordance, the accumulation of abscisic acid (ABA) plays an essential role in abiotic pressure signaling and transduction pathways, arbitrating many feedbacks.

Advancement in molecular biology, genomics, proteomics and metabolomics has furnished perception into plant gene regulatory unit system, which particularly comprises of inducible genes (environmental elements and developmental cues), expression programming and regulatory factors (cis-element and trans-element), corresponding biochemical pathways and various signal factors. Genetic studies revealed that strain tolerance is mainly quantitative trait loci (QTLs), making genetic development choices difficult.

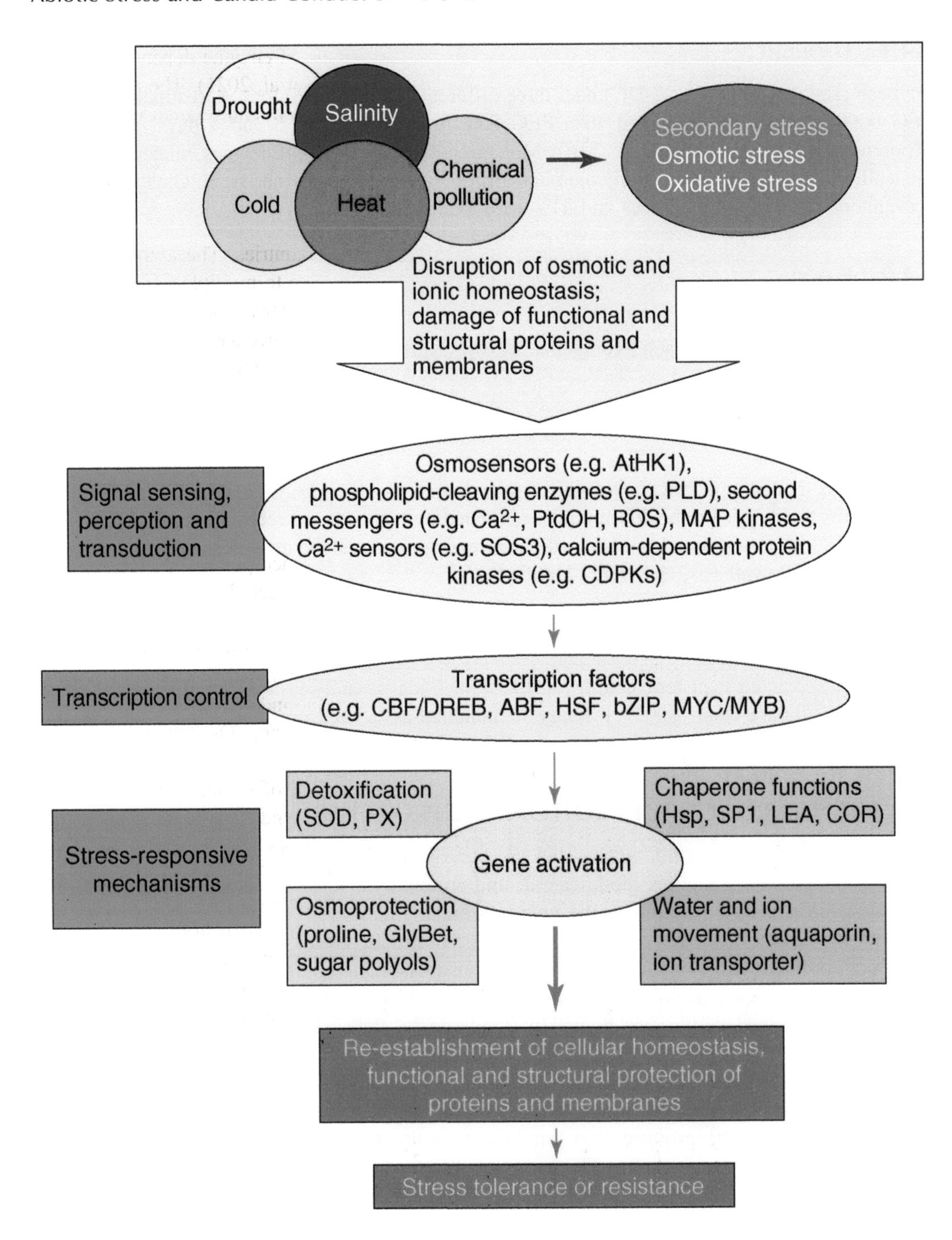

FIGURE 3.1 The complexities from signal sensing to responsive mechanism unwound (Reprinted with permission from Wang et al. 2003).

3.4 PROTEOMIC APPROACHES TO ABIOTIC STRESSES

The proteomic approaches conducted for the abiotic stress conditions for the identification of protein ultimately twine the development of antistress strategies. Variety of crops were subjected to different abiotic conditions for controlled duration and the study included 2-DE proteomic approach. Some of the studies are listed as follows.

3.4.1 Drought

Soybean crop was studied by 2-DE under three different extreme conditions as treatment with 10% PEG 6000, stopped watering and 10% PEG 6000 treatment for 4, 5 and 4 days, respectively (Toorchi et al. 2009; Alam et al. 2010; Mohammadi et al. 2012). Enzymes related to Lignin biosynthesis, small G-protein family members, osmolytes and transmembrane H2O channels, ROS scavengers, molecular chaperones and TFs were identified.

3.4.2 Flood

Wheat was submerged in water for 2 days and the study for proteins involved was conducted by using 2-DE proteomic approach (Kong et al. 2010). Cytoskeleton proteins, disease/defense-related proteins, etc. were identified.

3.4.3 Salinity

Cowpea was treated with NaCl and 2-DE approach was adopted (de Abreu et al. 2014). Twenty-two proteins that happen to be differentially regulated by both salt and recovery were identified by LC-ESI-MS/MS. Ca++ signal proteins, ethylene receptors, metabolic enzymes, enzymes involved in ATP and ETP synthesis, etc. were identified.

3.4.4 Cold

Maize was subjected to a temperature of 10°C for 7 days, analysis was done using the 2-DE approach (Kollipara et al. 2002). Primary metabolism-associated enzymes, signal transduction molecules, defence-related proteins, etc. were identified.

3.5 HIGH-THROUGHPUT PROTEOMIC APPROACHES

The high-throughput proteomic approaches are contributing immensely to the structure identification of the proteins, function, modifications and protein dynamics. As shown in Figure 3.2, the route to MS/MS is after performing the liquid chromatography.

Synthesis of crucial metabolic proteins, indulged in production of osmo-protectants and regulatory proteins having role in pathways of signal transduction and kinases, are needed to respond to the abiotic strain. These techniques help to draft new transcripts, and a steady degree of strain adaptation is attained within few hours. In standard, the transcriptional regulation of genes is directly controlled by using a network of TFs and TF binding sites (TFBS). TFs are proteins with a DNA area that attaches to the cis-performing region in the promoter of a target gene. They set off (activators) or repress (repressors) the activity of the RNA polymerase, hence regulating gene expression. TFs may be grouped into families according to their DNA-binding domain. The presence or absence of transcription elements, activators and suppressors regulating transcription of goal genes often entails an entire cascade of signaling processes, decided through a type of tissue, developmental stage or climate circumstance.

Genes induced by abiotic stress are predominantly partitioned into two gatherings regarding their protein items. First type are those whose coding regions straightforwardly present the plant cells resistance against ecological pressure, as late embryogenesis abundant (LEA) protein, against freezing protein, osmotic administrative protein, chemicals for integrating betaine, proline and different osmo-regulators. The second one includes different gatherings of qualities, whose coding regions assume a significant job in controlling the expression of genes and signal transduction, for example, the transcriptional factors. At any rate, four distinctive regulons can be recognized, two ABA free (1 and 2) and two ABA subordinate (3 and 4): (1) the CBF/DREB regulon; (2) the NAC

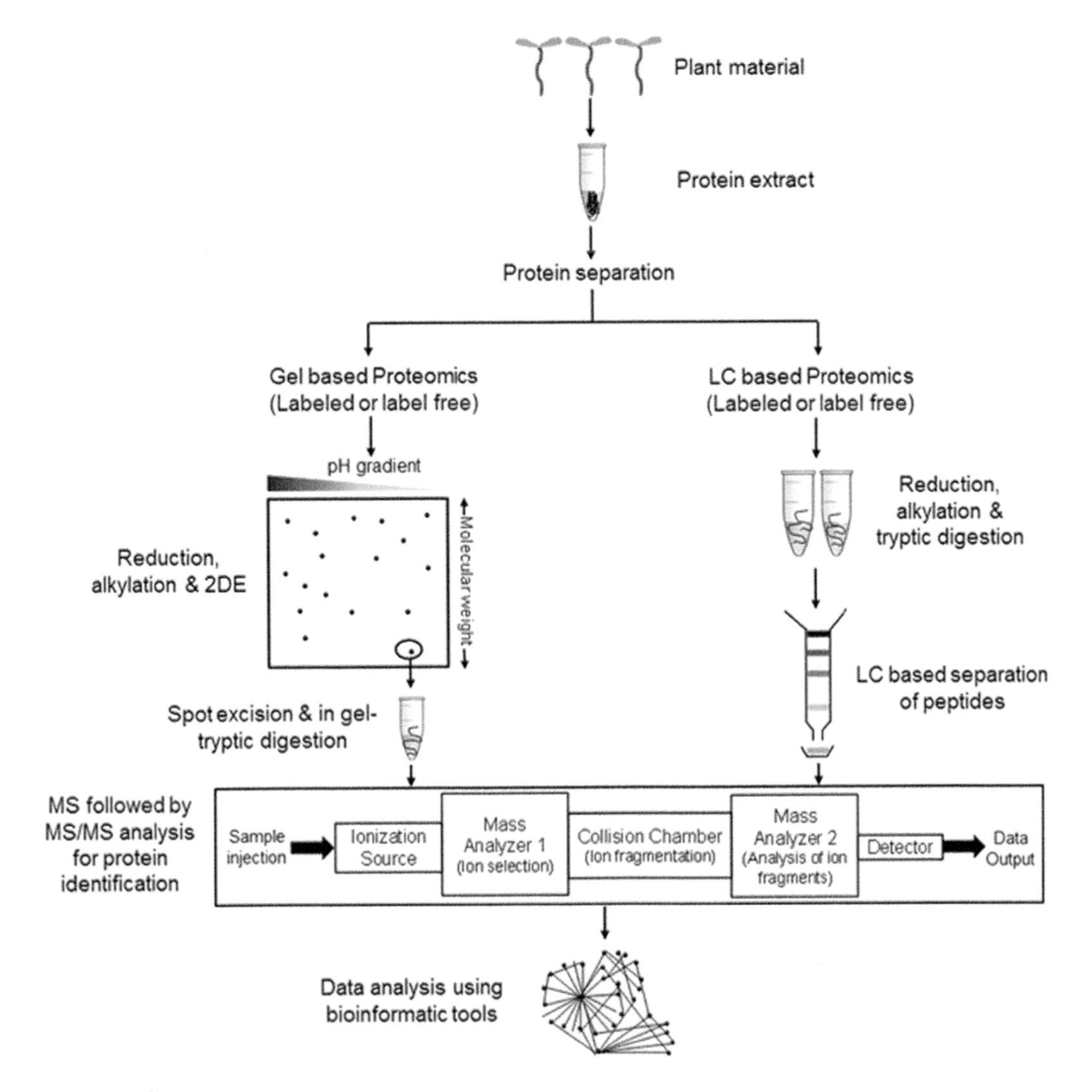

FIGURE 3.2 The pathway to proteomic approach (Reprinted from Ghosh and Xu 2014).

(NAM, ATAF and CUC) and ZF-HD (zinc-finger homeodomain) regulon; (3) the AREB/ABF (ABA-responsive component restricting protein/ABA-restricting variable) regulon; and (4) the MYC (myelocytomatosis oncogene)/MYB (myeloblastosis oncogene) regulon.

Stress flagging manages ionic imbalance, water transport, metabolic reconstruction, re-organizing gene expression, ionic homeostasis, water homeostasis and cell balance. The significance of understanding abiotic stress and the complex pathways related to coping with it will assist us in expanding our capacity to improve pressure obstruction in crops and accomplish horticultural firmness and nourishment security for a developing populace.

3.6 STRESS SENSING

Plants sense distinctive environmental stresses. During drought and salt distress, there is distinct disparity between dry spell and salt stress, which can be recognized by observing the primary and auxiliary pressure signals. Primary consequence signs for salt pressure are osmotic and ionic or metal toxicity in cells. The vital sign for a dry spell is hyperosmotic stress, principal reason for

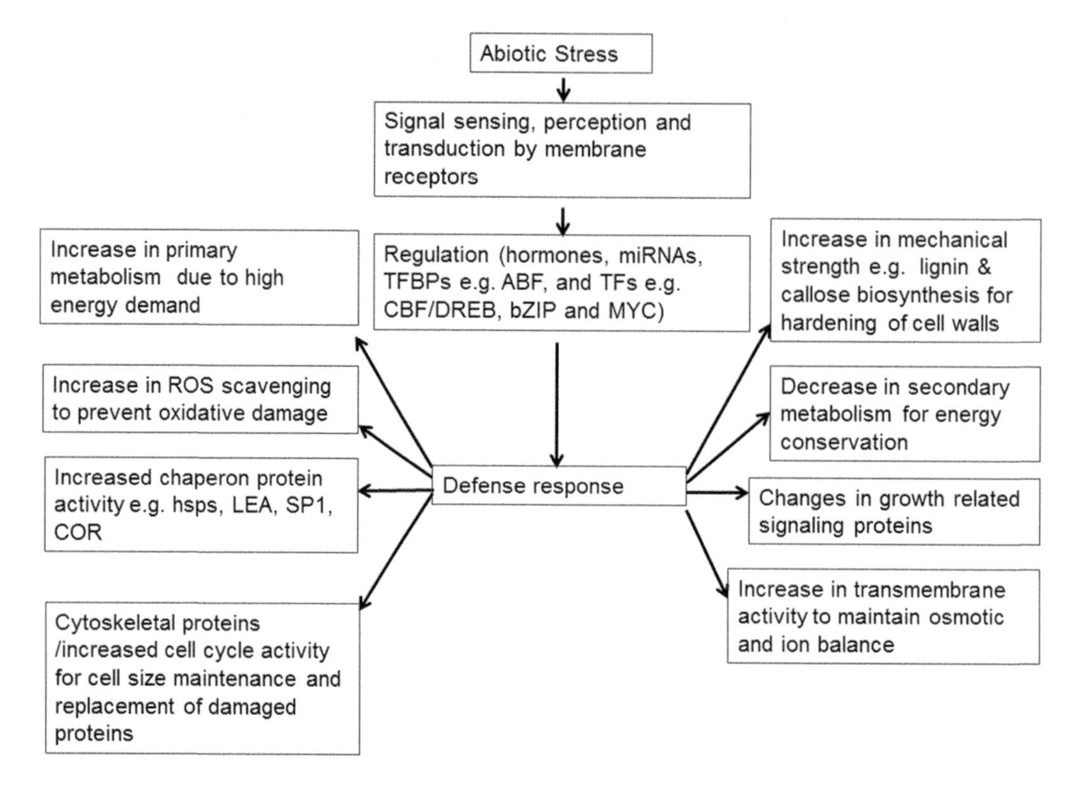

FIGURE 3.3 Changing parameters due to generated Abiotic stress (Reprinted from Onaga and Wydra 2016).

ABA amassing, which evokes numerous versatile reactions in plants. Hypo-osmotic pressure may not be a significant regulator for plant cells. Auxiliary signs for both stresses are perplexing. Like oxidative stress, it causes harm to cell components including proteins, layer lipids, nucleic acids and metabolic impairment. The two burdens have extraordinary signals triggered. Ecological stresses adversely affect plants causing hereditary, morphological and physiological changes (Figure 3.3).

Stress sensing is an intricate pathway. General techniques embraced by plants are escape, evasion and resistance altogether for the plant to endure. Both frigid and sultry distresses can alter the liquidity of phosphatide films of plant cells. This change might be detected by receptors such as kinases (RLKs) (Goff et al. 2007), G-protein-coupled receptors and ion channels. This chain of primary signaling receptors activates secondary intracellular flagging molecules that include Ca^{2+}, lipids, inositol phosphate, ABA, ROS and phytohormones (Osakabe et al. 2013; Zhu 2016), at last inciting changes in gene articulation, protein creation and metabolic pathways to improve plant stress resilience. The plant RLKs reflect diversity of their extracellular domains due to their need to evolve rapidly, in order to survive the ever-changing population of ligands produced by pathogens (Shiu et al. 2001). They comprise of an extracellular zone that may work in binding ligands or protein–protein co-operations, a transmembrane space and an intracellular kinase space. The two-part sensor-reaction controller frameworks, including histidine kinases first found in prokaryotes for the impression of different natural signals, also exist in eukaryotes, including plants. When the extracellular sensor area recognizes a signal, the cytoplasmic histidine build-up gets auto-phosphorylated. The phosphoryl moiety then is passed to an aspartate receiver in a reaction regulator that may comprise some of the sensor protein or a different protein. The sensors may couple with a downstream mitogen-activated protein kinase (MAPK) course or straightforwardly phosphorylate explicit focuses to start cell reactions. After getting a signal the cells regularly use numerous phosphoproteins to transduce and enhance the expression. Protein phosphorylation and dephosphorylation are the most well-known intracellular flagging modes. They manage a broad scope of cell procedures, such as compound enactment, macromolecules, protein restriction and debasement. Optional signs (i.e., hormones and second delegates: inositol phosphates and ROS) can start another passage of flagging occasions, varying from the essential flagging in existence. Chaperones likewise detect sultry distress. The misfolded proteins would discharge heat stress TFs released from chaperones, permitting heat stress interpretation components for initiating the best susceptible genes. It is by large accepted that plant cells sense outer ecological boosts by different sensors, which are restricted on the plasma layer, in the cytosol or internal organelles. A lot of passages are adopted simultaneously to bring about the biochemical response against stress under redox control, which supervises the post-translational modifications such as ubiquitination and phosphorylation and transcriptional activation simultaneously, which results in gene expression and production of proteins such as osmo-protectants and antioxidant proteins, which are responsible for direct action in providing stress tolerance. Signal transduction arrangements for the cold, dry season and salt stress can be separated into three significant sign types: (I) osmotic/oxidative stress flagging that utilizes MAPK modules, includes ROS searching catalysts and osmolytes, (II) Ca2+ dependent stress response that leads to the enactment of LEA proteins- type genes, for example, DRE/CRT (drought-responsive/C-repeat) class of genes), includes the creation of stress-responsive proteins for the most of unclear capacities and (III) Ca2+ subordinate salt overly sensitive (SOS) flagging that manages ionic homeostasis (Xiong et al. 2002). It includes the SOS pathway, which is explicit to salt and ionic stress (Shinozaki et al. 2003).

The prodigious plant reaction to abiotic stress includes the regulators of the microenvironment and their suite of biochemical and molecular systems (Figure 3.4). The investigation of the

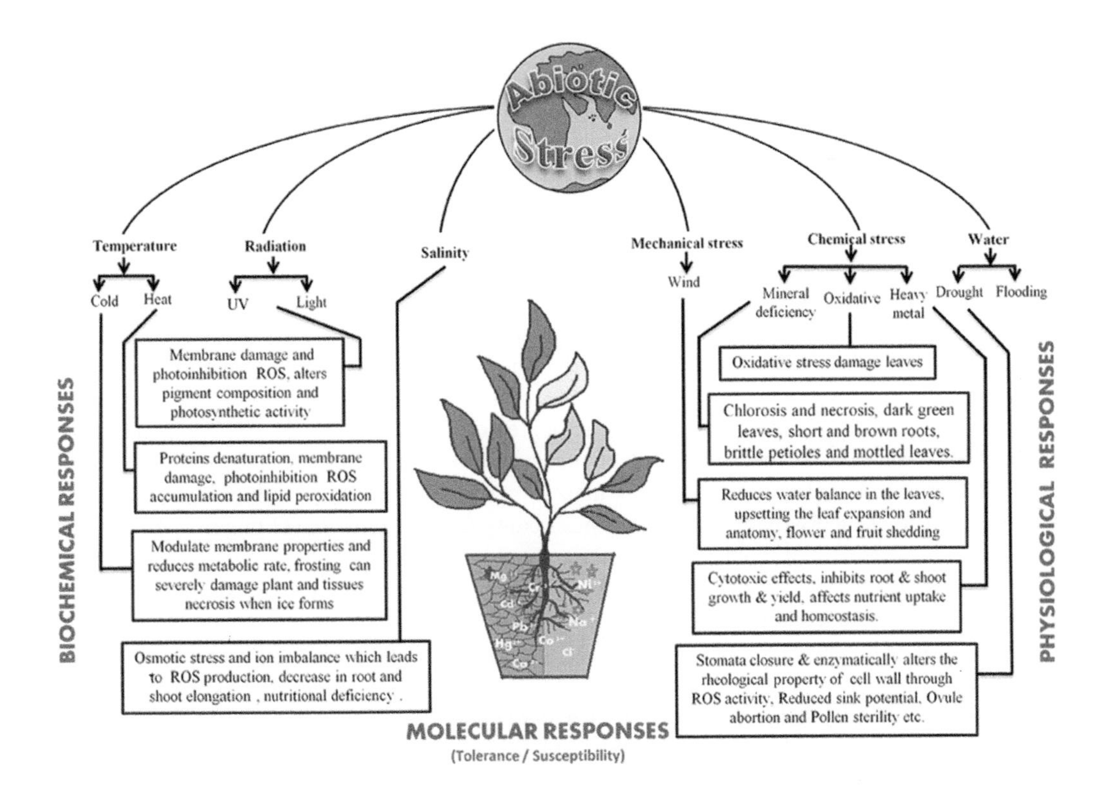

FIGURE 3.4 Abiotic stress and biochemical and physiological responses (Reprinted from Tripathi et al. 2015).

elements of stress-inducible genes is a significant instrument apart from exclusively comprehending the sub-atomic components of stress resistance and the reactions of higher plants, also to improve the stress resilience of harvests by gene regulation. Many genes are associated with abiotic stress reactions (Shinozaki et al. 2003; Öktem et al. 2008). Numerous dry spell inducible qualities are additionally actuated by salt and cold stress, which recommends comparative systems of stress reactions. These genes are arranged into three significant gatherings: (1) those that encode items that legitimately ensure plant cells against stresses, for example, heat stress proteins (HSPs) or chaperones, LEA proteins (Battaglia and Covarrubias 2013), osmo-protectants, liquid catalyst proteins, detoxification chemicals and free-radical scroungers (Wang and Jiao 2000); (2) those associated with flagging falls and in transcriptional control, for example, MAPK, Calmodulin-dependent protein kinase (CDPK) (Ludwig et al. 2004) and SOS kinase (Zhu et al. 2001), phospholipases (Sagar and Singh 2021) and transcriptional factors (Cho et al. 2000; Uno et al. 2000) (3) those engaged with water and ion up take and transport, namely, aquaporins and ionic transporters (Jing et al. 2023).

The LEA proteins are expressed in response to the abiotic stress experienced in the vegetative tissues by plants, bacteria and some invertebrates. These proteins prepare the host to be proactive and help in the damage control mechanisms. These hydrophilic proteins are expressed in a wide range from ferns to angiosperms. These proteins have been classified in seven groups based on their amino acid sequence similarity and on the presence of distinctive conserved motifs. These proteins are suggested to have a relevant role in the plant response to this unfavorable environmental condition (Marina et al. 2013).

The gene articulation portfolio in the plants of Solanaceae showed that count of differentially articulated genes was greater when faced with heat and cold stress as compared to any other abiotic stresses (Tolosa et al. 2020). Regulation of certain TFs, Dof (DNA-binding one finger), WRKY (WRKYs TFs are known to contain the highly conserved amino acid sequence WRKYGQK and the zinc-finger-like motifs Cys(2)-His(2) or Cys(2)-HisCys, and bind to the TTGAC(C/T) W-box *cis*-element in the promoter of their target genes), MYB (MYB DNA-binding domain contains approximately 52 amino acid residues and a diverse C-terminal region which controls the regulation of the protein activities), NAC (originated from three TFs as NAM (no apical meristem, Petunia), ATAF1–2 (*Arabidopsis thaliana* activating factor), and CUC2 (cup-shaped cotyledon, *Arabidopsis*), sharing the same DNA-binding domain), bZIP (Basic leucine zipper), ERF (Ethylene Responsive Factor), ARF (Auxin response factors) and HSF (Heat shock factors) are reported to be widely associated with abiotic and biotic stress response in the plants. These genes were further subcategorized for their specialized role in different stress conditions (Figure 3.5).

It specifies that plants are more adaptable to cold and heat pressure. These basic genes have been additionally dissected to distinguish their part in various signal transduction pathways. Enzymes of the three pathways, viz., MAP kinase, Jasmonic acid and Abscisic acid pathways, have shown comparability with the basic genes. The degree of articulation works to give some insight for the marker genes as needed for the plant species to adapt to different kinds of strains. These genes that are conserved may go about as marker genes for their part in abiotic strains just as various signal transduction pathways. The genes coding C3HC4-type RING finger family protein have been accounted for in cool, dry seasons and saltiness stress in rice (Rabbani et al. 2003). Arginine decarboxylase 2 (ADC2) is a significant catalyst answerable for the synthesis of polyamine under pressure conditions. It is liable for low temperature and drying out pressure in Poncirus trifoliate (Wang et al. 2011). It has been accounted for that during abiotic stresses, diverse flagging pathways facilitate the expression of several genes. The outflow of ADC2 gene is instigated by jasmonic acid and ABA pathway in *A. thaliana* (Perez-Amador et al. 2002). The gene coding for TFs such as zinc-finger (C3HC4-type RING finger) family protein, nuclear TF Y subunit B-2 and WRKY were found to have differential articulation under abiotic stresses. The articulation of these TFs is managed by different flagging pathways. WRKY gene family has likewise been proposed to assume a significant part in the guiding of transcription related with

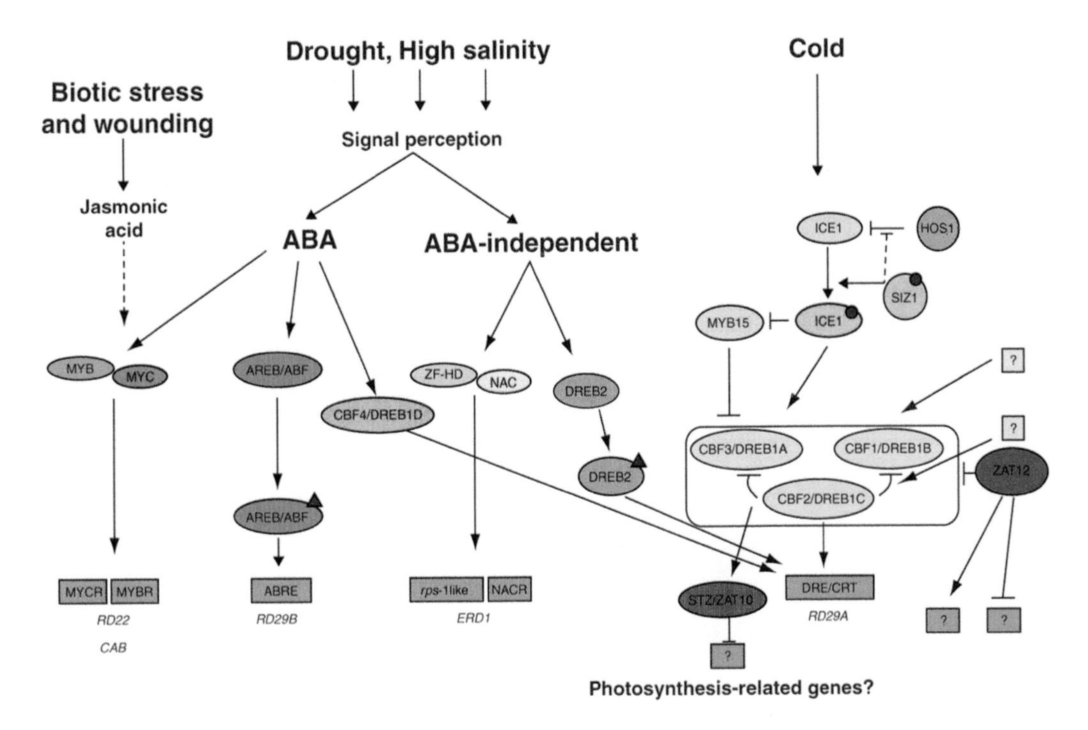

FIGURE 3.5 Abiotic stress and related factors (Reprinted with permission from Saibo et al. 2009).

pressure reactions. The genes coding enzymes of sign pathway are additionally associated with reaction toward abiotic stresses in different significant plant species. These genes can be additionally approved for their accessibility and reaction for stress resilience in other neglected plants (Dhawan et al. 2013).

The study of gene expression in abiotic stress conditions can also help in improving the plant growth and scale up the production of biomass. Several standard approaches have been designed that are based on altering – plant–microbe interactions; cell wall biosynthesis; and phytohormone levels. Simultaneously, using approaches based on functional genomics and genetics, a significant number of genes have been found that play a major role in tolerance against abiotic stress (Joshi et al. 2018).

3.7 GENETIC ENGINEERING AND STRESS TOLERANCE

Genetic engineering has also come up with an answer for producing better stress-tolerant plants involving enhancement of osmo-protectants. Glycine-betaine (GB) amasses in a series of organisms under abiotic stress and has been researched on in extraordinary detail. Plants that have been known to aggregate GB normally have been accounted to develop well under dry spell and saline climate. Gathering of GB in transgenic apple communicating pressure controller gene, Osmyb4, was connected to improved dry spell and cold resilience. Synthesis of GB is done either by oxidation of choline or N-methylation of glycine by three known pathways. In plants, enzyme choline monooxygenase (CMO) first transforms choline into betaine aldehyde and afterward an enzyme dependent on NAD^+, BADH- betaine aldehyde dehydrogenase produces GB. These proteins are primarily found in chloroplast stroma and their activity is expanded because of salt pressure (Giri et al. 2011). Other osmo-protectants targeted using this approach include proline, mannitol and trehalose.

REFERENCES

Ahmad, G., Khan, A.A., Mohamed, H.I., 2021. Impact of the low and high concentrations of fly ash amended soil on growth, physiological response, and yield of pumpkin (*Cucurbita moschata* Duch. Ex Poiret L.). *Environ. Sci. Pollut. Res.* 28, 17068–17083.

Alam, I., Sharmin, S.A., Kim, K.H., Yang, J.K., Choi, M.S., Lee, B.-H., 2010. Proteome analysis of soybean roots subjected to short-termdrought stress. *Plant Soil.* 333, 491–505.

Ali, W., Isayenkov, S.V., Zhao, F.J., Maathuis, F.J., 2009. Arsenite transport in plants. *Cell. Mol. Life Sci.* 66, 2329–2339.

Arif, N., Yadav, V., Singh, S., Singh, S., Ahmad, P., Mishra, R.K., Sharma, S., Tripathi, D.K., Dubey, N.K., Chauhan, D.K., 2016. Influence of high and low levels of plant-beneficial heavy metal ions on plant growth and development. *Front. Environ. Sci.* 4, 69.

Arjona-García, C., Blancas, J., Beltrán-Rodríguez, L., López Binnqüist, C., Colín Bahena, H., Moreno-Calles, A.I., Sierra-Huelsz, J.A., López-Medellín, X., 2021. How does urbanization affect perceptions and traditional knowledge of medicinal plants? *J. Ethnobiol. Ethnomed.* 17, 48.

Atamaleki, A., Yazdanbakhsh, A., Fakhri, Y., Salem, A., Ghorbanian, M., Mousavi Khaneghah, A., 2021. A systematic review and meta-analysis to investigate the correlation vegetable irrigation with wastewater and concentration of potentially toxic elements (ptes): A case study of spinach (*Spinacia oleracea*) and radish (*Raphanus raphanistrum* subsp. *sativus*). *Biological Trace Elem. Res.* 199, 792–799.

Battaglia, M., Covarrubias, A.A., 2013. Late Embryogenesis Abundant (LEA) proteins in legumes. *Front. Plant Sci.* 4, 190.

Bhargava, A., Srivastava, S., 2013. *Quinoa: Botany, Production and Uses.* CABI, Oxfordshire, UK.

Bhargava, A., Srivastava, S., 2014. Transgenic approaches for phytoextraction of heavy metals. In: Ahmad, P., Wani, M.R., Azooz, M.M., Tran, L.P. (Eds.) *Improvement of Crops in the Era of Climatic Changes.* Springer, New York, pp. 57–80.

Bhattacharjee, R.B., Singh, A., Mukhopadhyay, S.N., 2008. Use of nitrogen-fixing bacteria as biofertiliser for non-legumes: Prospects and challenges. *Appl. Microbiol. Biotechnol.* 80, 199–209.

Chen, Y., Sun, S.K., Tang, Z., Liu, G., Moore, K.L., Maathuis, F., Miller, A.J., McGrath, S.P., Zhao, F.J., 2017. The Nodulin 26-like intrinsic membrane protein OsNIP3;2 is involved in arsenite uptake by lateral roots in rice. *J. Exp. Bot.* 68, 3007–3016.

Cho, R.J., Campbell, M.J., 2000. Transcription, genomes, function. *Trends Genet.* 16, 409–415.

de Abreu, C.E., Araújo, G.S., Monteiro-Moreira, A.C., Costa, J.H., Leite H.B., Moreno, F.B., Prisco, J.T., Gomes-Filho, E., 2014. Proteomic analysis of salt stress and recovery in leaves of *Vigna unguiculata* cultivars differing in salt tolerance. *Plant Cell Rep.* 33, 1289–1306.

Dhawan, C., Kharb, P., Sharma, R., Uppal S., Aggarwal R.K., 2013. Development of male-specific SCAR marker in date palm (*Phoenix dactylifera* L.). *Tree Genet. Genom.* 9, 1143–1150.

Ferreira, D., Figueiredo, J., Laureano, G., Machado, A., Arrabaça, J.D., Duarte, B., Figueiredo, A., Matos, A.R., 2021. Membrane remodelling and triacylglycerol accumulation in drought stress resistance: The case study of soybean phospholipases A. *Plant Physiol. Biochem.* 169, 9–21.

Ghosh, D., Xu, J., 2014. Abiotic stress responses in plant roots: A proteomics perspective. *Front. Plant Sci.* 5, 6.

Giri, C., Ochieng, E., Tieszen, L.L., Zhu, Z., Singh, A., Loveland, T., Masek, J., Duke, N., 2011. Status and distribution of mangrove forests of the world using earth observation satellite data. *Global Ecol. Biogeog.* 20, 154–159.

Goff, K.E., Ramonell, K.M., 2007. The role and regulation of receptor-like kinases in plant defense. *Gene Regul. Syst. Biol.* 26, 167–175.

Gupta, S., Satpati, S., Nayek, S., Garai, D., 2010. Effect of wastewater irrigation on vegetables in relation to bioaccumulation of heavy metals and biochemical changes. *Environ. Monit. Assess.* 165, 169–177.

Jing, W., Li, Y., Zhang, A., Zhou, X., Gao, J., Ma, N., 2023. Aquaporin, beyond a transporter. *Horticultural Plant J.* 9, 29–34.

Joshi, R., Singla-Pareek, S.L., Pareek, A., 2018. Engineering abiotic stress response in plants for biomass production. *J. Biol. Chem.* 293, 5035–5043.

Kollipara, K.P., Saab, I.N., Wych, R.D., Lauer, M.J., Singletary, G.W., 2002. Expression profiling of reciprocal maize hybrids divergent for cold germination and desiccation tolerance. *Plant Physiol.* 129, 974–992.

Kong, A., Thorleifsson, G., Gudbjartsson, D.F., Masson, G., Sigurdsson, A., Jonasdottir, A., Walters, G.B., Jonasdottir, A., Gylfason, A., Kristinsson, K.T., Gudjonsson, S.A., Frigge, M,L,, Helgason, A., Thorsteinsdottir, U., Stefansson, K., 2010. Fine-scale recombination rate differences between sexes, populations and individuals. *Nature* 467, 1099–1103.

Låsztity, R., Låsztity, B., Hidvégi, M., Simon-Sarkadi, I., 1992. Effect of fertilizers on the yield, protein content and amino acid composition of winter cereals. *Periodica Polytechnica Ser. Chem. Eng.* 86, 25–41.

Letshwenyo, M.W., Mokokwe, G., 2020. Accumulation of heavy metals and bacteriological indicators in spinach irrigated with further treated secondary wastewater. *Heliyon* 6, e05241.

Li, W., Yao, A., Zhi, H., Kaur, K., Zhu, Y.C., Jia, M., Zhao, H., Wang, Q., Jin, S., Zhao, G., Xiong, Z.Q., Zhang, Y.Q., 2016. Angelman syndrome protein Ube3a regulates synaptic growth and endocytosis by inhibiting BMP signaling in *Drosophila. PLoS Genet.* 12, e1006062.

Ludwig, W., Strunk, O., Westram, R., Richter, L., Meier, H., Yadhukumar, Buchner, A., Lai, T., Steppi, S., Jobb, G., Förster, W., Brettske, I., Gerber, S., Ginhart, A.W., Gross, O., Grumann, S., Hermann, S., Jost, R., König, A., Liss, T., Lüssmann, R., May, M., Nonhoff, B., Reichel, B., Strehlow, R., Stamatakis, A., Stuckmann, N., Vilbig, A., Lenke, M., Ludwig, T., Bode, A., Schleifer, K.H., 2004. ARB: a software environment for sequence data. *Nucleic Acids Res.* 32, 1363–1371.

Miles, L.S., Breitbart, S.T., Wagner, H.H., Johnson, M.T.J., 2019. Urbanization shapes the ecology and evolution of plant-arthropod herbivore interactions. *Front. Ecol. Evol.* 7, 310.

Mishra, K., Sanwal, E., Tandon, P.K., Gupta, K., 2015. Metabolic effects of pesticide effluents on *Nicotiana tabacum* and *Vigna radiata* plants. *Intern. J. Environ.* 4, 87–94.

Mohammadi, P.P., Moieni, A., Hiraga, S., Komatsu, S., 2012. Organ-specific proteomic analysis of drought-stressed soybean seedlings. *J. Prot.* 75, 1906–1923.

Öktem, H.A., Eyidoðan, F., Demirba, D., Bayraç, A.T., Öz, M.T., Özgür, E., Selçuk, F., Yücel, M., 2008. Antioxidant responses of lentil to cold and drought stress. *J. Plant Biochem. Biotechnol.* 17, 15–21.

Onaga, G., Wydra, K., 2016. Advances in plant tolerance to abiotic stresses. In: Abdurakhmonov, I.Y. (Ed.), *Plant Genomics*. InTechOpen, London, pp. 229–272.

Osakabe, Y., Yamaguchi-Shinozaki, K., Shinozaki, K., Tran, L.S., 2013. Sensing the environment: key roles of membrane-localized kinases in plant perception and response to abiotic stress. *J. Exp. Bot.* 64, 445–458.

Perez-Amador, M.A., Leon, J., Green, P.J., Carbonell, J., 2002. Induction of the arginine decarboxylase ADC2 gene provides evidence for the involvement of polyamines in the wound response in Arabidopsis. *Plant Physiol.* 130, 1454–1463.

Purbajanti, E.D., Slamet, W., Fuskhah, E., Rosyida, 2019. Effects of organic and inorganic fertilizers on growth, activity of nitrate reductase and chlorophyll contents of peanuts (*Arachis hypogaea* L.). *IOP Conf. Ser.: Earth Environ. Sci.* 250, 012048.

Rabbani, M.A., Maruyama, K., Abe, H., Khan, M.A., Katsura, K., Ito, Y., Yoshiwara, K., Seki, M., Shinozaki, K., Yamaguchi-Shinozaki, K., 2003. Monitoring expression profiles of rice genes under cold, drought, and high-salinity stresses and abscisic acid application using cDNA microarray and RNA gel-blot analyses. *Plant Physiol.* 133, 1755–1767.

Sagar, S., Singh, A., 2021. Emerging role of phospholipase C mediated lipid signaling in abiotic stress tolerance and development in plants. *PLant Cell Rep.* 40(11): 2123–2133.

Saibo, N.J., Lourenço, T., Oliveira, M.M., 2009. Transcription factors and regulation of photosynthetic and related metabolism under environmental stresses. *Ann. Bot.* 103, 609–623.

Shinozaki, K., Yamaguchi-Shinozaki, K., Seki, M., 2003. Regulatory network of gene expression in the drought and cold stress responses. *Curr. Opin. Plant Biol.* 6, 410–417.

Shiu, S.H., Bleecker, A.B., 2001. Receptor-like kinases from *Arabidopsis* form a monophyletic gene family related to animal receptor kinases. *Proc. Natl. Acad. Sci. USA* 98, 10763–10768.

Srivastava, S., Bhargava, A., 2015. Genetic diversity and heavy metal stress in plants. In: Ahuja, M.R., Jain, S.M., (Eds.) *Sustainable Development and Biodiversity*. Springer International Publishing, Switzerland, pp. 241–270.

Sule, R.O., Condon, L., Gomes, A.V., 2022. A common feature of pesticides: oxidative stress-the role of oxidative stress in pesticide-induced toxicity. *Oxid. Med. Cell Longev.* 19, 5563759.

Tohamey, S., Ebrahim, S.A., 2015. Inducing resistance against leaf rust disease of wheat by some micro-elements and tilt fungicide. *Plant Pathol. J.* 14(4): 175–181.

Tolosa, L.N., Zhang, Z., 2020. The role of major transcription factors in solanaceous food crops under different stress conditions: Current and future perspectives. *Plants* 9, 56.

Toorchi, M., Yukawa, K., Nouri, M.Z., Komatsu, S., 2009. Proteomics approach for identifying osmotic-stress-related proteins in soybean roots. *Peptides* 30, 2108–2117.

Tripathi, A., Goswami, K., Sanan-Mishra, N., 2015. Role of bioinformatics in establishing microRNAs as modulators of abiotic stress responses: the new revolution. *Front. Physiol.* 6, 286.

Uno, Y., Furihata, T., Abe, H., Yoshida, R., Shinozaki, K., Yamaguchi-Shinozaki, K., 2000. *Arabidopsis* basic leucine zipper transcription factors involved in an abscisic acid-dependent signal transduction pathway under drought and high-salinity conditions. *Proc. Natl. Acad. Sci. USA* 97, 11632–11637.

Wang, S.Y., Jiao, H., 2000. Scavenging capacity of berry crops on superoxide radicals, hydrogen peroxide, hydroxyl radicals, and singlet oxygen. *J. Agric. Food Chem.* 48, 5677–5684.

Wang, W., Vinocur, B., Altman, A., 2003. Plant responses to drought, salinity and extreme temperatures: towards genetic engineering for stress tolerance. *Planta* 218, 1–14.

Wang, J., Sun, P.P., Chen, C.L., Wang, Y., Fu, X.Z., Liu, J.H., 2011. An arginine decarboxylase gene PtADC from *Poncirus trifoliata* confers abiotic stress tolerance and promotes primary root growth in *Arabidopsis*. *J. Exp. Bot.* 62, 2899–2914.

Xiong, L., Lee, H., Ishitani, M., Tanaka, Y., Stevenson, B., Koiwa, H., Bressan, R.A., Hasegawa, P.M., Zhu, J.K. 2002. Repression of stress-responsive genes by FIERY2, a novel transcriptional regulator in *Arabidopsis*. *Proc. Natl. Acad. Sci. USA* 99, 10899–108904.

Yang, H.C., Rosen, B.P., 2016. New mechanisms of bacterial arsenic resistance. *Biomed. J.* 39, 5–13.

Yang, H.C., Fu, H.L., Lin, Y.F., Rosen, B.P., 2012. Pathways of arsenic uptake and efflux. *Curr. Top. Membr.* 69, 325–358.

Zhang, P., Ma, G., Wang, C., Lu, H., Li, S., Xie, Y., Ma, D., Zhu, Y., Guo, T., 2017. Effect of irrigation and nitrogen application on grain amino acid composition and protein quality in winter wheat. *PLoS One*, 12, e0178494.

Zhu, J.K., 2016. Abiotic stress signaling and responses in plants. *Cell* 167, 313–324.

Zhu, Y.G., Smith, S.E., Smith, F.A., 2001. Zinc (Zn)-phosphorus (P) interactions in two cultivars of spring wheat (*Triticum aestivum* L.) differing in P uptake efficiency. *Ann. Bot.* 88, 941–945.

Section II

Microbial and Medical Biotechnology

4 Long Non-coding RNA(s)

Is the Junk Worth It?

Arushi Mishra
Bioinformatics Laboratory, Department of Biological Sciences, Birla Institute of Technology and Science (BITS) Pilani, Rajasthan, India

CONTENTS

4.1 INTRODUCTION

Gene regulation is enigmatically controlled procedure, which is performed smoothly due to well-versed coordination between functional and structural organization of the eukaryotic genome not only at transcriptional but also at posttranscriptional level through epigenetic modifications. With the

DOI: 10.1201/9781003324706-6

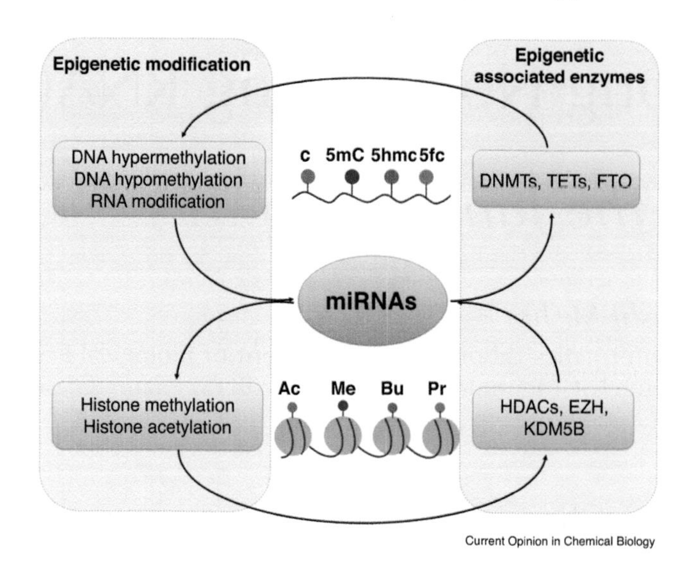

FIGURE 4.1 The miRNA-epigenetic feedback loop (Reprinted with permission from Yao et al. 2019).

progressing studies, crosstalk between regulatory processes of RNA and epigenetics has been identified, with increasing evidences of RNA playing regulatory roles in the epigenetic processes (Figure 4.1) (Holoch and Moazed 2015; Yao et al. 2019). With technological advancement in molecular and cellular characterization in the past decade, the abscise "junk" region of genome has now been characterized as an essential modulator of the timing and occurrence of gene expression of innumerable genes. After carefully curating, the ENCODE project demonstrated that about 80% of the non-coding genome interacts with regulatory proteins, thereby enabling their functions (ENCODE 2012). Thus, it was clarified that the noncoding part of genome is transcribed into noncoding RNA (ncRNA) including both the small and long noncoding RNAs (lncRNAs). Henceforth, the importance of lncRNA in different signaling pathways and metabolic processes has been discussed along with views on whether and why it is absolutely essential to study this class of noncoding RNAs?

4.2 HISTORICAL OVERVIEW

Historically, the human genome has been roughly separated into coding and noncoding regions. Structural ncRNAs, such as tRNAs and rRNAs, and the protein-coding genes have been found to be essential for different cellular functions. Initially, it was thought that most of the non-coding genome is junk DNA, without any legit function. As we know, gene regulation occurs at two major steps which are transcription and translation, thus proteins act as the key functional regulators (Morris and Mattick 2014). Jacob, Meselson and Brenner discovered messenger RNAs (mRNA) in the 1960s, who later also showed that they have major functions within cells (Brenner et al. 1961). Later on, other structural RNA classes were also discovered, such as rRNAs, tRNAs and heteronuclear RNAs (hnRNAs) (Warner et al. 1966; Kim et al. 1974). While the international human genome sequencing program was close to obtaining the first complete sequence, researchers encountered an unexpected difficulty which was – how to differentiate between translated and untranslated genes and finally improved sequencing technologies solved this problem. Several international research programs, such as the Functional Annotation of the Mammalian Genome (FANTOM) project, showed that most of the RNA transcribed from mammalian genome do not code for proteins. Thus, thousands of non-coding RNAs, including short (200 nt) RNAs, were identified. Furthermore, alternate splicing has been found as a common event, in not only coding but also non-coding sections of the genome (Bardou et al. 2014). As an inference of this, the definition of a gene was revised that non-coding

RNAs may not simply be junk DNA or transcriptional noise (ENCODE Project Consortium 2012; Gerstein et al. 2007; Kapranov et al. 2007) and later next generation sequencing (NGS) helped to differentiate between the expression pattern of coding and non-coding RNAs (Dinger, Pang, et al. 2008; Mortazavi et al. 2008). Even more evidence confirming the existence of noncoding transcription in the intergenic regions came from correlations with chromatin and histone modifications, such as DNase1 hypersensitivity; H3K9ac, H3K4me3 and H3K36me3; or binding of transcription factors (TFs) at the loci and their expression (Guttman et al. 2009, 2011; van Bakel et al. 2010; ENCODE Project Consortium 2012). The very first functional lncRNA that was discovered was H19 (Brannan et al. 1990) which was found to be abundantly expressed in many mesoderm- and endoderm-derived tissues. It was also shown to regulate a network of co-expressed imprinted genes that supervise fetal and post-natal development (Gabory et al. 2010). Another lncRNA described was X-inactive specific transcript (XIST), transcribed only from the inactive X chromosome in mammalian female cells (Brown et al. 1991; Penny et al. 1996).

4.3 CLASSIFICATION OF lncRNA

The different modes of deriving lncRNAs include,

- Protein-coding gene with structural damage can transform into lncRNA
- Abreast non-transcribed regions can generate lncRNA by chromosomal rearrangements
- Retrotransposition of a copy of noncoding gene can form lncRNA
- Tandem duplication events can lead to development of lncRNA with adjacent repeats.
- Inserting a transposon can also generate a functional lncRNA
- Mitochondrial genome

Based on the location where they originate, they can be classified as – intergenic lncRNA (lincRNA), intronic lncRNA (sense/antisense), exonic lncRNA and overlapping lncRNA (Kung et al. 2013). The classification of lncRNAs is described in Figure 4.2.

4.3.1 lncRNAs

This lncRNA subclass can be further subdivided into several categories depending on the lncRNA position relative to the associated protein-coding gene (Marques et al. 2013).

Sense lncRNAs: they imbricate with protein-coding regions. While most of the lncRNAs do not encode proteins; few have both coding and non-coding potential. For instance, the intragenic lncRNA SRA (steroid receptor RNA activator) was among the first lncRNAs found to encode a small peptide (Chooniedass-Kothari et al. 2004; Kapranov et al. 2005; Denoeud et al. 2007; Djebali et al. 2012).

Antisense lncRNAs: also known as natural antisense transcripts (NATs), also extends over with protein-coding regions, but contradictory to sense lncRNA, they are transcribed from the opposite strand (Schwartz et al. 2008). However, sense or antisense lncRNAs can imbricate with both exonic or intronic regions. NATs exist in virtually every living species, including eukaryotes, prokaryotes and even viruses (Britto-Kido et al. 2013). Nonetheless, no evolutionarily conserved NAT has been indexed (Katayama et al. 2005; Wahlestedt 2013).

Bidirectional lncRNAs: the transcription start site is present less than 1 kb away from its neighboring protein-coding gene. They are usually transcribed in the opposite direction (Knauss and Sun 2013).

Intronic lncRNAs: they define true lncRNA as they are the predominant non-coding RNAs in the human genome and include nothing but intronic RNAs (TINs), thus representing more than 70% of all non-coding RNAs (except rRNAs) in human cells (Nakaya et al. 2007), and partially intronic RNAs (PINs), which are generally unspliced (Louro et al. 2009; St Laurent et al. 2015).

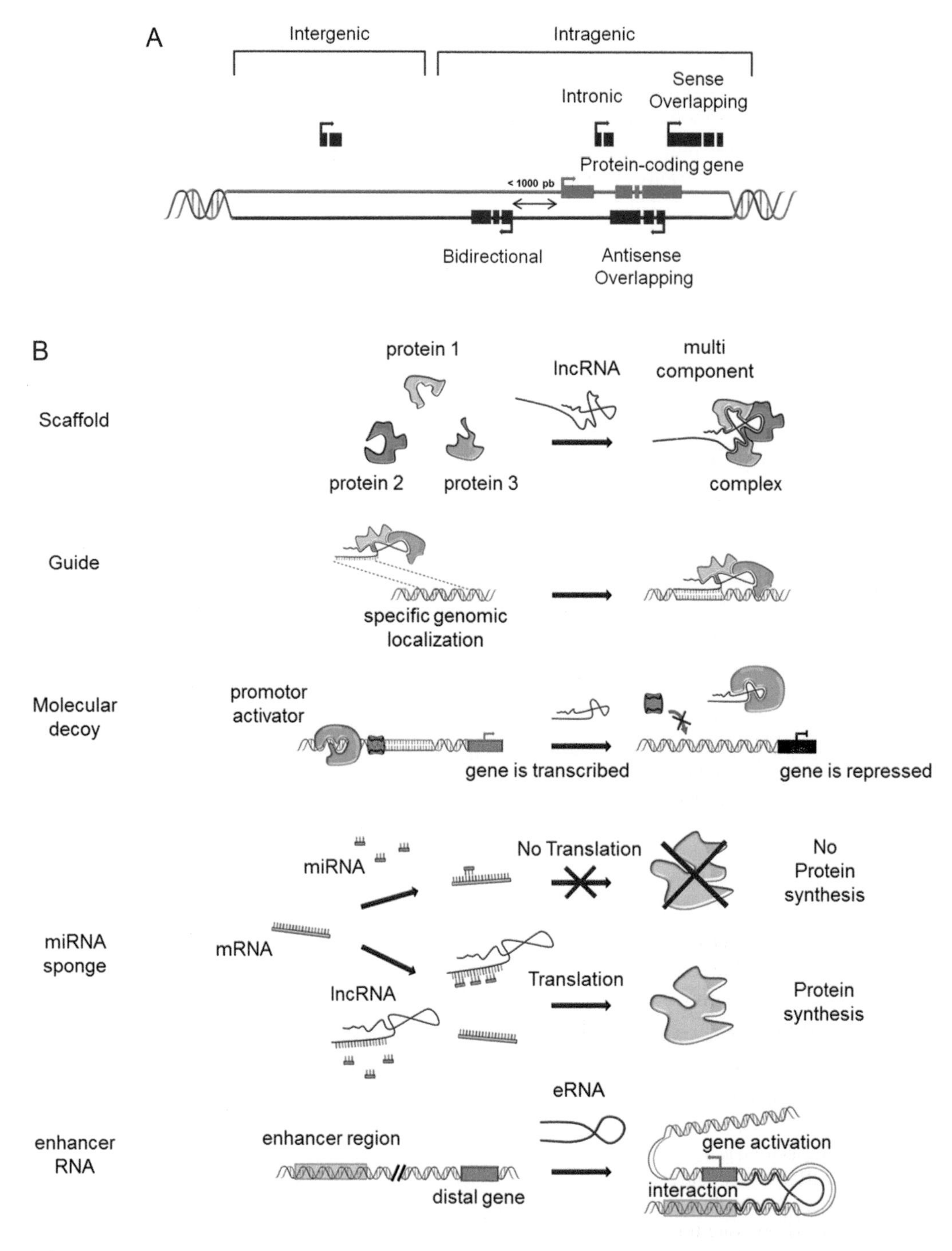

FIGURE 4.2 General classification of lncRNAs and mechanisms of action. (A) Sense-long non-coding RNAs (lncRNAs) overlap with a protein-coding gene on the same DNA strand and antisense lncRNAs overlap on the opposite strand. Bidirectional lncRNAs are transcripts in which the transcription start site is located at less than 1 kb away from the neighboring protein-coding gene. Intronic lncRNAs are located within an intron of a protein-coding gene, while intergenic lncRNAs (or lincRNAs) are located more than 5 kb away from a protein-coding region. Intragenic lncRNAs include sense and antisense overlapping, bidirectional and intronic lncRNAs. (B) lncRNAs can act via different mechanisms of actions: "Scaffold": lncRNAs can serve as adaptors by bringing together multiple components, such as proteins, RNAs or DNAs, to form

4.3.2 Intergenic lncRNAs

The basic difference between intragenic and intergenic is that the long intergenic (or intervening) lncRNAs, called lincRNAs, do not overlap with protein-coding regions (introns or exons). As a result, their expression profiling and its effects are simpler to interpret (Hangauer et al. 2013). Many functional lincRNAs display stability comparable to mRNAs (Clark et al. 2012), active transcription marks such as the K4H36 chromatin signature at their promoter (Guttman et al. 2009), and cell-specific expression profiles (Khalil et al. 2009; Jia et al. 2010). LincRNAs are not at all evolutionarily conserved (Necsulea et al. 2014). Most lincRNAs overlay with repetitive elements, such as SINEs or LINEs (Ulitsky et al. 2013; Kelley and Rinn 2012). These repetitive elements help lincRNAs by homologous base pairing with other RNAs (Gong and Maquat 2011; Carrieri et al. 2012). This subclass further includes vlincRNAs or macroRNAs which are very long intergenic non-coding RNAs. Their length can vary from 50 kb to 1 Mb and they could be derived from pseudogenes (St Laurent et al. 2015; Milligan and Lipovich 2014; Lazorthes et al. 2015).

4.4 FUNCTIONAL ASPECTS OF lncRNA

lncRNA were earlier thought to be transcriptional noise but they were later demonstrated to be functional molecules and it was shown that just like proteins they also have roles in cellular functions. Various mechanisms such as proliferation and apoptosis, development and differentiation, X-chromosome inactivation and genomic imprinting have been found to be manifested by these lncRNAs (Lee et al. 2013; Akhade et al. 2017). They have also been found to play important roles in human diseases such as coronary artery disease, amyotrophic lateral sclerosis, Alzheimer's disease as well as cancer with either oncogenic or tumor-suppressive role, since they can mediate their effects in cis or trans manner by directly binding to DNA, RNA or proteins. But the mechanism by which they govern these disease processes still remains hidden and is being investigated. Thus, understanding lncRNA is of utmost importance to study human diseases and their biology. The different mechanisms of action by which lncRNAs function are listed in Figure 4.2.

4.4.1 Transcriptional Regulation

In the context of transcription regulation, the role of lncRNA is multifaceted. The intergenic lncRNA has been divided into two major categories, on the basis of chromatin marks at their promoters: those emanating from enhancer regions or those transcribed from promoter-like lncRNA loci (Marques et al. 2013). Studies have shown that lncRNA can act as scaffolds,

ribonucleoprotein complexes [e.g., HOTAIR (Hox antisense intergenic RNA), ANRIL (Antisense non-coding RNA in the INK4 locus), MALAT1 (Metastasis associated lung adenocarcinoma transcript 1), TUG1 (Taurine up-regulated gene 1), NEAT1 (Nuclear paraspeckle assembly transcript 1)]. 'Guide': via standard base pairing, lncRNAs can guide ribonucleoprotein complexes to a specific genomic location [e.g., HOTTIP (HOXA transcript at the distal tip), XIST (X specific inactive transcript), AIR (Antisense Igf2r RNA), KCNQ1ot1 (Kcnq1 opposite transcript 1), lincRNA-p21). "Molecular decoy": lncRNAs can bind to protein complexes and prevent their interaction with their natural targets (e.g., GAS5 (Growth arrest-specific 5), PANDA (P21 associated ncRNA DNA damage activated), NEAT1 (Nuclear paraspeckle assembly transcript 1)). "miRNA sponge": ceRNAs (competing endogenous RNAs) can bind to and sequester miRNA, leading to the active transcription of their mRNA targets (e.g., MD1 (Myoblast Differentiation 1), H19, ROR (Regulator of Reprogramming)]. "Enhancer": eRNAs (enhancer RNAs) are derived from enhancer regions and regulate neighboring gene expression in cis and trans, being often implicated in creating chromatin loops for stabilization and activation of the promoter region of their target gene [e.g., lncRNA-CSR (class switch recombination), HOTTIP (HOXA transcript at the distal tip)] (Reprinted with permission from Bouckenheimer et al. 2016).

recruiting coactivators and transcription complexes to facilitate transcription (Hung et al. 2011). On the other hand, they have also been shown to repress transcription, by directly blocking access, acting as decoys or competing with the TF itself (Willingham et al. 2005). Some studies have also shown the role of lncRNA directly on Pol II-mediated transcription as well as acting as coactivators. The example of some of these lncRNA that modulate transcription through direct or indirect means includes Gas5, SRA and NRON (Lanz et al. 1999, Schnell et al. 2004). H19 is one of the highly expressed lncRNA in cancer and it mediates (1) SP1-TGFBR2 (TF-gene) interaction through let-7b and miR-200b, (2) ETS1-TGFBR2 (TF-gene) interaction through miR-29a and miR-200b, and (3) STAT3-KLF11 (TF-gene) interaction through miR-17. Such regulatory triplets can be used to predict the potential function of lncRNA and miRNAs in cancer. This critical regulation could play a pivotal role in understanding the molecular mechanisms of cancer initiation and progression and could serve as biomarkers for cancer diagnosis, drug development and therapeutic strategy development (Aimin Li et al. 2020).

4.4.2 Posttranscriptional Regulation

RNA processing is composed of several distinct steps involving several RBPs. After transcription, these RBPs facilitate the overall turnover of the RNA, its stability, degradation and localization. These events together determine the dynamic and steady-state levels of mRNA. lncRNA have been shown to participate along with RBPs in modulating at least some of these posttranscriptional mechanisms. It is notable that one of the most characterized lncRNA, NEAT1 is localized to paraspeckles and is an essential structural component of this membrane-less organelle. MALAT1 has been shown to localize with nuclear speckles. These organelles have important roles in RNA processing events (Tripathi et al. 2010; Tsuiji et al. 2011). SRSF1 (SR splicing factor 1) can interact with MALAT1 through its RRM domain (RNA Recognition Motifs). It is required for efficient localization of SRSF1 as well as several other splicing factors to nuclear speckles. SR protein interactions with MALAT1 in the nuclear speckles alter their phosphorylation status, thereby leading to alterations in alternative splicing patterns (Tripathi et al. 2010). Apart from the facilitation of correct localization, lncRNA are also known to act as sponges thus intervening with miRNA and mRNA interactions.

4.4.3 Post-Translational Regulation

Although evidence supporting the role of lncRNA in post-translational regulation is limited, nevertheless there are recent studies that show that the lncRNA-p21 can associate with polysome bound β-catenin (Ebralidze et al. 2008; Yoon et al. 2012). There have also been other studies showing lncRNA interaction with polysomes suggesting that the lncRNA are capable of fine-tuning gene expression through all stages of transcription and translation. lncRNA can also interact directly with the proteins and regulate their stability by preventing protein ubiquitination and degradation. Liu et al. found that MT1JP acts as a crucial factor in cellular transformation through modulation of p53 translation by binding and stabilization of the RNA binding protein TIAR (Liu et al. 2016). LoNA, a nucleolar-specific lncRNA, has been detected to reduce the production of rRNA ribosome biosynthesis by binding to nucleolin through its 5' region and sequestering it, thereby suppressing rRNA transcription and diminishing fibrillarin activity by recruiting 3' end thereby reducing rRNA methylation (Li et al. 2018).

4.4.4 Interactions with RBPS

About 7% of all the proteins are RBPs that regulate the functioning and fate of a broad range of transcripts thereby contributing to cellular homeostasis. Recent proteomics and transcriptomics studies have revealed that lncRNA are often included in the RBP-based complexes. mRNA decay is an essential regulatory mechanism in gene expression. Long ncRNA regulates the mRNA decay by interacting with the RBPs and thereby modulates important cellular functions and pathological events. NORAD (non-coding RNA activated by DNA damage) is one such example of lncRNA, which is highly conserved and abundantly expressed in the mammalian tissues as a result of tissue damage. This lncRNA functions as a multivalent binding platform for the PUMILIO (PUM) family of RBPs (Lee et al. 2016).

4.4.5 Chromatin Maintenance

Chromatin is a nucleoprotein complex that is a very dynamic entity in the cell that governs gene expression. The chromatin states are defined by the proteins that are part of it, and the various epigenetic factors that interact with chromatin. The Xist lncRNA is one of the pioneer examples under this category (Loda and Heard. 2019). Several lncRNA have been shown to be part of the chromatin, both active as well as inactive chromatin. They act as scaffolds providing a binding site for the activator and/or repressor complexes and through cis interactions can regulate gene expression. A well-known example under this category is the lncRNA HOTAIR (Khalil et al. 2009). HOTAIR and chromatin remodeling complexes such as Polycomb repressive complexes 1 and 2 (PRC1 and PRC2) (Rinn et al. 2007) mediate mono-ubiquitinylation of Lysine 119 of Histone 2A (H2AK119ub) (Wang et al. 2004) and di- and trimethylation of Histone H3 on lysine 27 (H3K27me2 and H3K27me3) (Cao et al. 2004), respectively; as well as Lysine Specific Demethylase 1 (LSD1)/CoREST which demethylates mono- and di-methylated Histone 3 at Lysine 4 (H3K4) and G9a histone methyltransferase which catalyzes Histone 3 Lysine 9 di- and trimethylation (H3K9me2 and H3K9me3). This RNA–protein interaction between HOTAIR and PRC2 is essential for proper localization of PRC2 across the HOXD cluster. Infact, deletion of HOTAIR results in the mislocalization of PRC2 from the HOXD cluster, in trans, and concomitantly activates HOXD genes (Shi et al. 2004). Thus, lncRNA can act as recruiters of epigenetic regulators to the chromatin and also as tethering junctions holding the chromatin in a particular state.

4.5 DUAL CONTRASTING ROLE PLAY BY lncRNA

Functional activity of long non-coding RNAs is not limited to the physiological events in cellular processes, it has a very important and regulatory role in several signaling pathways which are directly or indirectly involved in cancer or tumorigenesis. lncRNA functions through either as a regulatory component by interaction with proteins, such as receptors, TFs as well as by interacting with RNA and lipids. lncRNA are known to modulate the signaling transduction by directly associating with receptors, several protein kinases, TFs and with various downstream signaling molecules. They can enhance the phosphorylation events, or downregulate the signaling pathways by altering enzymatic activity of kinases. To date, various researchers have investigated the role of lncRNA in cancer progression and have identified that this occurs by altering signaling pathways. Association of lncRNA Lnc-DC with STAT3 prevents the SHP1-mediated dephosphorylation in JAK-STAT signaling (Wang et al. 2014). The various lncRNAs with their role and mechanism are mentioned in Table 4.1. Understanding the functional activity and regulatory role of lncRNA in signaling pathways should create the direction for therapeutic targets in cancer treatment. Cellular activities that are a result of various complex signaling cascade which are activated by different extra- or intracellular signals, Researchers have identified that lncRNA could be interacting with all of the steps of these complex signaling pathways (Lin and Yang 2018). A representation of the balance between oncogenic and tumor-suppressor lncRNAs can be seen in Figure 4.3.

TABLE 4.1
lncRNAs Involved in Various Cancer Types

lncRNA	Role	Type of Cancer	Mechanism of Action	Biological Activity
H19	Oncogenic	NSCLC	Activation of STAT3 via sponging miR-17	Cell growth, migration and invasion
LINC00152	Oncogenic	NSCLC	Activation of EGFR/PI3K/Akt signaling pathway	Cell proliferation
SOX21-AS1	Oncogenic	Colorectal cancer	Activation of MYO6 via sponging miR-145	Cell growth, proliferation and invasion
NEAT1	Oncogenic	Glioblastoma	Activation of c-Met via sponging miR-449b-5p	Cell proliferation, invasion and migration
	Oncogenic	Glioblastoma	Activation of SOX2 via sponging miR-132	Cell invasion and migration
MALAT1	Oncogenic	Breast cancer	Activation of CDK4/E2F1 pathway via sponging miR-124	Cell proliferation and cell cycle
	Tumor suppressor	Breast cancer	pTEN-miRNA-MALAT1 16 axis	Tumor suppression
GAS5	Tumor suppressor	TNBC (Triple-negative Breast cancer)	miR-196a-5p/ forkhead box protein O1(FOXO)/PI3K (phosphatidylinositol 3-kinase)/ AKT axis	Tumor suppression
LET	Tumor suppressor	nasopharyngeal carcinoma	EZH2- mediated H3K27 histone trimethylation of LET promoter	Tumor suppression

- Receptor dimerization (Homo or Hetero-dimer) could be regulated by lncRNA.
- Recruitment of protein kinases to receptors could be enhanced because of association of lncRNA for activation/inactivation.
- Sometimes lncRNA itself acts as a Second Messenger in signaling cascades
- By associating with Kinases, lncRNA modifies effectors molecules through post-translational mechanisms in a Kinase dependent manner.
- lncRNA regulates the entry of downstream signaling components, especially TFs entry into nucleus by associating and modulating the TF binding to cognate sites.

4.6 ASSOCIATION WITH SIGNALING CASCADES

4.6.1 lncRNA and Wnt-Signaling

Wnt Signaling is conserved from embryo to adult and plays a crucial role in physiology. Cellular proliferation, growth and differentiation are driven by this signaling during embryogenesis whereas in adults, it is involved in maintaining tissue homeostasis. There are three major Wnt-signaling pathways, i) canonical (β-catenin dependent), ii) noncanonical (Wnt/PCP) and iii) Wnt/Ca2+ signaling. Among them, β-catenin mediated signaling has been more extensively studied. In the absence of Wnt, cytoplasmic level of β-catenin is decreased by ubiquitin-mediated proteasomal degradation. Two kinases GSK3/CK1 are bound to β-catenin with the help of scaffold protein Axin and APC. Multiple phosphorylation of β-catenin 11 on serine and threonine residues by kinases

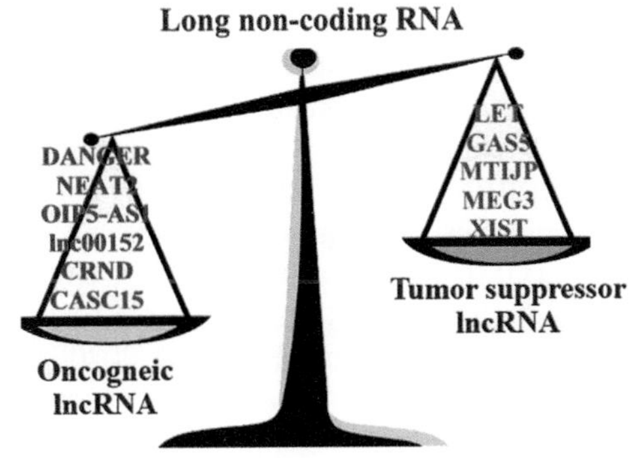

FIGURE 4.3 The figure demonstrates the dual role of lncRNAs. The oncogenic LncRNAs play dominant role in activation of signaling pathways and the tumor-suppressor lncRNAs functioning majorly in epigenetic regulation.

lead to degradation. In the nucleus, transcription of downstream genes is kept off by Groucho protein. When Wnt binds to its receptors, β-catenin is stabilized and enters into the nucleus, and along with co-activator TCF, it starts the transcription of several growth-related Wnt target genes. Recently, it has been shown that lncRNA interacts with a variety of signaling pathways, Wnt signaling is one of them. The perturbation of Wnt signaling by lncRNA in different cancers is summarized below. lncRNA is directly involved in the regulation of the intermediate component of Wnt signaling (for example β-catenin) and sometimes also suppresses the signaling resulting in cancer progression. In colorectal cancer, lncRNA CCAT-2 helps in the recruitment of TCF3 (a co-activator of Wnt-targeted gene transcription) to the promoter, directly associating with it resulting in activation of Wnt signaling. Another lncRNA RBM-AS1 acts as a transcriptional activator of Wnt-targeted genes by binding with β-catenin and recruiting it to the promoter (Javed et al. 2020). In Renal cancer cell carcinoma, lncRNA CRND modifies PI3K/AKT/GSK3-beta signaling pathway in an unknown manner, resulting in up-regulation of β-catenin level in cells which leads to activation of Wnt signaling (Yang et al. 2018). In hepatocellular carcinoma, lncRNA TCF7 recruits SWI/SNF chromatin remodeling complex to the promoter of TCF co-activator of Wnt-targeted gene expression leading to activation of Wnt-signaling (Wang et al. 2015). lncRNA CASC11 stabilizes hnRNP-K (a poly-C binding protein family), which protects β-catenin from GSK3 by interacting with Axin and GSK3, leading to the upregulation of βcatenin signaling (Hu et al. 2018). In NSCLC cells, lncRNA AK126698 (Fu et al. 2016) and in colorectal cancer, lncRNA ADAMTS9-A1 (Li et al. 2020) both suppress the Wnt signaling by targeting and inhibiting Wnt receptor Frizzled 8 and by downregulating β-catenin, respectively.

4.6.2 lncRNA and TGF-β Signaling

TGF-beta/Smad signaling pathway works through the TGF-β superfamily ligands (TGF β-1, 2,3,5; BMP, Activin), which bind to type-II receptor, allow the type-I dimerization with type-II receptor and are phosphorylated. Following this, activated type-I receptor phosphorylates Smad2 and Smad3 which form complexes with Smad4 and enter into the nucleus thus regulating gene expression. Cellular growth, differentiation, cellular homeostasis and several other cellular activities are maintained by TGF β signaling. 12 Recent research has revealed that lncRNA can cause the up and downregulation of TGF β signaling, which is summarized below. lncRNA can recruit the phosphatases as a result of lncRNA-mediated dephosphorylation of intermediate components of signaling cascade. In hepatic cells, lncRNA MALAT1 recruits a phosphatase, PPM1A to the phosphorylated Smad2/Smad3 together with other proteins such as SETD2. PPM1A dephosphorylate Smad2/Smad3, so there is a reduction of Smad4 activation leading to termination of signaling (Zhang et al. 2020). lncRNA interfere with the recruitment of TFs on promoters of TGF-β target gene through epigenetic modification. lncRNA TGFB2-AS1 keeps the transcription of TGF-β targeted genes, turned off, by H3K12 methylation. LncRNA binds to the 3' terminal end of the

EED adaptor of Polycomb Repressive Complex (PRC2) and recruits it for methylation (Papoutsoglou et al. 2019). Involvement of lncRNA CASC9 has an important role in development of colorectal cancer, wherein the association of lncRNA CASC9 with its binding protein CPSF3, stabilizes the ligands, TGF-β2 and TRET, thus resulting in hyperactivation of signaling, resulting in more accumulation of Smad3 phosphorylation (Luo et al. 2019). lncRNA acts as an activator of almost all the components of TGF-beta signaling cascade causing progression of cancer. In hepatocellular carcinoma, lncRNA MALAT1 upregulates TGF-β concentration, NORAD, KR19 up-regulate TGFBR1 and TGFBR2; lncRNA34A up-regulate Smad4; KRT19, lncRNA LFAR1 up-regulate Smad2/3 complex, and all of these up-regulate TGF-β signaling (Han et al. 2020).

4.6.3 lncRNA and JAK-STAT Pathway

JAK-STAT signaling is driven by cytokine receptors as well as by hormone signaling. JAK (Janus kinase) is a part of the receptor, activated upon binding of cytokines to its specific receptor. The activated JAK kinase then phosphorylates the cytosolic part of receptor which is the binding site of SH2 domain-containing protein STAT (signal transducer and activator of transcription), the bound STAT is also phosphorylated by JAK, leading to the dimerization of STAT which enters into the nucleus, regulating key processes like cell division, cell growth, differentiation and cell migration. The normal activation and inactivation of this signaling are maintained through dephosphorylation of receptors and JAK by SHP1 phosphatase, and ubiquitin-mediated destruction by SOCS (suppressor of cytokine signaling) protein. lncRNA binds and inactivates such inhibitors. lncRNA LINC00669 in Nasopharyngeal cell carcinoma (NPC) associates with SOCS protein and blocks its ubiquitin ligase activity so that STAT1 levels increase, resulting in longer activation of JAK-STAT signaling and nasopharyngeal carcinoma progression (Qing et al. 2020). lncRNA can up-regulate the component of JAK-STAT signaling cascade from ligand to intermediate components, leading to an increase in the phosphorylation event of kinases and STAT. lncRNA also acts as a double negative gate that is the repressor of JAK-STAT signaling. In NSCLC, lncRNA PART1 acts as a repressor of microRNA miR-635, which is a repressor of JAK1 and JAK3 gene transcription. Sponging mechanism of PART1 accelerates the JAK/STAT stabilization and phosphorylation, ultimately causing cancer progression (Zhu et al. 2019). A few lncRNA act as a negative regulator of JAK-STAT signaling by downregulating the STAT. In colorectal cancer, lncRNA RP11-468F2.5 binds to STAT5 and STAT6 resulting in inhibition of JAK/STAT-mediated cell proliferation and promotes apoptosis (Jiang et al. 2019).

4.6.4 lncRNA and PI3/AKT Pathway

Several cytokines and RTK signaling events activate and recruit the PI3K (Phosphatidyl-inositol3 Kinase). PI3K uses phospholipids (Phosphatidylinositol) as a substrate and generates PI-3 phosphates [PI (3,4) P2 and PI (3,4,5) P3] which act as docking sites of other proteins, such as protein kinase-B or Akt. lncRNA up-regulates expression of signaling components such as protein kinase-B, which promote the cellular signaling by phosphorylation events. lncRNA DANCR promotes the expression of PKB gene through phosphorylation of RXRA (thyroxine receptor) at 49/78 serine residues (Tang et al. 2018). In lung cancer, lncRNA OECC has been shown to associate with Akt genes, stabilizing the mRNA and promoting gene expression (Zhu et al. 2019). In gastric cancer, lncRNA AK023391 accelerates the PI3K/Akt pathway through an increase in phosphorylated PI3K levels (Huang et al. 2017).

4.6.5 lncRNA and MAPK/ERK Signaling

Another important signaling pathway required for cellular proliferation, gene expression, differentiation, mitosis and cell survival is MAPK (Mitogen-activated Protein Kinases)/ERK 14 (Extracellular signal-regulated kinase) signaling also known as RAS-RAF-MAPK signaling. MAPKs are serine threonine kinases that phosphorylate downstream proteins which are TFs. This phosphorylation event acts like an on/off switch for signaling. In case of MAPK signaling also, lncRNA plays a role in both up and downregulation. lncRNA can stabilize and help in maintaining the phosphorylated intermediate components of the signaling cascades, through upregulation of the phosphorylated MEK1/2 and C-RAF, which upregulates p38 activation by phosphorylation (Liu et al. 2019; Wang et al. 2018). lncRNA ENS-653 associates with ERK1/2 and ERK5 (Intermediate component of MAPK signaling cascade) and promotes phosphorylation, leading to progression of MAPK signaling in Papillary Thyroid Cancer (Song et al. 2019). In Breast cancer, lncRNA linc-ROR associates with dual specific kinase-DUASP7, which is a negative regulator of ERK phosphorylation, and blocks its activity thus resulting in upregulation of p-ERK (Peng et al. 2017). Some lncRNA function as an endogenous RNA which compete with other miRNA that are repressors of MAPK signaling, thus ultimately resulting in upregulation of MAPK signaling. In NSCLC, miRNA miR-181a blocks the phosphorylation of MAPK1 and MAP2K1. lncRNA SNHG12 inhibits miR-181a and upregulates the phosphorylated MAPK1 and MAP2K1 leading to activation of signaling (Wang et al. 2017).

4.7 lncRNA AS TUMOR SUPPRESSORS

lncRNA partake gene regulation both in cis and trans manner. Mature transcripts of lncRNA have been found to regulate the expression of nearby or far-off genes in a sequence-specific manner by regulating the structure and organization of their chromatin, by directly binding to their promoter or enhancer regions, or by aiding the development of complexes that support or mask active transcription (Johnson et al. 2012; Kopp et al. 2018). A number of lncRNA are delineated as mediators of epigenetic silencing because it has been found that they recruit complexes like Polycomb Repressive Complex 2 (PRC2), which is responsible for trimethylation of Lysine at 27th position on Histone 3 (H3K27me3), referred to as a repressive modification that silences the transcription of the genetic locus (Tsai et al. 2010; Hirata et al. 2015). Thus, lncRNA not only promote or activate signaling pathways, but they are also part of repressive complexes, acting as negative regulators of cancer progression. A few examples are listed below:

4.7.1 GAS5

Growth Arrest-Specific 5 (GAS5), a non-coding gene sequesters miR-21, thus functions as an endogenous sponge and as a result, reciprocal repression of both miR-21 and GAS5 occurs. This interaction occurs on the putative miR-21 binding site at exon 4 of GAS5 and RISC (RNA-induced silencing complex) (Zhang et al. 2013). GAS5 also inhibits TNBC (Triple-negative Breast cancer) cell invasion through miR-196a-5p/forkhead box protein O1(FOXO)/PI3K (phosphatidylinositol 3-kinase)/AKT axis that initiates with competitive binding of GAS5 to miR-196a-5p (Li et al. 2018). The mechanism has been demonstrated in Figure 4.4.

4.7.2 MT1JP

Reported as a tumor suppressor, MT1JP (Metallothionein 1J pseudogene), modulates p53 protein expression level, thereby regulating p53 related signaling pathway (Liu et al. 2016). Reduced MT1JP expression leads to significant lymphatic metastasis and advanced TNM (Tumor Node Metastasis) stage for gastric cancer patients. In-vitro and in-vivo studies also show the suppressive role of MT1JP on the migration and invasion of gastric cancer (Xu et al. 2018; Zhang et al. 2018).

4.7.3 LET

As the name suggests, Low Expression in Tumor (LET) lncRNA is usually down-regulated in most of the cancer types. It is negatively regulated by Hypoxia-induced deacetylase 3 (HDAC3) thus,

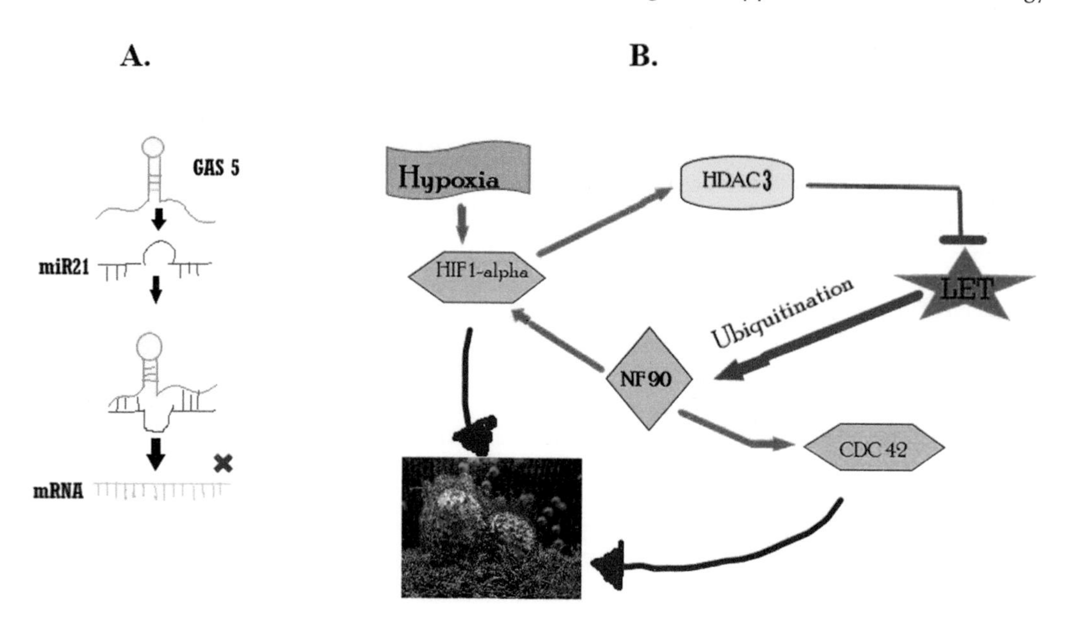

FIGURE 4.4 lncRNAs involved in Invasion and Migration (A) GAS5 repressing miR-21 by binding and sequestering it. (B) Negative regulation of lncRNA LET by HDAC3.

promoting cellular invasion by nuclear factor 90 (NF90) stabilization which is an outcome of its downregulation (Sun et al. 2015) as demonstrated in Figure 4.4. In nasopharyngeal carcinoma cells, its repression has been linked to EZH2- mediated H3K27 histone trimethylation of LET promoter, which leads to inhibition of cell proliferation and apoptosis (Yang et al. 2013).

4.7.4 MALAT1

Metastasis Associated Lung Adenocarcinoma Transcript 1 is both a tumor promoting as well as suppressing lncRNA. Its escalated expression is associated with relapse and metastatic progression in breast cancer while on the other hand, few recent studies provide insight about its metastasis-suppressing activity. Mechanistically, it prevents the association of prometastatic TF TEA domain family member 2 (TEAD) with its co-activator YAP (Yes-associated protein) (Kim et al. 2018). Kim et al. reported that somatic knockout of MALAT1 specifically induced metastasis. Kwok et al. suggested that PTEN-miRNA-MALAT1 16 axis may promote tumorigenesis and reported the first evidence proving that MALAT1 possesses novel tumor-suppressive capabilities in breast cancer (Kwok et al. 2018).

4.7.5 MEG3

Known as Maternally expressed gene 3, this lncRNA belongs to the DLK1-MEG3 imprinted locus. MEG3 expression inhibition has been found in various types of human tumors and tumor cell lines including those derived from brain, bladder, bone marrow, breast, cervix, colon, liver, lung, meninges and prostate (Zhou et al. 2012; Braconi et al. 2011; Wang et al. 2012). MEG3 expression is lost or significantly decreased in 25% of neuroblastomas (Astuti et al. 2005), 81% of hepatocellular cancers (Braconi et al. 2011), and 82% of gliomas. Recently, in meningiomas also MEG3 expression was examined and it was found that loss of MEG3 expression is strongly linked with tumor grade. It was detected in 1 of 11 grade II, 4 of 9 grade I, and none of seven grade III meningiomas (Zhang et al. 2010a).

4.7.6 XIST

XIST is considered to be a potential tumor suppressor in cancer (Xing et al. 2018). It was found by a pathway screening based on gene set enrichment analysis (GSEA) that XIST activates moesin (MSN) on the X chromosome which leads to activation of c-Met signaling pathway. Loss of c-Met signaling decreases the transmigration abilities of Breast cancer cells and suppresses the breast-to-brain metastatic abilities in vivo (Xu et al. 2018).

4.8 lncRNA AS THERAPEUTIC TARGETS

Oncogenic and suppressive functions of lncRNA have been shown to have a deterministic role in cancer. Due to its involvement in various cellular processes and its functional outcome, lncRNA are an attractive therapeutic target especially against cancer. Their restricted appearance and ability to interact with proteins, lipids as well as with RNA makes them amenable to being targeted by the existing anticancer drugs. In Papillary Thyroid Carcinoma (PTC), lncRNA BANCR regulate EMT by up and downregulation of E-cadherin and N-cadherin and regulate RAF/MEK/ERK pathway by increasing the phosphorylation event of intermediates. Using inhibitors of BANCR 17 would be a therapeutic target in PTC as well as various disorders caused by upregulation of the pathway (Wang et al. 2018). lncRNA acts as a dual regulator in cancer and the mechanism of regulation in most of the cases occurs through genetic regulation and/or epigenetic regulation. So, targeting lncRNA at all these levels can be considered from a therapeutic perspective.

4.8.1 Nucleic Acid Modulators of lncRNA

Small interfering RNA (siRNA) forms a complex with RISC (RNA-Induced Silencing Complex) and binds to its target region and degrade it. siRNA can be applied to target the lncRNA as a therapeutic. In Esophageal squamous cell carcinoma knockdown of lncRNA CASC9 by siRNA downregulates the metastasis (Liang et al. 2018). Targeting lncRNA MALAT1 by siRNA is very effective in metastasis suppression in human bladder cancer and lung adenocarcinoma (Matsui and Corey 2017). A single-stranded oligonucleotide is synthesized by 15–25 bases which are complementary to the RNA, known as Antisense Oligonucleotides (ASO). It can bind to the RNA and degrade it by recruiting RNase H. Nowadays the use of ASOs to target the lncRNA is showing very effective results. In breast cancer, use of MALAT1 ASO has shown reduction of metastasis and slower tumor formation (Arun et al. 2016). Prokaryotic organisms used the CRISPR system as an immune defence system to degrade foreign genetic material. CRISPR with CAS9 nuclease binds the target side and degrades it. Now this system is used widely as a genome editing tool in mammalian cells and can be applied to target the lncRNA. Recently, it has been shown that metastasis can be controlled by CRISPR/Cas9 mediated inhibition of lncRNA LOC389641 in pancreatic duct carcinoma (Goyal et al. 2017). In several signaling pathways, the regulatory effect of lncRNA has been shown to occur through the sponging of miRNA. Interaction of lncRNA and miRNA regulates EMT in various cancers, in lung cancer miRNA217 interacts with MALAT1 and triggers its decay, causing decrease of EMT (Lu et al. 2015). Hence, targeting this crosstalk of lncRNA and miRNA would be an important therapeutic target.

4.8.2 Small Molecule Modulators of lncRNA

lncRNA interaction with the target is governed by structural adaptations. This can be targeted by various small molecules to disrupt the function of lncRNA. Sponging of miRNA-186 by lncRNA HULC causes progression of Ewing sarcoma. Small molecule YK-4-279 inhibits such sponging activity of lncRNA HULC (Mercatelli et al. 18 2020). Several plant-derived natural compounds are found to be effective regulators of lncRNA. Some studies have shown that lncRNA MALAT1,

NEAT11, H19 and PVT1 are affected by natural compounds. In multiple myeloma cells, NEAT1 promotes cell proliferation, migration and invasion. Treatment of resveratrol causes the down-regulation of NEAT activity (Geng et al. 2018).

4.9 FUTURE PROSPECTS

In this overview, a glimpse of lncRNA and their role in cancer have been dealt with at length. These non-coding RNAs play a very crucial role in other disorders as well, essentially neurological disorders like Alzheimer's and Schizophrenia. Since these classes of non-coding RNAs are found to be associated with RBPs as well as seen to be causing modifications in the chromatin structure, they pave a crucial path for several signaling pathways yet to be explored. By the recent discoveries it's been clear that even being translationally inactive, their far-reaching effects are worth studying in-depth to bring out their dual role more precisely and prominently in development and diseases. Moreover, studying lncRNAs would be of utmost help to reveal and define the regulatory networks, molecular markers and characteristics of human cancers. Since they interestingly portray a dual role of oncogenic and tumor suppressor, digging deep into their contrasting relationship might be helpful in creating valuable insights into the prevention of cancer as well as other neurological disorders.

REFERENCES

Akhade, V.S., Pal, D., Kanduri, C., 2017. Long noncoding RNA: Genome organization and mechanism of action. *Adv. Exp. Med. Biol.* 1008, 47–74.

Arun, G., Diermeier, S., Akerman, M., Chang, K.C., Wilkinson, J.E., Hearn, S., Kim, Y., MacLeod, A.R., Krainer, A.R., Norton, L., Brogi, E., Egeblad, M., Spector, D.L., 2016. Differentiation of mammary tumors and reduction in metastasis upon Malat1 LncRNA loss. *Genes Dev.* 30, 34–51.

Astuti, D., Latif, F., Wagner, K., Gentle, D., Cooper, W.N., Catchpoole, D., Grundy, R., Ferguson-Smith, A.C., Maher, E.R., 2005. Epigenetic alteration at the DLK1-GTL2 imprinted domain in human neoplasia: Analysis of neuroblastoma, phaeochromocytoma and Wilms' tumour. *Brit. J. Cancer* 92, 1574–1580.

Bardou, F., Ariel, F., Simpson, C.G., Romero-Barrios, N., Laporte, P., Balzergue, S., Brown, J.W., Crespi, M., 2014. Long noncoding RNA modulates alternative splicing regulators in *Arabidopsis*. *Dev. Cell* 30, 166–176.

Bouckenheimer, J., Assou, S., Riquier, S., Hou, C., Philippe, N., Sansac, C., Lavabre-Bertrand, T., Commes, T., Lemaître, J.M., Boureux, A., De Vos, J., 2016. Long non-coding RNAs in human early embryonic development and their potential in ART. *Hum. Reprod. Update* 23, 19–40.

Braconi, C., Henry, J.C., Kogure, T., Schmittgen, T., Patel, T., 2011. The role of microRNAs in human liver cancers. *Sem. Oncol.* 38, 752–763.

Brannan, C.I., Dees, E.C., Ingram, R.S., Tilghman, S.M., 1990. The product of the H19 gene may function as an RNA. *Mol. Cell. Biol.* 10, 28–36.

Brenner, S., Jacob, F., Meselson, M., 1961. An unstable intermediate carrying information from genes to ribosomes for protein synthesis. *Nature* 190, 576–581.

Britto-Kido, S.deA., Ferreira Neto, J.R., Pandolfi, V., Marcelino-Guimarães, F.C., Nepomuceno, A.L., Vilela Abdelnoor, R., Benko-Iseppon, A.M., Kido, E.A., 2013. Natural antisense transcripts in plants: A review and identification in soybean infected with *Phakopsora pachyrhizi* SuperSAGE library. *Sci. World J.* 2013, 219798.

Brown, C.J., Lafreniere, R.G., Powers, V.E., Sebastio, G., Ballabio, A., Pettigrew, A.L., Ledbetter, D.H., Levy, E., Craig, I.W., Willard, H.F., 1991. Localization of the X inactivation centre on the human X chromosome in Xq13. *Nature* 349, 82–84.

Cao, R., Zhang, Y., 2004. The functions of E(Z)/EZH2-mediated methylation of lysine 27 in histone H3. *Curr. Opin. Genet. Dev.* 14, 155–164.

Carrieri, C., Cimatti, L., Biagioli, M., Beugnet, A., Zucchelli, S., Fedele, S., Pesce, E., Ferrer, I., Collavin, L., Santoro, C., Forrest, A.R., Carninci, P., Biffo, S., Stupka, E., Gustincich, S. 2012. Long non-coding antisense RNA controls Uchl1 translation through an embedded SINEB2 repeat. *Nature* 491, 454–457.

Chooniedass-Kothari, S., Emberley, E., Hamedani, M.K., Troup, S., Wang, X., Czosnek, A., Hube, F., Mutawe, M., Watson, P.H., Leygue, E., 2004. The steroid receptor RNA activator is the first functional RNA encoding a protein. *FEBS Lett.* 566, 43–47.

Clark, M. B., Johnston, R. L., Inostroza-Ponta, M., Fox, A. H., Fortini, E., Moscato, P., Dinger, M. E., Mattick, J. S., 2012. Genome-wide analysis of long noncoding RNA stability. *Genome Res.* 22, 885–898.

Denoeud, F., Kapranov, P., Ucla, C., Frankish, A., Castelo, R., Drenkow, J., Lagarde, J., Alioto, T., Manzano, C., Chrast, J., Dike, S., Wyss, C., Henrichsen, C.N., Holroyd, N., Dickson, M.C., Taylor, R., Hance, Z., Foissac, S., Myers, R.M., Rogers, J., … Reymond, A., 2007. Prominent use of distal 5′ transcription start sites and discovery of a large number of additional exons in ENCODE regions. *Genome Res.*, 17, 746–759.

Dinger, M.E., Pang, K.C., Mercer, T.R., Mattick, J.S., 2008. Differentiating protein-coding and noncoding RNA: Challenges and ambiguities. *PLoS Comp. Biol.* 4(11), e1000176.

Djebali, S., Davis, C.A., Merkel, A., Dobin, A., Lassmann, T., Mortazavi, A., … Gingeras, T.R., 2012. Landscape of transcription in human cells. *Nature* 489, 101–108.

Ebralidze, A.K., Guibal, F.C., Steidl, U., Zhang, P., Lee, S., Bartholdy, B., Jorda, M.A., Petkova V., Rosenbauer, F., Huang, G., Dayaram, T., 2008. PU. 1 expression is modulated by the balance of functional sense and antisense RNAs regulated by a shared cis-regulatory element. *Genes Dev.* 22, 2085–2092.

ENCODE Project Consortium 2012. An integrated encyclopedia of DNA elements in the human genome. *Nature* 489, 57–74.

Fu, X., Li, H., Liu, C., Hu, B., Li, T., Wang, Y., 2016. Long noncoding RNA AK126698 inhibits proliferation and migration of non-small cell lung cancer cells by targeting Frizzled-8 and suppressing Wnt/β-catenin signaling pathway. *Onco Targets Ther.* 9, 3815–3827.

Gabory, A., Jammes, H., Dandolo, L., 2010. The H19 locus: role of an imprinted non-coding RNA in growth and development. *BioEssays* 32, 473–480.

Geng, W., Guo, X., Zhang, L., Ma, Y., Wang, L., Liu, Z., Ji, H., Xiong, Y., 2018. Resveratrol inhibits proliferation, migration and invasion of multiple myeloma cells via NEAT1-mediated Wnt/β-catenin signaling pathway. *Biomed. Pharmacother.* 107, 484–494.

Gerstein, M.B., Bruce, C., Rozowsky, J.S., Zheng, D., Du, J., Korbel, J.O., Emanuelsson, O., Zhang, Z.D., Weissman, S., Snyder, M., 2007. What is a gene, post-ENCODE? History and updated definition. *Genome Res.* 17, 669–681.

Gong, C., Maquat, L.E., 2011. LncRNAs transactivate STAU1-mediated mRNA decay by duplexing with 3‘ UTRs via Alu Elements. *Nature* 470, 284–290.

Goyal, A., Myacheva, K., Groß, M., Klingenberg, M., Arque, D.B., Diederichs, S., 2017. Challenges of CRISPR/Cas9 applications for long non-coding RNA genes. *Nucl. Acids Res.* 45, 1–13.

Guttman, M., Amit, I., Garber, M., French, C., Lin, M.F., Feldser, D., Huarte, M., Zuk, O., Carey, B.W., Cassady, J.P., Cabili, M.N., Jaenisch, R., Mikkelsen, T.S., Jacks, T., Hacohen, N., Bernstein, B.E., Kellis, M., Regev, A., Rinn, J.L., Lander, E.S., 2009. Chromatin signature reveals over a thousand highly conserved large non-coding RNAs in mammals. *Nature* 458, 223–227.

Guttman, M., Donaghey, J., Carey, B.W., Garber, M., Grenier, J.K., Munson, G., Young, G., Lucas, A.B., Ach, R., Bruhn, L., Yang, X., Amit, I., Meissner, A., Regev, A., Rinn, J.L., Root, D.E., Lander, E.S., 2011. LincRNAs act in the circuitry controlling pluripotency and differentiation. *Nature* 477, 295–300.

Han, M., Liao, Z., Liu, F., Chen, X., Zhang, B., 2020. Modulation of the TGF-beta signaling pathway by long noncoding RNA in hepatocellular carcinoma. *Biomark. Res.* 8, 1–11.

Hangauer, M.J., Vaughn, I.W., McManus, M.T., 2013. Pervasive transcription of the human genome produces thousands of previously unidentified long intergenic noncoding RNAs. *PLoS Genet.* 9, e1003569.

Hirata, H., Hinoda, Y., Shahryari, V., Deng, G., Nakajima, K., Tabatabai, Z.L., Ishii, N., Dahiya, R., 2015. Long noncoding RNA MALAT1 promotes aggressive renal cell carcinoma through Ezh2 and interacts with miR-205. *Cancer Res.* 75, 1322–1331.

Holoch, D., Moazed, D., 2015. RNA-mediated epigenetic regulation of gene expression. *Nat Rev Genet.* 16, 71–84.

Hu, X.Y., Hou, P.F., Li, T.T., Quan, H.Y., Li, M.L., Lin, T., Liu, J.J., Bai, J., Zheng, J.N., 2018. The roles of Wnt/β-catenin signaling pathway related LncRNA in cancer. *Int. J. Biol. Sci.* 14, 2003–2011.

Huang, Y., Zhang, J., Hou, L., Wang, G., Liu, H., Zhang, R., Chen, X., Zhu, J., 2017. LncRNA AK023391 promotes tumorigenesis and invasion of gastric cancer through activation of the PI3K/Akt signaling pathway. *J. Exp. Clin. Cancer Res.* 36, 1–14.

Hung, T., Wang, Y., Lin, M.F., Koegel, A.K., Kotake, Y., Grant, G.D., Horlings, H.M., Shah, N., Umbricht, C., Wang, P., Wang, Y., 2011. Extensive and coordinated transcription of noncoding RNAs within cell-cycle promoters. *Nat. Genet.* 43, 621–629.

Javed, Z., Khan, K., Sadia, H., Raza, S., Salehi, B., Sharifi-Rad, J., Cho, W.C., 2020. LncRNA and Wnt signaling in colorectal cancer. *Cancer Cell Int.* 20, 1–10.

Jia, H., Osak, M., Bogu, G.K., Stanton, L.W., Johnson, R., Lipovich, L., 2010. Genome-wide computational identification and manual annotation of human long noncoding RNA genes. *RNA* 16, 1478–1487.

Jiang, L., Zhao, X.H., Mao, Y.L., Wang, J.F., Zheng, H.J., You, Q.S., 2019. Long non-coding RNA RP11-48E2.5 curtails colorectal cancer cell proliferation and stimulate apoptosis via JAK/STAT signaling pathway targeting STAT5 and STAT6. *J. Exp. Clin. Cancer Res.* 38, 1–16.

Johnson, R., Derrien, T., Bussotti, G., Tanzer, A., Djebali, S., Tilgner, H., et al., 2012. The GENCODE v7 catalog of human long noncoding RNAs: Analysis of their gene structure, evolution, and expression. *Genome Res.* 22, 1775–1789.

Kapranov, P., Drenkow, J., Cheng, J., Long, J., Helt, G., Dike, S., Gingeras, T.R., 2005. Examples of the complex architecture of the human transcriptome revealed by RACE and high-density tiling arrays. *Genome Res.* 15, 987–997.

Kapranov, P., Cheng, J., Dike, S., Nix, D.A., Duttagupta, R., Willingham, A.T., et al., 2007. RNA maps reveal new RNA classes and a possible function for pervasive transcription. *Science* 316, 1484–1488.

Katayama, S., Tomaru, Y., Kasukawa, T., Waki, K., Nakanishi, M., Nakamura, M., …FANTOM Consortium, 2005. Antisense transcription in the mammalian transcriptome. *Science* 309, 1564–1566.

Kelley, D., Rinn, J., 2012. Transposable elements reveal a stem cell-specific class of long noncoding RNAs. *Genome Biol.* 13, R107.

Khalil, A.M., Guttman, M., Huarte, M., Garber, M., Raj, A., Rivea Morales, D., et al., 2009. Many human large intergenic noncoding RNAs associate with chromatin-modifying complexes and affect gene expression. *Proc. Natl. Acad. Sci. USA* 106, 11667–11672.

Kim, J., Piao, H.L., Kim, B.J., Yao, F., Han, Z., Wang, Y., et al., 2018. Long noncoding RNA MALAT1 suppresses breast cancer metastasis. *Nat. Genet.* 50, 1705–1715.

Kim, S.H., Suddath, F.L., Quigley, G.J., McPherson, A., Sussman, J.L., Wang, A.H., Seeman, N.C., Rich, A., 1974. Three-dimensional tertiary structure of yeast phenylalanine transfer RNA. *Science* 185, 435–440.

Knauss, J.L., Sun, T., 2013. Regulatory mechanisms of long noncoding RNAs in vertebrate central nervous system development and function. *Neuroscience* 235, 200–214.

Kopp, F., Mendell, J.T., 2018. Functional classification and experimental dissection of long noncoding RNAs. *Cell* 172, 393–407.

Kung, J.T., Colognori, D., Lee, J.T., 2013. Long noncoding RNAs: past, present, and future. *Genetics* 193, 651–669.

Kwok, Z.H., Roche, V., Chew, X.H., Fadieieva, A., Tay, Y., 2018. A non-canonical tumor suppressive role for the long non-coding RNA MALAT1 in colon and breast cancers. *Int. J. Cancer* 143, 668–678.

Lanz, R.B., McKenna, N.J., Onate, S.A., Albrecht, U., Wong, J., Tsai, S.Y., Tsai, M.J., O'Malley, B.W., 1999. A steroid receptor coactivator, SRA, functions as an RNA and is present in an SRC-1 complex. *Cell* 97, 17–27.

Lazorthes, S., Vallot, C., Briois, S., Aguirrebengoa, M., Thuret, J.Y., St Laurent, G., Rougeulle, C., Kapranov, P., Mann, C., Trouche, D., Nicolas, E., 2015. Erratum: A vlincRNA participates in senescence maintenance by relieving H2AZ-mediated repression at the INK4 locus. *Nat. Commun.* 6, 6918.

Lee, J.T., Bartolomei, M.S., 2013. X-inactivation, imprinting, and long noncoding RNAs in health and disease. *Cell* 152, 1308–1323.

Lee, S., Kopp, F., Chang, T.C., Sataluri, A., Chen, B., Sivakumar, S., Yu, H., Xie, Y., Mendell, J.T., 2016. Noncoding RNA NORAD regulates genomic stability by sequestering PUMILIO proteins. *Cell* 164, 69–80.

Li, A., Mallik, S., Luo, H., Jia, P., Lee, D.F., Zhao, Z., 2020. H19, a long non-coding RNA, mediates transcription factors and target genes through interference of MicroRNAs in pan-cancer. *Mol. Ther. Nucl. Acids* 21, 180–191.

Li, D., Zhang, J., Wang, M., Li, X., Gong, H., Tang, H., Chen, L., Wan, L., Liu, Q., 2018. Activity dependent LoNA regulates translation by coordinating rRNA transcription and methylation. *Nat. Comm.* 9, 1726.

Liang, Y., Chen, X., Wu, Y., Li, J., Zhang, S., Wang, K., Guan, X., Yang, K., Bai, Y., 2018. LncRNA CASC9 promotes esophageal squamous cell carcinoma metastasis through upregulating LAMC2 expression by interacting with the CREB-binding protein. *Cell Death Diff.* 25, 1980–1995.

Lin, C., Yang, L., 2018. Long noncoding RNA in cancer: Wiring signaling circuitry. *Trends Cell Biol.* 28, 287–301.

Liu, B., Zhao, H., Zhang, L., Shi, X., 2019. Silencing of long-non-coding RNA ANCR suppresses the migration and invasion of osteosarcoma cells by activating the p38 MAPK signaling pathway. *BMC Cancer* 19, 1–9.

Liu, L., Yue, H., Liu, Q., Yuan, J., Li, J., Wei, G., Chen, X., Lu, Y., Guo, M., Luo, J., Chen, R., 2016. LncRNA MT1JP functions as a tumor suppressor by interacting with TIAR to modulate the p53 pathway. *Oncotarget* 7, 15787–15800.

Loda, A., Heard, E., 2019. Xist RNA in action: Past, present, and future. *PLoS Genet.* 15, e1008333.

Louro, R., Smirnova, A.S., Verjovski-Almeida, S., 2009. Long intronic noncoding RNA transcription: expression noise or expression choice? *Genomics* 93, 291–298.

Lu, L., Luo, F., Liu, Y., Liu, X., Shi, L., Lu, X., Liu, Q., 2015. Posttranscriptional silencing of the LncRNA MALAT1 by miR-217 inhibits the epithelial-mesenchymal transition via enhancer of zeste homolog 2 in the malignant transformation of HBE cells induced by cigarette smoke extract. *Toxicol. Appl. Pharmacol.* 289, 276–285.

Luo, K., Geng, J., Zhang, Q., Xu, Y., Zhou, X., Huang, Z., Shi, K.Q., Pan, C., Wu, J., 2019. LncRNA CASC9 interacts with CPSF3 to regulate TGF- beta signaling in colorectal cancer. *J. Exp. Clin. Cancer Res.* 38, 1–16.

Marques, A.C., Hughes, J., Graham, B., Kowalczyk, M.S., Higgs, D.R., Ponting, C.P., 2013. Chromatin signatures at transcriptional start sites separate two equally populated yet distinct classes of intergenic long noncoding RNAs. *Genome Biol.* 14, R131.

Matsui, M., Corey, R.D., 2017. Noncoding RNAs as drug targets. *Nat. Rev. Drug. Discov.* 16, 167–179.

Mercatelli, N., Fortini, D., Palombo, R., Paronetto, M.P., 2020. Small molecule inhibition of Ewing sarcoma cell growth via targeting the long non-coding RNA HULC. *Cancer Lett.* 469, 111–123.

Milligan, M.J., Lipovich, L., 2014. Pseudogene-derived lncRNAs: Emerging regulators of gene expression. *Front. Genet.* 5, 476.

Morris, K.V., Mattick, J.S., 2014. The rise of regulatory RNA. *Nat. Rev. Genet.* 15, 423–437.

Mortazavi, A., Williams, B.A., McCue, K., Schaeffer, L., Wold, B., 2008. Mapping and quantifying mammalian transcriptomes by RNA-Seq. *Nat. Methods* 5, 621–628.

Nakaya, H.I., Amaral, P.P., Louro, R., Lopes, A., Fachel, A.A., Moreira, Y.B., El-Jundi, T.A., da Silva, A.M., Reis, E.M., Verjovski-Almeida, S., 2007. Genome mapping and expression analyses of human intronic noncoding RNAs reveal tissue-specific patterns and enrichment in genes related to regulation of transcription. *Genome Biol.* 8, R43.

Necsulea, A., Soumillon, M., Warnefors, M., Liechti, A., Daish, T., Zeller, U., Baker, J.C., Grützner, F., Kaessmann, H., 2014. The evolution of lncRNA repertoires and expression patterns in tetrapods. *Nature* 505, 635–640.

Papoutsoglou, P., Tsubakihara, Y., Caja, L., Morén, A., Pallis, P., Ameur, A., Heldin, C.H., Moustakas, A., 2019. The TGFB2-AS1 LncRNA regulates TGF-beta signaling by modulating corepressor activity. *Cell Rep.* 28, 3182–3198.

Peng, W.X., Huang, J.G., Yang, L., Gong, A.H., Mo, Y.Y., 2017. Linc-RoR promotes MAPK/ERK signaling and confers estrogen-independent growth of breast cancer. *Mol. Cancer.* 16, 1–11.

Penny, G.D., Kay, G.F., Sheardown, S.A., Rastan, S., Brockdorff, N., 1996. Requirement for Xist in X chromosome inactivation. *Nature* 379, 131–137.

Qing, X., Tan, G.L., Liu, H.W., Li, W., Ai, J.G., Xiong, S.S., Yang, M.Q., Wang, T.S., 2020. LINC00669 insulate the JAK/STAT suppressor SOCS1 to promote nasopharyngeal cancer cell proliferation and invasion. *J. Exp. Clin. Cancer Res.* 39, 1–16.

Rinn, J.L., Kertesz, M., Wang, J.K., Squazzo, S.L., Xu, X., Brugmann, S.A., Goodnough, L.H., Helms, J.A., Farnham, P.J., Segal, E., Chang, H.Y., 2007. Functional demarcation of active and silent chromatin domains in human HOX loci by noncoding RNAs. *Cell.* 129, 1311–1323.

Schnell, J.R., Dyson, H.J., Wright, P.E., 2004. Structure, dynamics, and catalytic function of dihydrofolate reductase. *Annu. Rev. Biophys. Biomol. Struct.* 33, 119–140.

Schwartz, J.C., Younger, S.T., Nguyen, N.-B., Hardy, D.B., Monia, B.P., Corey, D.R., Janowski, B.A., 2008. Antisense transcripts are targets for activating small RNAs. *Nat. Struct. Mol. Biol.* 15, 842–848.

Shi, Y., Lan, F., Matson, C., Mulligan, P., Whetstine, J.R., Cole, P.A., Casero, R.A., Shi, Y., 2004. Histone demethylation mediated by the nuclear amine oxidase homolog LSD1. *Cell* 119, 941–953.

Song, B., Li, R., Zuo, Z., Tan, J., Liu, L., Ding, D., Lu, Y., Hou, D., 2019. LncRNA ENST00000539653 acts as an oncogenic factor via MAPK signaling in papillary thyroid cancer. *BMC Cancer* 19, 2–12.

St Laurent, G., Shtokalo, D., Tackett, M.R., Yang, Z., Eremina, T., Wahlestedt, C., Urcuqui-Inchima, S., Seilheimer, B., McCaffrey, T.A., Kapranov, P., 2015. Intronic RNAs constitute the major fraction of the non-coding RNA in mammalian cells. *BMC Genom.* 13, 504.

Sun, Q., Liu, H., Li, L., Zhang, S., Liu, K., Liu, Y., Yang, C., 2015. Long noncoding RNA-LET, which is repressed by EZH2, inhibits cell proliferation and induces apoptosis of nasopharyngeal carcinoma cell. *Med. Oncol.* 32, 226.

Tang, J., Zhong, G., Zhang, H., Yu, B., Wei, F., Luo, L., Kang, Y., Wu, J., Jiang, J., Li, Y., Wu, S., Jia, Y., Liang, X., Bi, A. 2018. LncRNA DANCR upregulates PI3K/AK signaling through activating serine phosphorylation of RXRA. *Cell Death Dis.* 9, 1–12.

Tripathi, V., Ellis, J.D., Shen, Z., Song, D.Y., Pan, Q., Watt, A.T., Freier, S.M., Bennett, C.F., Sharma, A., Bubulya, P.A., Blencowe, B.J., Prasanth, S.G., Prasanth, K.V., 2010. The nuclear-retained noncoding RNA MALAT1 regulates alternative splicing by modulating SR splicing factor phosphorylation. *Mol Cell.* 39, 925–938.

Tsai, M.C., Manor, O., Wan, Y., Mosammaparast, N., Wang, J.K., Lan, F., Shi, Y., Segal, E., Chang, H.Y., 2010. Long noncoding RNA as modular scaffold of histone modification complexes. *Science* 329, 689–693.

Tsuiji, H., Yoshimoto, R., Hasegawa, Y., Furuno, M., Yoshida, M., Nakagawa, S., 2011. Competition between a noncoding exon and introns: Gomafu contains tandem UACUAAC repeats and associates with splicing factor-1. *Genes Cells.* 16, 479–490.

Ulitsky, I., Bartel, D.P., 2013. lincRNAs: genomics, evolution, and mechanisms. *Cell* 154, 26–46.

van Bakel, H., Nislow, C., Blencowe, B.J., Hughes, T.R. 2010. Most 'dark matter' transcripts are associated with known genes. *PLoS Biol.* 8, e1000371.

Wahlestedt, M., Norddahl, G.L., Sten, G., Ugale, A., Frisk, M.A., Mattsson, R., Deierborg, T., Sigvardsson, M., Bryder, D., 2013. An epigenetic component of hematopoietic stem cell aging amenable to reprogramming into a young state. *Blood* 121, 4257–4264.

Wang, H., Wang, L., Erdjument-Bromage, H., Vidal, M., Tempst, P., Jones, R.S., Zhang Y. 2004. Role of histone H2A ubiquitination in Polycomb silencing. *Nature* 431, 873–878.

Wang, P., Chen, D., Ma, H., Li, Y., 2017. LncRNA SNHG12 contributes to multidrug resistance through activating the MAPK/Slug pathway by sponging miR-181a in non-small cell lung cancer. *Oncotarget.* 8, 1–16.

Wang, P., Xue, Y., Han, Y., Lin, L., Wu, C., Xu, S., Jiang, Z., Xu, J., Liu, Q., Cao, X. 2014. The STAT3-binding long noncoding RNA lnc-DC controls human dendritic cell differentiation. *Science* 344, 310–313.

Wang, Y., Gu, J., Lin, X., Yan, W., Yang, W., Wu, G., 2018. LncRNA BANCR promotes EMT in PTC via the Raf/MEK/ERK signaling pathway. *Oncol. Lett.* 15, 5865–5870.

Wang, P., Ren, Z., Sun, P., 2012. Overexpression of the long non-coding RNA MEG3 impairs in vitro glioma cell proliferation. *J. Cell. Biochem.* 113, 1868–1874.

Wang, Y., He, L., Du, Y., Zhu, P., Huang, G., Luo, J., Yan, X., Ye, B., Li, C., Xia, P., Zhang, G., Tian, Y., Chen, R., Fan, Z., 2015. The long noncoding RNA LncTCF7 promote self-renewal of human liver cancer stem cells through activation of Wnt signaling. *Cell Stem Cell.* 16, 413–425.

Warner, J.R., Soeiro, R., Birnboim, H.C., Girard, M., Darnell, J.E., 1966. Rapidly labeled HeLa cell nuclear RNA. *J. Mol. Biol.* 19, 349–361.

Willingham, A.T., Orth, A.P., Batalov, S., Peters, E.C., Wen, B.G., Aza-Blanc, P., Hogenesch, J.B., Schultz, P.G., 2005. A strategy for probing the function of noncoding RNAs finds a repressor of NFAT. *Science.* 309, 1570–1573.

Xing, F., Liu, Y., Wu, S.Y., Wu, K., Sharma, S., Mo, Y.Y., et al., 2018. Loss of XIST in breast cancer activates MSN-c-Met and reprograms microglia via exosomal miRNA to promote brain metastasis. *Cancer Res.* 78, 4316–4330.

Xu, Y., Zhang, G., Zou, C., Zhang, H., Gong, Z., Wang, W., Ma, G., Jiang, P., Zhang, W., 2018. LncRNA MT1JP suppresses gastric cancer cell proliferation and migration through MT1JP/MiR-214- 3p/RUNX3 axis. *Cell Physiol. Biochem.* 46, 2445–2459.

Yang, F., Huo, X.S., Yuan, S.X., Zhang, L., Zhou, W.P., Wang, F., Sun, S.H., 2013. Repression of the long noncoding RNA-LET by histone deacetylase 3 contributes to hypoxia-mediated metastasis. *Mol Cell.* 49, 1083–1096.

Yang, G., Shen, T., Yi, X., Zhang, Z., Tang, C., Wang, L., Zhou, Y., Zhou, W., 2018. Crosstalk between long non-coding RNA and Wnt/beta-catenin signaling in cancer. *J. Cell Mol. Med.* 22, 2062–2070.

Yao, Q., Chen, Y., Zhou, X., 2019. The roles of microRNAs in epigenetic regulation. *Curr. Opin. Chem. Biol.* 51, 11–17.

Yoon, J.H., Abdelmohsen, K., Srikantan, S., Yang, X., Martindale, J.L., De, S., Huarte, M., Zhan, M., Becker, K.G., Gorospe, M., 2012. LincRNA-p21 suppresses target mRNA translation. *Mol Cell.* 47, 648–655.

Zhang, G., Li, S., Lu, J., Ge, Y., Wang, Q., Ma, G., et al., 2018. LncRNA MT1JP functions as a ceRNA in regulating FBXW7 through competitively binding to miR-92a3p in gastric cancer. *Mol Cancer.* 17, 87.

Zhang, X., Gejman, R., Mahta, A., Zhong, Y., Rice, K.A., Zhou, Y., Cheunsuchon, P., Louis, D. N., Klibanski, A., 2010a. *Maternally Expressed Gene 3*, an imprinted noncoding RNA gene, is associated with meningioma pathogenesis and progression. *Cancer Res.* 70, 2350–2358.

Zhang, J., Han, C., Song, K., Chen, W., Ungerleider, N., Yao, L., Ma, W., Wu, T., 2020. The long noncoding RNA MALAT1 regulates TGF-beta signaling through formation of a LncRNA-protein complex with Smads, SETD2 and PPM1A in hepatic cells. *PLoS One.* 1371, 1–16.

Zhang, Z., Zhu, Z., Watabe, K., Zhang, X., Bai, C., Xu, M., Wu, F., Mo, Y.Y., 2013. Negative regulation of LncRNA GAS5 by miR-21. *Cell Death Diff.* 20, 1558–1568.

Zhou, Y., Zhang, X., Klibanski, A., 2012. MEG3 noncoding RNA: a tumor suppressor. *J. Mol. Endocrinol.* 48, R45–R53.

Zhu, D., Yu, Y., Wang, W., Wu, K., Liu, D., Yang, Y., Zhang, C., Qi, Y., Zhao, S., 2019. Long noncoding RNA PART1 promotes progression of non-small cell lung cancer cells via JAK-STAT signaling pathway. *Cancer Med.* 8, 6064–6481.

5 Exosomes as a Therapeutic Tool Against Infectious Diseases

Nishant Sharma
Translational Health Science and Technology Institute, Faridabad, Haryana, India
The Public Health Research Institute at the International Centre for Public Health (ICPH), Newark, NJ, USA

Nevadita Sharma
B.S. Anangpuria Institute of Technology and Management, Faridabad, Haryana, India
The Public Health Research Institute at the International Centre for Public Health (ICPH), Newark, NJ, USA

CONTENTS

5.1 INTRODUCTION

The name "membrane fragments" or "exosomes", used for the first time in 1987 by Rose Johnstone, was initially given for the lipid-layered spherical structures secreted outside the eukaryotic cells. However, besides exosomes, there exist other classes of extracellular vesicles (EVs) such as microvesicles, ectosomes, apoptotic bodies and inclusion bodies, each having distinct mechanism of biogenesis, release, composition and function (Zhang et al. 2019, 2020) (Table 5.1). Since 1997 the term exosome has been used for a multienzyme ribonuclease complex involved in RNA processing (Mitchell et al. 1997). Exosomes are in fact bilayer extracellular membrane microvesicles secreted by most eukaryotic cells having a diameter of 30–200 nm and a density range of 1.13–1.19 g/mL (Théry et al. 2002; Gurunathan et al. 2019; Kowalczyk et al. 2022). The packaged content of exosomes comprises of lipids, nucleic acids and proteins. A vast proportion of these proteins are reported to be associated with the plasma membrane, cytosol and lipid metabolism (Zhang et al. 2019). With advancement of bioanalytical tools, centrifugation techniques and deep knowledge about these structures, better protocols have been designed for isolation and

DOI: 10.1201/9781003324706-7

TABLE 5.1
Comparison of Exosomes, Microvesicles and Apoptotic Bodies

Empty Cell	Exosomes (EXO)	Microvesicles (MVs)	Apoptotic Body (ABs)	Literature
Diameter	40–160 nm	100–1000 nm	1–5 μm	Doyle and Wang (2019); Dai et al. (2020)
Content	Proteins, lipids, nucleic acids, enzymes	Proteins, lipids, nucleic acids	Content like cytosol – intact cellular organelles, chromatin, small amounts of glycosylated proteins	Doyle and Wang (2019); Zhang et al. (2019)
Function	Cell transport, protein release, storage, cell-to-cell communication, immune stimulation, support for myelin formation in the nervous system, neurite growth, tissue regeneration	Cell transport, cell-to-cell communication, blood coagulation, adhesion	–	Doyle and Wang (2019); Cheng et al. (2020)
Origin	Multivesicular body (MVB)	Cell membrane	Cell membrane	Bai et al. (2020); Zhang et al. (2019)
Location	All cell types	Erythrocytes, thrombocytes, lymphocytes, vascular endothelial cells	Dying cells that release apoptotic bodies into the extracellular space	Doyle and Wang (2019); Zhang et al. (2019)
Possibilities	Use of neurodegenerative diseases biomarkers as carriers. Better understanding of exosome function could lead to the development of new cancer therapies	A better understanding of MV function could lead to the development of new cancer therapies	–	Doyle and Warg (2019)

Source: Reprinted From Kowalczyk et al. 2022.

purification of exosomes (Ayala-Mar et al. 2019). Though many researchers have identified the biochemical composition of EVs, many research groups are investigating and trying to identify specific biomarkers for EVs, exosomes and other vesicular structures (Zaborowski et al. 2015; Deng and Miller 2019; Kowal et al. 2016). The common observation of these studies establishes that different vesicular structures have different content due to their dissimilar routes of formation as well as environmental stress. The proteomic analysis has also revealed that despite being from same source, vesicular content varies depending on the method of isolation and storage (Abels and Breakefield 2016; Hessvik and Llorente 2018). Despite all these variations, there is a common consensus and understanding that EVs and exosomes play a crucial role in cell–cell communication, cancer progression and cargo delivery (Abels and Breakefield 2016; Minciacchi 2015; Zhang et al. 2019). These findings have led to increasing research being conducted toward the following:

i. understanding the cell–cell interactions and cellular trafficking mediated via. EVs;
ii. identifying potential biomarkers for early and better diagnosis of cancer and other intracellular infections; and
iii. modifying vesicular content and use them as transport vehicles for therapeutics.

5.2 BIOGENESIS OF EXOSOMES

The initial endosomes originate through invagination of the cell membrane to form an early vesicular structure (Figure 5.1). The cellular cargo starts to accumulate in this structure resulting in the development of early sorting endosomes (ESEs). These structures acquire further diverse cellular components with the help of transport proteins and evolve into late-sorting endosomes (LSEs). LSEs develop into multivesicular bodies (MVBs) by indentation and later fuse with the

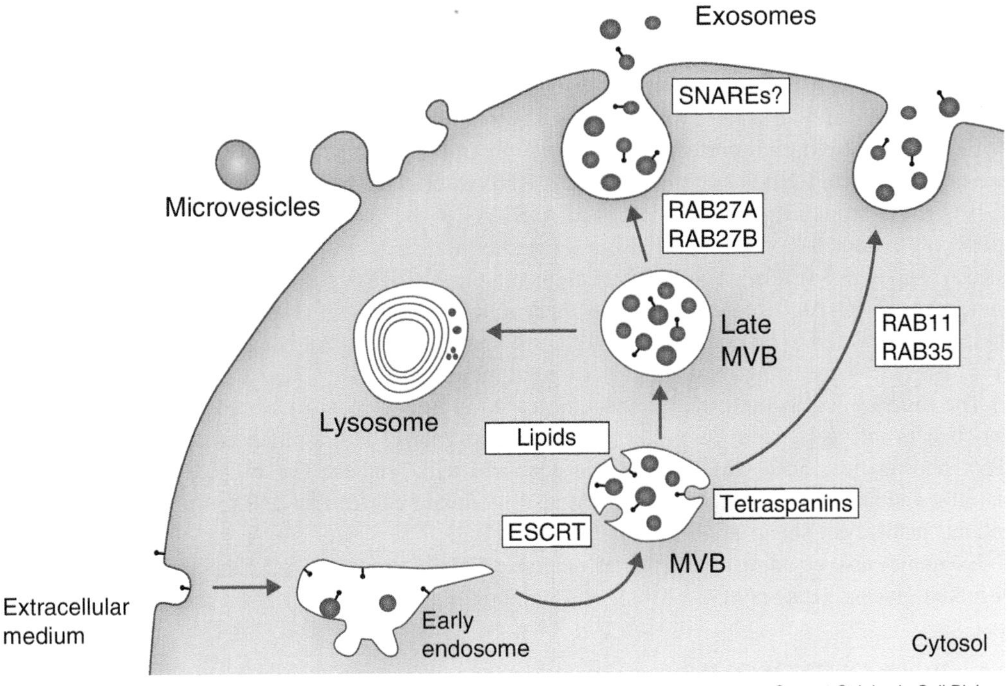

FIGURE 5.1 Exosome biogenesis and secretion (Reprinted with permission from Kowal et al. 2014).

cell membrane. During this process, some proteins get incorporated into the invaginating membrane as well as some of the cytosolic components getting engulfed and packaged (Zhang et al. 2019, 2020). These events cause the substances internal to the cells to be shunt outside in the form of vesicles. These vesicles are termed as exosomes.

However, the content of exosome, mechanism and the key molecules involved in the formations have not been understood well and is being investigated by many research groups. Currently, there are two different hypotheses proposed for the biogenesis of exosomes, namely, (i) ESCRT dependent and (ii) ESCRT independent. Recent evidence shows that the formation of MVB depends on "endosomal sorting complex required for transport (ESCRT)" function. ESCRT is a transport protein machinery which comprise of four different protein ESCRTs (0, I, II and III) that work in tandem for facilitating the formation of MVB, vesicle budding and protein sorting (Henne et al. 2011; Hurley 2015). This hypothesis is supported by ESCRT components being identified in exosomes isolated from various sources. Interestingly, recent scientific research indicates the existence of an alternative pathway for sorting exosomal content into MVBs without involving ESCRT machinery. This ESCRT-independent mechanism relies on raft-based microdomains for sorting the cargo within the endosomal membrane. These microdomains are thought to be highly enriched in sphingomyelinases (SMs) which later form ceramides (Airola and Hannun 2013). This alternate mechanism is ceramide dependent and relies on the properties of lipids in exosome biogenesis. In some cases, proteins, such as tetraspanins, also reported to be involved in exosome biogenesis and protein sorting (Perez-Hernandez et al. 2013).

5.3 COMPOSITION OF EXOSOMES

Exosomes are typically a group of secreted EVs rich in proteins, lipids, DNA, mRNA, noncoding RNAs and glycoconjugates (Figure 5.2). Proteomic content of exosome consists of proteins with various functions, such as tetraspanins (such as CD9, CD63, CD81 and CD82) which are involved in cell penetration and invasion; heat shock proteins (such as HSP70 and HSP90) which are involved in stress response; MVB formation proteins like Alix and TSG101; as well as proteins involved in membrane transport and fusion (Rab2, Rab7, flotillin and annexins). Besides these, some proteins are also reported to be involved in exosome biogenesis like Alix and TSG101 while others are reported to be unique and can be used as exosomal marker proteins such as TSG101 and HSP70 (Table 5.2).

Exosomes have been reported to harbor not only proteins but also various types of nucleic acids like microRNA (miRNA) and long noncoding RNA (lncRNA) (Mathivanan et al. 2012; Kołat et al. 2019). RNA sequencing analysis showed miRNAs to be the most abundant type of RNA in exosomes extracted from human plasma (Huang et al. 2013). Besides miRNAs, ribosomal RNA (rRNA) and lncRNA form the next most abundant class of RNA present in exosomes followed by some transfer RNAs (tRNAs) and small nuclear RNAs (snRNA) (Huang et al. 2013). Some of these were reported to be consistent with the exosomes isolated from cancer patient samples (Xie et al. 2019) and are being exploited as potential biomarkers.

The third critical component of exosome cargo is lipids. In most of the cases, exosomes are enriched in raft-associated lipids like cholesterol (primarily B lymphocytes), phosphatidylserine (PS), phosphatidic acid, SM, arachidonic acid and leukotrienes. These lipids not only provide stability and structure to the exosomes but also modulate cholesterol and prostaglandin-mediated cellular pathways (Zhang et al. 2019).

Exosomes also contain sugars such as mannose, polylactosamine, α-2,6 sialic acid and complex N-linked glycans (Batista et al. 2011).

5.4 ROLE OF EXOSOMES

As heterogeneous as the group of exosomes in terms of size and contents, their functions are far more diverse. With the expansion of the knowledge about these structures, their biological

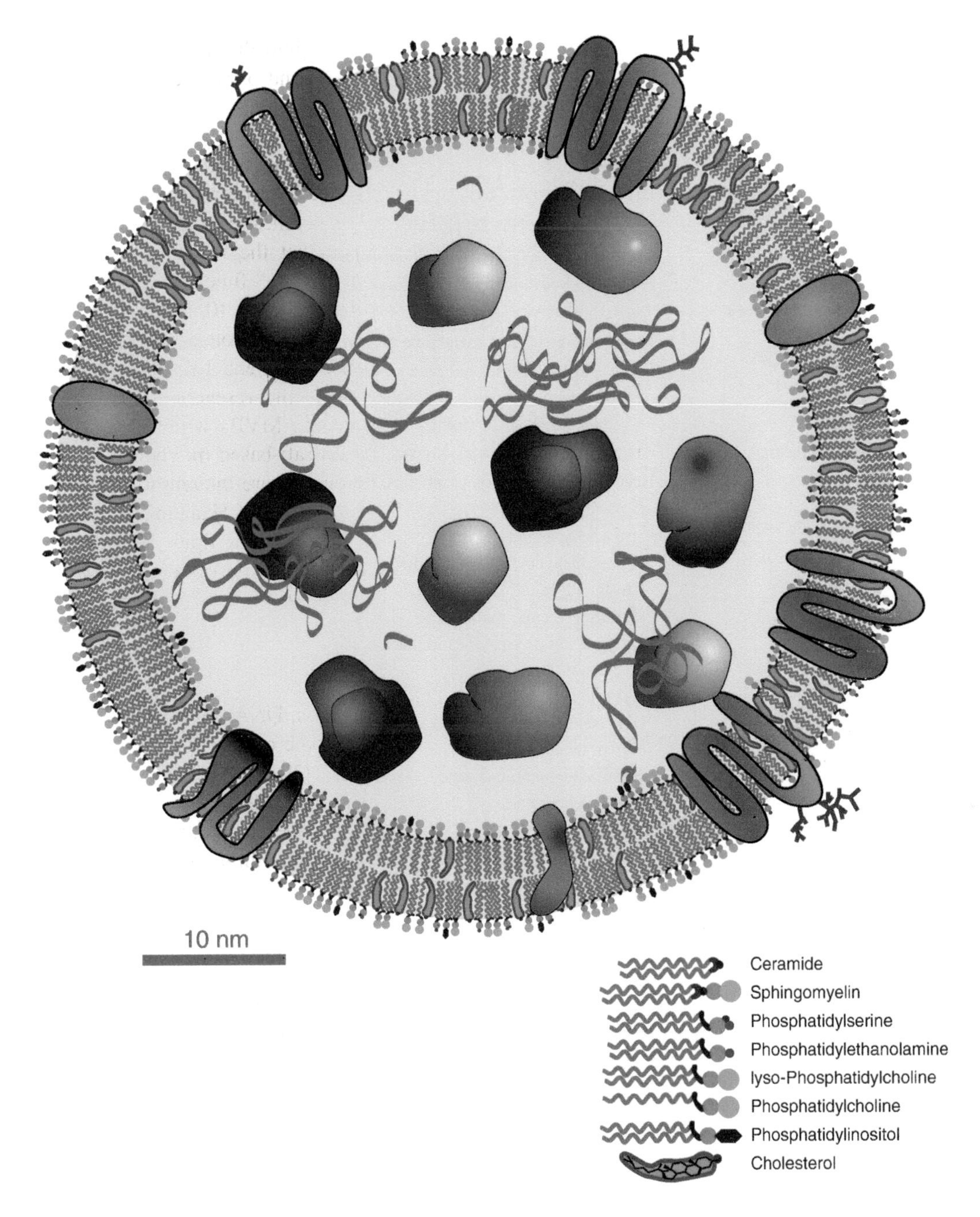

FIGURE 5.2 Structure of mid-size exosome depicting various components (Reprinted with permission from Vlassov et al. 2012).

importance is becoming more significant. The biological function depends on their cell/tissue of origin. Most important roles of exosomes are in cell–cell communication and immunomodulation.

Exosomes are basically a package of effector molecules enclosed in lipid membrane which acts as a delivery vehicle for this packaged cargo. These vehicles deliver their cargo by distinct mechanisms and pathways which in some cases are specific for certain cell types. This further adds complexity to the role of exosomes in cell–cell communication. Commonly, cell communication

TABLE 5.2
Different Types of Proteins Found in Exosomes

Protein Category and Description	Examples
Tetraspanins	CD9, CD63, CD81, CD82, CD37, CD53
Heat shock proteins (HSP)	HSP90, HSP70, HSP27, HSP60
Cell adhesion	Integrins, Lactadherin, Intercellular Adhesion Molecule 1
Antigen presentation	Human leukocyte antigen class I and II/peptide complexes
Multivesicular body Biogenesis	Tsg101, Alix, Vps, Rab proteins
Membrane transport	Lysosomal-associated membrane protein 1/2, CD13, PG regulatory-like protein
Signaling proteins	GTPase HRas, Ras-related protein, furloss, extracellular signal-regulated kinase, Src homology 2 domain phosphatase, GDP dissociation inhibitor, Syntenin-1, 14-3-3 Proteins, Transforming protein RhoA
Cytoskeleton components	Actins, Cofilin-1, Moesin, Myosin, Tubulins, Erzin, Radixin, Vimentin
Transcription and protein synthesis	Histone1, 2, 3, Ribosomal proteins, Ubiquitin, major vault protein, Complement factor 3
Metabolic enzymes	Fatty acid synthase Glyceraldehyde-3-phosphate dehydrogenase Phosphoglycerate kinase 1 Phosphoglycerate mutase 1 Pyruvate kinase isozymes M1/M2 ATP citrate lyase ATPase Glucose-6-phosphate isomerase Peroxiredoxin 1 Aspartate aminotransferase Aldehyde reductase
Trafficking and membrane fusion	Ras-related protein 5, 7 Annexins I, II, IV, V, VI Synaptosomal-associated protein Dynamin, Syntaxin-3
Antiapoptosis	Alix, Thioredoxine, Peroxidase
Growth factors and cytokine	Tumor Necrosis Factor (TNF)-α, TNF Receptors, Transforming growth factor-β
Death receptors	FasL, TNF-related apoptosis-inducing ligand
Iron transport	Transferrin receptor
References	Simpson et al. (2008); Natasha et al. (2014); Yellon and Davidson (2014); Choi et al. (2015)

Source: Reprinted from Zhang et al. 2019.

with adjacent cells involves direct contact through gap junctions or protein/protein interactions between cell surface proteins. On the other hand, communication with distant cells is mediated by secreted factors like hormones and cytokines which initiate signal transduction and pass the message to the receptor cells (Camussi et al. 2010; Record 2014; Turturici et al. 2014).

It has been reported that exosomes containing cell-specific biological effector molecules such as proteins, lipids and nucleic acids interact and modulate target cells via the following mechanisms:

i. stimulating the target cells directly via surface-bound ligands;
ii. transferring activated receptors to target cells; and
iii. reprogramming target cells by delivering functional active proteins, lipids and nucleic acids (RNAs).

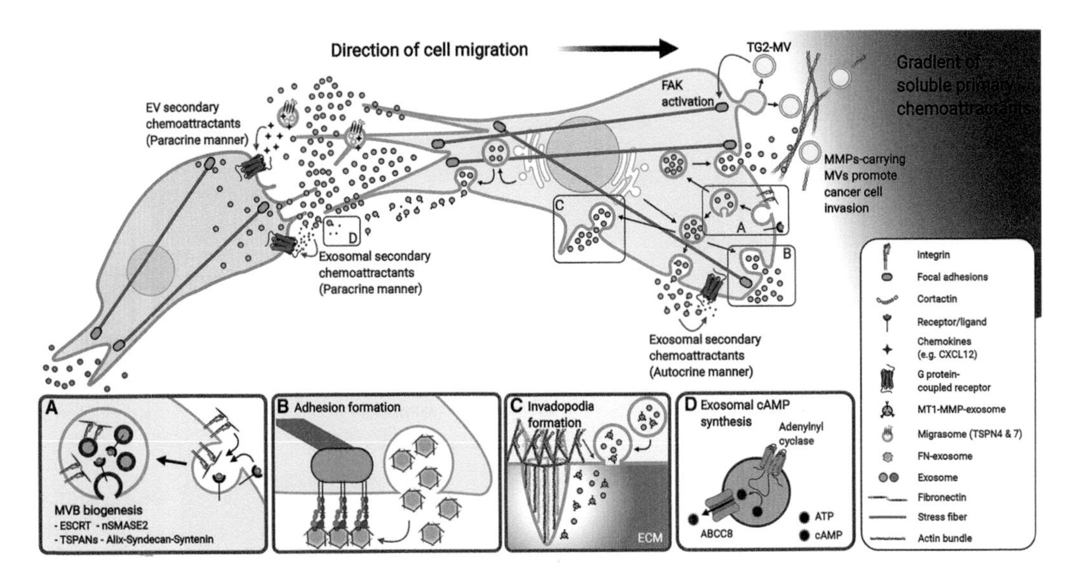

FIGURE 5.3 Role of exosomes in directional migration of cells (Reprinted with permission from Sung et al. 2021).

Therefore, exosomes serve as a means not only to communicate with neighboring cells but also with the cells that are distantly located (Whiteside 2016).

Exosomes play a crucial role in immunoregulation and are reported to be involved in antigen presentation, immune activation and suppression and immune tolerance (Castro et al. 2014). Exosomes originating from immune and non-immune cells have been reported to play roles such as immune regulation, antigen presentation and apoptosis depending on the cells from which they have been originated (Colombo et al. 2014; Théry et al. 2009). Exosomes derived from dendritic cells (DCs) harbor many DC components and also express molecules (such as CD40, CD80, CD86, integrins, MHC-I and MHC-II) crucial for DCs immune functions. It has been reported that exosomes derived from DCs express transmembrane FasL, TNF and TRAIL on their surface and mediate the apoptosis of tumor cells (Munich et al. 2012). EVs derived from B lymphocytes also reported to modulate adaptive immune responses by promoting secretion of IL-5 and IL-13 and T-cell proliferative responses (Gutiérrez-Vázquez et al. 2013). The study of miRNAs in the exosomes derived from human breast milk showed the abundance of some of them like miR-181a and miR-155 during initial months of lactation and their decrease in the later months (Kosaka et al. 2010). The abundance of these miRNAs having significant immunoregulatory function indicates their role in the development of the infant's immune system.

Exosomes, along with other EVs, are also known to be involved in the migration of *Dictyostelium* cells since they carry the adenylyl cyclase that synthesizes cAMP, a well-known chemoattractant (Figure 5.3). The adenylyl cyclase (ACA)-containing EVs are released as specific transporters which aid in the formation of head-to-tail arrays of migrating cells and promote both autocrine and paracrine migration (Sung et al. 2021). Additionally, it has also been reported that the platelets also secrete exosomes, whose main function is in the coagulation reaction, but they are also presumed to be involved in the inflammatory response (Puhm et al. 2021). Besides these, exosomes are studied and found to be involved in targeting specific cells (mediated via cell surface proteins) and influencing various immune pathways (Zhang et al. 2019). This feature further substantiates their importance in understanding, diagnosing and curing many diseases and infections.

Exosomes have also been documented as quite significant in tumorigenesis via modulation and restructuring of the cellular microenvironment resulting in formation of a metastatic niche, along with the attenuation/modulation of tumor immune responses (Figure 5.4). In such instances,

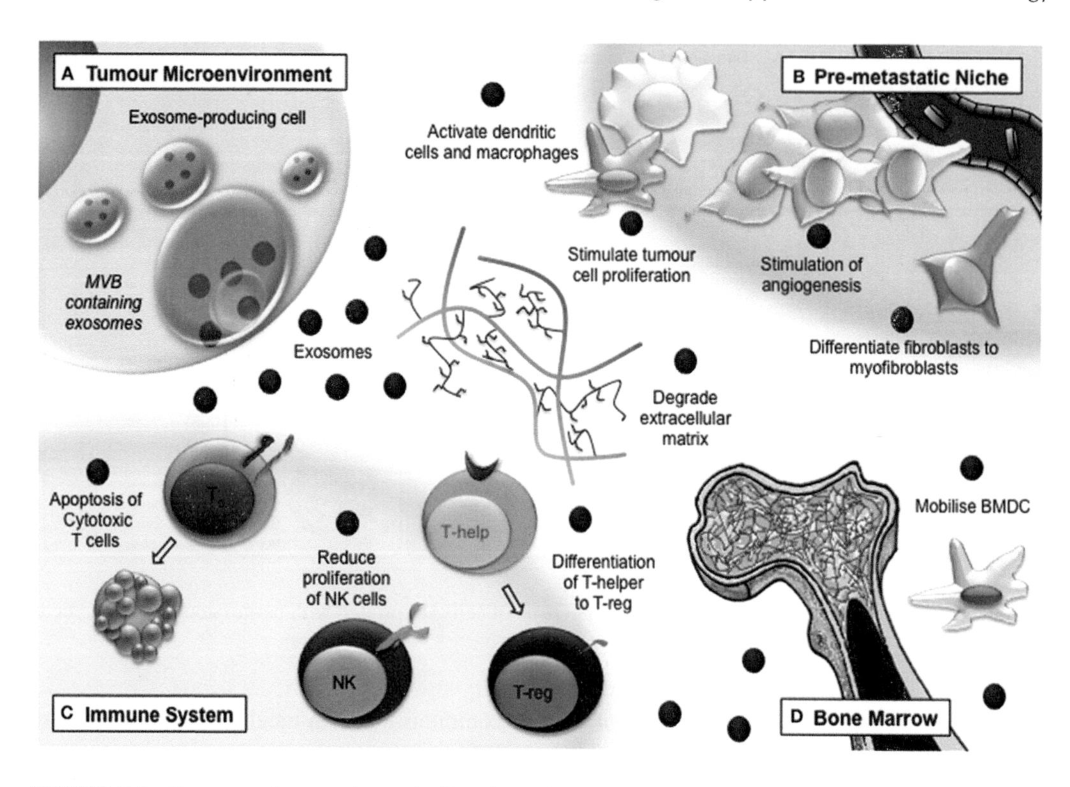

FIGURE 5.4 Exosomes in tumorigenesis (Reprinted from Tickner et al. 2014).

exosomes are known to induce microenvironmental alterations in tissue leading to tumor formation, reducing the anti-tumor immune responses, migration of cancerous cells by avoiding immune detection and attachment to secondary sites, thus aiding in establishment of metastatic growth.

5.5 APPLICATIONS OF EXOSOMES

5.5.1 Exosomes as Diagnostics Tools: Biomarkers

Over the years, various research groups have reported exosomes in almost all body fluids (such as blood, urine, saliva and breast milk). Exosomes derived from these body fluids possess a distinct profile which is the characteristic of their originating source (Zhang et al. 2019). Thus, exosomes harboring cell/tissue-specific cargos will reflect the cellular processes and the state of originating source and therefore can be utilized as biomarkers for detection of various cellular conditions or diseases (Kalluri et al. 2020). Currently, major diseases that are being the target of exosome-mediated diagnosis include cancer (Figure 5.4) and diseases affecting the central nervous system (CNS) (Li et al. 2017; Liu et al. 2019). The reason being the fact that exosomes can be extracted from fluids which required minimum invasion of the body and these exosomes harbor all types of biomolecules which can be used for diagnosis. For instance, some studies indicated that the nucleic acids packed in the exosomes can give insight into the cancer-inducing genetic mutations in the parental tissue and thus can be used for detection of cancer (Zhu et al. 2020). The presence of specific miRNAs in fluids like saliva also provides insight into early diagnosis of the cancer. For instance, increased level of circulating exosomal miR-21 is reported to be correlated with pancreatic, colorectal, colon, breast, liver, ovarian and cancers, and its increased level in urine-derived exosomes is associated with bladder and prostate cancer (Campos et al. 2021; Shi 2016). Besides cancer, exosome content also helps in identifying other underlying infections. For example, the presence of certain miRNAs in breast milk was found to be correlated with underlying breast infections such as HIV-1 (Zahoor et al. 2010). Not only RNAs, but other biomolecules are also been reported to be present in exosomes and can be used as a diagnostic tool. For example, urinary exosomes of prostate cancer patients were reported to harbor cancer signatures such as prostate-specific antigen (Greening et al. 2015). The discoveries of multiple microRNAs as well as other biomolecules will enhance the diagnostic and prognostic potential of exosome by providing set of biomarkers specific for disease and cellular stress.

5.5.2 Exosomes as Drug Delivery Vehicles

Exosomes by themselves or as drug delivery vehicles are being actively explored as novel and better therapeutic agents. Exosomes are nanoscale membranous structures with ability to recognize and target specific cells or tissues (Zhang et al. 2019). They can mediate delivery of active payloads such as nucleic acids, proteins and modulate recipient cells. Also, exosomes provide a great advantage as they can be isolated from a recipient and provide biocompatibility with minimal clearance by the phagocytic system (Ha et al. 2016; Zhang et al. 2019). This provides an excellent solution to the problem of immunogenicity and can be harnessed for delivering therapeutic agent(s) without getting rapidly cleared from the host. It can also facilitate delivery of the therapeutic agent (s) across the blood-brain barrier. With this approach, variety of therapeutic materials like short interfering-RNA (siRNA), recombinant proteins and anti-inflammatory drugs can be packaged and delivered for diverse applications (Ha et al. 2016). The various approaches being utilized includes:

i. extracting exosomes from donor cells and loading them with therapeutic agents;
ii. loading donor cells with therapeutic agent(s) such that, it gets packed inside exosomes during exosome generation from donor cells; and
iii. transfecting donor cells with recombinant DNA (expressing therapeutic agent) such that it gets packed in the exosomes (Bunggulawa et al. 2018; Liu and Su 2019).

These exosomes are then used to deliver the therapeutic agents at the target site. The fact that the exosome can deliver therapeutic agents like siRNA which are prone to degradation makes them even better choice of delivery agent. This approach has been used to deliver siRNA to silence MAPK genes in lymphocytes (Wahlgren et al. 2012) and in one animal study gene silencing was used to treat Huntington's disease (Dar et al. 2021), indicating their vast potential to target and treat specific disease. Mesenchymal stem cells (MSCs) are multipotent cells with immune response largely mediated through paracrine factors including exosomes. This property can be exploited to develop cell-free therapy using exosomes derived from MSCs (Ma et al. 2020). Besides these, in a promising study, it has been also reported that exosomes derived from genetically modified macrophages (transfected with a plasmid DNA coding for enzyme catalase) could be utilized for curing Parkinson's disease. The study reported substantial anti-inflammatory and neuroprotective effects in mice model with neuroinflammation (Haney et al. 2013). Another approach being investigated is using the DC-derived exosomes and packaging them with tumor peptides. These exosomes can be used in a cell-free approach toward stimulating the immune system and mounting a tumor-specific immune response. These DCs-derived exosomes are being investigated for human trials and are being complemented with other approaches to develop vaccines against specific tumors (Näslund et al. 2013).

5.6 CAVEAT AND FUTURE ASPECTS

Exosome research is a budding field with most of the advancement being done over the past decade. However, despite of all the advantages the field has its own set of limitations. Most of the studies involving identification of biomarkers were done on a small sample size and the true potential need to be validated in clinical scenarios across the whole population. Majority of the drug loading and delivery research has been investigated in animal models. Though many of them have moved to clinical trials, but still it is a long way from being approved and launched into market for therapeutics. Another issue is the approval and ethical clearance of these therapies. As many of the approaches involve human samples and their genetic modifications, a proper set of guidelines needs to be formulated while keeping in mind the true nature and potential of exosomes. However, the major hurdle is the scalability of the approach to industrial scale. In order to utilize the true potential of exosome as vaccine therapeutics, the production and isolation of exosome need to be considerably amplified (Fu et al. 2020; Kennedy et al. 2021). Though, from the time of first report of exosome to the current age, lots of advancement have been made but the efficiency of the process need to be further improved. Still, this field is rapidly evolving and holds the key to not only understand underlying mechanism of cell–cell communication but also to enable us with tools to diagnose and also treat diseases like cancer, Parkinson and other cellular infections.

REFERENCES

Abels, E.R., Breakefield, X.O., 2016. Introduction to extracellular vesicles: Biogenesis, RNA cargo selection, content, release, and uptake. *Cell Mol. Neurobiol.* 36, 301–312.

Airola, M.V., Hannun, Y.A., 2013. Sphingolipid metabolism and neutral sphingomyelinases. *Handb. Exp. Pharmacol.* 215, 57–76.

Ayala-Mar, S., Donoso-Quezada, J., Gallo-Villanueva, R.C., Perez-Gonzalez, V.H., González-Valdez, J., 2019. Recent advances and challenges in the recovery and purification of cellular exosomes. *Electrophoresis* 40, 3036–3049.

Bai, Y., Guo, J., Liu, Z., Li, Y., Jin, S., Wang, T., 2020. The role of exosomes in the female reproductive system and breast cancers. *OTT* 13, 12567–12586.

Batista, B.S., Eng, W.S., Pilobello, K.T., Hendricks-Muñoz, K.D., Mahal, L.K., 2011. Identification of a conserved glycan signature for microvesicles. *J. Proteome Res.* 10, 4624–4633.

Bunggulawa, E.J., Wang, W., Yin, T., Wang, N., Durcan, C., Wang, Y., Wang, G., 2018. Recent advancements in the use of exosomes as drug delivery systems. *J. Nanobiotechnol.* 16, 81.

Campos, A., Sharma, S., Obermair, A., Salomon, C., 2021. Extracellular vesicle-associated miRNAs and chemoresistance: a systematic review. *Cancers* 13, 4608.

Camussi, G., Deregibus, M.C., Bruno, S., Cantaluppi, V., Biancone, L., 2010. Exosomes/microvesicles as a mechanism of cell-to-cell communication. *Kidney Int.* 78, 838–848.

Castro, B.M., Prieto, M., Silva, L.C., 2014. Ceramide: A simple sphingolipid with unique biophysical properties. *Prog. Lipid Res.* 54, 53–67.

Cheng, L., Zhang, K., Qing, Y., Li, D., Cui, M., Jin, P., Xu, T., 2020. Proteomic and lipidomic analysis of exosomes derived from ovarian cancer cells and ovarian surface epithelial cells. *J. Ovarian Res.* 13, 9.

Choi, D.S., Kim, D.K., Kim, Y.K., Gho, Y.S., 2015. Proteomics of extracellular vesicles: exosomes and ectosomes. *Mass Spectrom. Rev.* 34, 474–490.

Colombo, M., Raposo, G., Thery, C., 2014. Biogenesis, secretion, and intercellular interactions of exosomes and other extracellular vesicles. *Annu. Rev. Cell Dev. Biol.* 30, 255–289.

Dai, J., Su, Y., Zhong, S., Cong, L., Liu, B., Yang, J., Tao, Y., He, Z., Chen, C., Jiang, Y., 2020. Exosomes: Key players in cancer and potential therapeutic strategy. *Signal Transduct. Target Ther.* 5, 145.

Dar, G.H., Mendes, C.C., Kuan, W.L., Speciale, A.A., Conceição, M., Görgens, A., et al. 2021. GAPDH controls extracellular vesicle biogenesis and enhances the therapeutic potential of EV mediated siRNA delivery to the brain. *Nature Comm.* 12, 6666.

Deng, F., Miller, J., 2019. A review on protein markers of exosome from different bio-resources and the antibodies used for characterization. *J. Histotechnol.* 42, 226–239.

Doyle, L., Wang, M., 2019. Overview of extracellular vesicles, their origin, composition, purpose, and methods for exosome isolation and analysis. *Cells* 8, 727.

Fu, S., Wang, Y., Xia, X., Zheng, J.C., 2020. Exosome engineering: Current progress in cargo loading and targeted delivery. *NanoImpact* 20, 100261.

Greening, D.W., Gopal, S.K., Xu, R., Simpson, R.J., Chen, W., 2015. Exosomes and their roles in immune regulation and cancer. *Sem. Cell Develop. Biol.* 40, 72–81.

Gurunathan, S., Kang, M.-H., Jeyaraj, M., Qasim, M., Kim, J.-H., 2019. Review of the isolation, characterization, biological function, and multifarious therapeutic approaches of exosomes. *Cells* 8, 307.

Gutiérrez-Vázquez, C., Villarroya-Beltri, C., Mittelbrunn, M., Sánchez-Madrid, F., 2013. Transfer of extracellular vesicles during immune cell-cell interactions. *Immunol Rev.* 251, 125–142.
Ha, D., Yang, N., Nadithe, V., 2016. Exosomes as therapeutic drug carriers and delivery vehicles across biological membranes: Current perspectives and future challenges. *Acta Pharm. Sin. B.* 6, 287–296.
Haney, M.J., Zhao, Y., Harrison, E.B., Mahajan, V., Ahmed, S., He, Z., et al., 2013. Specific transfection of inflamed brain by macrophages: A new therapeutic strategy for neurodegenerative diseases. *PLoS One* 8, e61852.
Henne, W.M., Buchkovich, N.J., Emr, S.D., 2011. The ESCRT pathway. *Dev. Cell.* 21, 77–91.
Hessvik, N.P., Llorente, A., 2018. Current knowledge on exosome biogenesis and release. *Cell. Mol. Life Sci.* 75, 193–208.
Huang, X., Yuan, T., Tschannen, M., Sun, Z., Jacob, H., Du, M., et al., 2013. Characterization of human plasma-derived exosomal RNAs by deep sequencing. *BMC Genomics* 14, 319.
Hurley, J.H., 2015. ESCRTs are everywhere. *EMBO J.* 34, 2398–2407.
Kalluri, R., LeBleu, V.S., 2020. The biology, function, and biomedical applications of exosomes. *Science* 367, eaau6977.
Kennedy, T.L., Russell, A.J., Riley, P., 2021. Experimental limitations of extracellular vesicle-based therapies for the treatment of myocardial infarction. *Trends Cardiovasc. Med.* 31, 405–415.
Kołat, D., Hammouz, R., Bednarek, A.K., Płuciennik, E., 2019. Exosomes as carriers transporting long non-coding RNAs: Molecular characteristics and their function in cancer. *Mol. Med. Rep.* 20, 851–862.
Kosaka, N., Izumi, H., Sekine, K., Ochiya, T., 2010. microRNA as a new immune-regulatory agent in breast milk. *Silence* 1, 7.
Kowal, J., Tkach, M., Théry, C., 2014. Biogenesis and secretion of exosomes. *Curr. Opin. Cell Biol.* 29, 116–125.
Kowal, J., Arras, G., Colombo, M., Jouve, M., Morath, J.P., Primdal-Bengtson, B., Dingli, F., Loew, D., Tkach, M., Théry, C., 2016. Proteomic comparison defines novel markers to characterize heterogeneous populations of extracellular vesicle subtypes. *Proc. Natl. Acad. Sci. (USA)* 113(8), E968–E977.
Kowalczyk, A., Wrzecinska, M., Czerniawska-Piatkowska, E., Kupczynski, R., 2022. Exosomes-spectacular role in reproduction. *Biomed. Pharmacother.* 148, 112752.
Li, W., Li, C., Zhou, T., Liu, X., Liu, X., Li, X., Chen, D., 2017. Role of exosomal proteins in cancer diagnosis. *Mol Cancer* 16, 145.
Liu, C., Su, C., 2019. Design strategies and application progress of therapeutic exosomes. *Theranostics* 9, 1015–1028.
Liu, W., Bai, X., Zhang, A., Huang, J., Xu, S., Zhang, J., 2019. Role of exosomes in central nervous system diseases. *Front. Mol. Neurosci.* 12, 240.
Ma, Z.J., Yang, J.J., Lu, Y.B., Liu, Z.Y., Wang, X.X., 2020. Mesenchymal stem cell-derived exosomes: Toward cell-free therapeutic strategies in regenerative medicine. *World J. Stem Cells* 12, 814–840.
Mathivanan, S., Fahner, C.J., Reid, G.E., Simpson, R.J., 2012. ExoCarta 2012: Database of exosomal proteins RNA and lipids. *Nucl. Acids Res.* 40, D1241–D1244.
Minciacchi, V.R., Freeman, M.R., Di Vizio, D., 2015. Extracellular vesicles in cancer: Exosomes, microvesicles and the emerging role of large oncosomes. *Semin. Cell Dev. Biol.* 40, 41–51.
Mitchell, P., Petfalski, E., Shevchenko, A., Mann, M., Tollervey, D., 1997. The exosome: A conserved eukaryotic RNA processing complex containing multiple 3'→5' exoribonucleases. *Cell* 91, 457–466.
Munich, S., Sobo-Vujanovic, A., Buchser, W.J., Beer-Stolz, D., Vujanovic, N.L., 2012. Dendritic cell exosomes directly kill tumor cells and activate natural killer cells via TNF superfamily ligands. *Oncoimmunology* 1, 1074–1083.
Näslund, T.I., Gehrmann, U., Qazi, K.R., Karlsson, M.C.I., Gabrielsson, S., 2013. *J. Immunol.* 190, 2712–2719.
Natasha, G., Gundogan, B., Tan, A., Farhatnia, Y., Wu, W., Rajadas, J., Seifalian, A.M., 2014. Exosomes as immunotheranostic nanoparticles. *Clin Ther.* 36, 820–829.
Perez-Hernandez, D., Gutierrez-Vazquez, C., Jorge, I., Lopez-Martin, S., Ursa, A., Sanchez-Madrid, F., Vazquez, J., Yanez-Mo, M., 2013. The intracellular interactome of tetraspanin-enriched microdomains reveals their function as sorting machineries toward exosomes. *J. Biol. Chem.* 288, 11649–11661.
Puhm, F., Boilard, E., Machlus, K.R., 2021. Platelet extracellular vesicles-beyond the blood. *Arterioscler. Thromb. Vasc. Biol.* 41, 87–96.
Record, M., 2014. Intercellular communication by exosomes in placenta: A possible role in cell fusion? *Placenta* 35, 297–302.
Shi, J., 2016. Considering exosomal miR-21 as a biomarker for cancer. *J. Clin. Med.* 5, 42.

Simpson, R.J., Jensen, S.S., Lim, J.W., 2008. Proteomic profiling of exosomes: Current perspectives. *Proteomics* 8, 4083–4099.

Sung, B.H., Parent, C.A., Weaver, A.M., 2021. Extracellular vesicles: Critical players during cell migration. *Develop. Cell* 56, 1861–1874.

Théry, C., Ostrowski, M., Segura, E., 2009. Membrane vesicles as conveyors of immune responses. *Nat. Rev. Immunol.* 9, 581–593.

Théry, C., Zitvogel, L., Amigorena, S., 2002. Exosomes: Composition, biogenesis and function. *Nat. Rev. Immunol* 2, 569–579.

Tickner, J.A., Urquhart, A.J., Stephenson, S.-A., Richard, D.J., O'Byrne, K.J., 2014. Functions and therapeutic roles of exosomes in cancer. *Front. Oncol.* 4, 127.

Turturici, G., Tinnirello, R., Sconzo, G., Geraci, F., 2014. Extracellular membrane vesicles as a mechanism of cell-to-cell communication: Advantages and disadvantages. *Am. J. Physiol. Cell Physiol.* 306, C621–C633.

Vlassov, A.V., Magdaleno, S., Setterquist, R., Conrad, R., 2012. Exosomes: Current knowledge of their composition, biological functions, and diagnostic and therapeutic potentials. *Biochim. et Biophys. Acta* 1820, 940–948.

Wahlgren, J., Karlson, T.D.L., Brisslert, M., Vaziri Sani, F., Telemo, E., Sunnerhagen, P., Valadi, H., 2012. Plasma exosomes can deliver exogenous short interfering RNA to monocytes and lymphocytes. *Nucleic Acids Res.* 40, e130.

Whiteside, T.L., 2016. Tumor-derived exosomes and their role in cancer progression. *Adv Clin. Chem.* 74, 103–141.

Xie, Y., Dang, W., Zhang, S., Yue, W., Yang, L., Zhai, X., Yan, Q., Lu, J. 2019. The role of exosomal noncoding RNAs in cancer. *Mol. Cancer* 18, 37.

Yellon, D.M., Davidson, S.M., 2014. Exosomes: nanoparticles involved in cardioprotection? *Circ. Res.* 114, 325–332.

Zaborowski, M.P., Balaj, L., Breakefield, X.O., Lai, C.P., 2015. Extracellular vesicles: Composition, biological relevance, and methods of study. *BioScience* 65, 783–797.

Zahoor, M.A., Yao, X.D., Henrick, B.M., Verschoor, C.P., Abimiku, A., Osawe, S., Rosenthal, K.L., 2010. Expression profiling of human milk derived exosomal microRNAs and their targets in HIV-1 infected mothers. *Sci. Rep.* 10, 12931.

Zhang, Y., Liu, Y., Liu, H., Tang, W.H., 2019. Exosomes: biogenesis, biologic function and clinical potential. *Cell. Biosci.* 9, 19.

Zhang, Y., Bi, J., Huang, J., Tang, Y., Du, S., Li, P., 2020. Exosome: a review of its classification, isolation techniques, storage, diagnostic and targeted therapy applications. *Int. J. Nanomed.* 15, 6917–6934.

Zhu, L., Sun, H.T., Wang, S., Huang, S.L., Zheng, Y., Wang, C.Q., et al., 2020. Isolation and characterization of exosomes for cancer research. *J. Hematol. Oncol.* 13, 152.

6 Plant Thionins

The Green Antimicrobial Agents

Raghvendra Pandey
Department of Botany, School of Life Sciences, Mahatma Gandhi Central University, Motihari, Bihar, India

Shilpi Srivastava
Amity Institute of Biotechnology, Amity University Uttar Pradesh, Lucknow Campus, Lucknow, India

CONTENTS

6.1 INTRODUCTION

Antibiotics, first discovered in 1928 and the word coined in 1942, are antimicrobial compounds produced by one organism that is antagonistic to the growth of other organisms. The discovery of antibiotics by Alexander Fleming revolutionized the medical arena and their utilization gained rapid strides in the succeeding decades. The structure of penicillin was elucidated in 1945 by D.C. Hodgkin for which she was awarded the Nobel Prize in 1964. The recognition of penicillin as a therapeutic agent is attributed to Heatley, Chain and Florey (Buynak 2004). The development of the first sulfonamide-based synthetic drug "Prontosil" in 1935 by Gerhard Domagk led to an "Era of Antibacterials" that has been a major factor in reducing mortality among human populations (Bhargava and Srivastava 2017). Then came the "golden era of discovery of antibiotics" wherein several antibiotics were developed between the 1950s and 1970s, along with many analogues that played a critical role in public health (Coates et al. 2002). Table 6.1 provides information on the different classes of antibiotics along with their mode of action. Antimicrobials were extensively used in surgery, organ transplant and chemotherapy (Williamson et al. 2017; Subramaniam and Girish 2020). Antibiotics find wide application in agriculture where they are administered for controlling plant pathogens. Applications in livestock farming include use for growth promotion, mass medication of animals and for improving feed conversion efficiency (Kuppusamy et al. 2018; Manyi-Loh et al. 2018; Roth et al. 2019). The use of antibiotics has increased tremendously with their sale expected to reach around 60 billion US$ in 2025 (Hamad 2010).

DOI: 10.1201/9781003324706-8

TABLE 6.1
Some Commonly Used Antibiotics and Their Modes of Resistance

Class of Antibiotic	Example	Modes of Resistance
Aminoglycosides	Gentamicin, Streptomycin, Spectinomycin	Acetylation, Altered target, Efflux, Nucleotidylation, Phosphorylation
β-Lactams	Ampicillin, Aztreonam, Cephamycin, Meropenem	Altered target, Efflux, Hydrolysis
Cationic peptides	Colistin	Altered target, Efflux
Glycopeptides	Vancomycin, Teicoplanin	Reprogramming of peptidoglycan biosynthesis
Macrolides	Azithromycin, Erythromycin	Altered target, Efflux, Glycosylation, Hydrolysis, Phosphorylation
Phenicols	Chloramphenicol	Acetylation, Altered target, Efflux
Quinolones	Ciprofloxacin	Acetylation, Altered target, Efflux
Rifamycins	Rifampin	ADP-ribosylation, Altered target, Efflux
Sulfonamides	Sulfamethoxazole	Altered target, Efflux
Tetracyclines	Minocycline, tigecycline	Altered target, Efflux, Monooxygenation

Source: Reprinted With Permission From Bhargava and Srivastava 2017.

6.2 ANTIBIOTIC RESISTANCE

Indiscriminate use of antibiotics in diverse sectors has resulted in the development of antibiotic resistance among the microorganisms which has become a global concern for human and animal health (Roth et al. 2019; Subramaniam and Girish 2020). The large-scale production and rampant use of antibiotics have led to higher concentrations of antibiotics in the environment, a condition known as "antibiotic pollution". The exposure to antibiotics selects for *de novo* acquired resistance and emergence of antibiotic-resistant organisms reduced the effectiveness of drugs directed for curing pathological conditions (Ter Kuile et al. 2016; Morehead and Scarbrough 2018). The situation has been aggravated by the extensive use of antibiotics for human therapeutics in many countries, especially the developing ones, due to easy availability without an appropriate prescription (Ayukekbong et al. 2017). The development of resistance in several Gram-negative bacteria having high prevalence of efflux pumps and rigid walls is a cause of concern since these attributes make them quite resilient against antimicrobial agents (Arias and Murray 2009; Fair and Tor 2014). Also, fewer pharmaceutical companies engaged in antibiotic discovery has led to less development and/or discovery of new antibiotics over the last 2–3 decades making the situation quite worrisome. With a negligible number of antibiotics in clinical trials, the future of mankind is at stake and urgent efforts are needed to develop novel anti-infective therapies.

6.3 ANTIMICROBIAL PEPTIDES (AMPS)

AMPs are a part of the nonspecific part of the innate immune system inherent in most lifeforms that have a major role to play in the host defense against pathogens (Li et al. 2021; Srivastava et al. 2021). These broad-spectrum antimicrobial agents, generally serving as the first line of defense, have been reported in viruses (Fischetti 2008; Parisien et al. 2008; Plotka et al. 2019), bacteria (Willey and Van Der Donk 2007; Simons et al. 2020), fungi (Wu et al. 2014; Haney et al. 2017), plants (Tam et al. 2015; Slavokhotova et al. 2017), invertebrates (Tassanakajon et al. 2015), fishes (Huang et al. 2007; Masso-Silva and Diamond 2014), amphibians (Xu and Lai 2015), reptiles (van Hoek 2014), birds (Cheng et al. 2015) and mammals (Selsted and Ouellette 2005) (Table 6.2). The AMPs have become the center of attraction with respect to antibiotic therapy and have sparked

TABLE 6.2
Distribution of AMPs Across the Living Kingdom and Their Biological Activities

Group	Organism	AMP	Biological Activity	Reference
Virus	Φ3626	Murein hydrolase	Antibacterial	Zimmer et al. (2002)
	ΦAB2	LysAB2	Antibacterial	Peng et al. (2017)
Bacteria	*Paenibacillus polymyxa*	Polymyxins	Antibacterial	Velkov et al. (2019)
	Enterobacteria	Microcins	Antibacterial	Duquesne et al. (2007)
Fungi	*Pseudoplectania nigrella*	Plectasin	Antibacterial	Mygind et al. (2005)
	Neosartorya fischeri	Antifungal protein 2 (NFAP)	Antifungal	Kovács et al. (2011)
	Coprinopsis cinerea	Copsin	Antibacterial	Essig et al. (2014)
Plants	*Pisum sativum*	Defensin 1	Antifungal	Lobo et al. (2007)
	Nicotiana alata	Defensin NaD1	Antifungal	Hayes et al. (2018)
	Oryza sativa	Defensin OsAFP1	Antifungal	Ochiai et al. (2018)
Arthropoda	*Panulirus argus*	Defensin	Antimicrobial	Montero-Alejo et al. (2012)
	Episesarma tetragonum	Crustin	Antibiofilm	Sivakamavalli et al. (2015)
Fishes	*Epinephelus coioides*	Defensin	Antiviral	Guo et al. (2012)
	Gadus morhua	Cathelicidin	Antibacterial; Antifungal	Broekman et al. (2011)
Amphibians	*Bufo gargarizans*	Buforin	Antibacterial; Antifungal	Cho et al. (2009)
	Odorrana livida	Cathelicidin	Antibacterial; Anti-inflammatory	Qi et al. (2019)
	Pelophylax nigromaculata	Cathelicidin	Antibacterial; Anti-inflammatory	Wang et al. (2021)
Reptiles	*Bungarus fasciatus*	Cathelicidin	Antibacterial; Antifungal	Wang et al. (2008)
	Hydrophis cyanocinctus	Cathelicidin	Antimicrobial; Anti-inflammatory	Wei et al. (2015)
	Alligator sinensis	Cathelicidin	Antimicrobial; Anti-inflammatory	Chen et al. (2017)
Birds	*Phasianus colchicus*	Cathelicidin Pc-CATH1,2,3	Antibacterial; Antifungal	Wang et al. (2011)
	Meleagris gallopavo	Cathelicidin CATH2	Antibacterial	Yacoub et al. (2016)
Mammals	*Homo sapiens*	Cathelicidin LL-37	Antibacterial; Antibiofilm	Duplantier 2013, Almaghrabi 2019)
	Bubalus bubalis	Cathelicidin CATH4	Antibacterial	Brahma et al. (2015)
	Sus scrofa	Protegrin-4	Antibacterial	Gour et al. (2019)

hopes for an effective alternative to conventional antibiotics. Apart from serving as antimicrobial agents, AMPs also exhibit a variety of biological functions such as stimulation of the immune regulation, angiogenesis, suppression of the inflammatory response, wound healing and antitumor activity (Roudi et al. 2017; Pfalzgraff et al. 2018; Mookherjee et al. 2020), which makes them ideal candidates for the development of novel therapeutics in the coming decades (Haney and Hancock 2013; Zhang et al. 2021).

AMPs are evolutionally conserved, gene-encoded entities that alter microbe metabolism both at cellular and membrane levels. These are constitutively expressed or go through rapid transcription when introduced in cells after microbial infection (Jain et al. 2022; Nawrot et al. 2014). The AMPs are produced as secondary metabolites and consist of short peptide sequences upto 50 amino acids. The AMPs vary in their sequence, secondary structural preferences and modes of action. Although majority of them are cationic (lysine and arginine) and positively charged (+2 to +11), anionic and hydrophobic AMPs are also known to exist (Epand and Vogel 1999; Harris et al. 2009; Cascales et al. 2018). Since the first reports of the extraction of AMPs from soil bacteria (Dubos 1939a,b), this field has seen rapid advances and about 3300 of them comprising from bacteria, archaea, protista, fungi, plants and animal kingdoms are mentioned in the AMP database. It is interesting to note that some of the AMPs like defensins and cyclotides are common between different classes of eukaryotes like plants and animals, some like heveins have been exclusively reported in plants, while some are specific to certain plant families (snakins) (Sarkar et al. 2021).

6.4 THIONINS

A number of AMPs have been isolated from plants, namely, thionins, cyclotides, snakins, lipid transfer proteins (LTPs), 2S albumins, hevein-type proteins, defensins, knottins and glycine-rich peptides (Benko-Iseppon et al. 2010; Hu et al. 2018). Since the isolation of the first thionin from wheat flour in 1942 by the research group led by Balls and collaborators, AMPs have been isolated from several plant parts and their expression determined to be either constitutive or induced (Table 6.3). Plant AMPs are basic, cysteine-rich polypeptides that permit formation of several disulfide bonds that impart stability during stress conditions thus facilitating in protection against chemical, thermal and proteolytic degradation (Jenssen et al. 2006; Hammami et al. 2009; Tam et al. 2015). The AMPs have molecular weight of around 2–5 kDa, 40–100 amino acids and have intermolecular s–s bonds (Höng et al. 2021). Apart from having a major role in regulating plant growth and development, these cysteine-rich peptides are majorly known for their role in plant defense systems (Campos et al. 2018).

Of the diverse plant AMPs, thionins have occupied a place of prominence and have carved a special niche for themselves as antimicrobial agents. Thionins, the first AMPs reported from plants, are small (45–47 amino acid long), basic, cysteine-rich AMPs having 3/4 disulfide bonds that display significant activity against several plant pathogens (Nawrot et al. 2014; Srivastava et al. 2021). The molecular weight of thionins is around 5 kDa. Initially, thionins were considered

TABLE 6.3
AMPs Discovered From Different Plant Parts

Plant Part	Plant	AMPs	References
Flower	*Brassica rapa*	Defensins	Park et al. (2002)
	Nicotiana alata	Defensins	Lay et al. (2003)
	Petunia hybrida	Defensins	Ghag et al. (2012)
	Sambucus nigra	Cyclotides	Álvarez et al. (2018)
Root	*Capsella bursa-pastoris*	Shepherins	Remuzgo et al. (2014)
	Heliophila coronopifolia	Defensins	Weiller et al. (2016)
Leaves	*Broussonetia papyrifera*	Hevein-like AMPs: PMAPI	Zhao et al. (2011)
	Solanum lycopersicum	Snakin-2	Herbel et al. (2017)
Fruits	*Capsicum annuum*	Thionin-like peptides	Taveira et al. (2014)
	Zizyphus jujube	Snakin-Z	Daneshmand et al. (2013)
Endosperm	*Triticum aestivum*	α-1-purothionin, PINA and PINB	Liu et al. (2000); Nawrot et al. (2014)
Bark	*Eucommia ulmoides*	EAFP1 and EAFP2	Huang et al. (2002)
	Pichia pastoris	Defensin	Wisniewskia et al. (2003)
Bulb/tubers	*Solanum tuberosum*	Snakin-2	Berrocal-Lobo et al. (2002)
	Tulipa gesneriana	Tu-AMP 1 and Tu-AMP 2	Fujimura et al. (2004)

Source: Reprinted From Srivastava et al. 2021.

as a family of homologous peptides isolated from wheat seeds but subsequent studies revealed their heterogeneous structure. Initially, a number of monocotyledonous species were initially reported to contain thionins especially in several members of family Poaceae (Bekes and Lasztity 1981; Bohlmann et al. 1988; Colilla et al. 1990; Castagnaro et al. 1994), but recent researches have confirmed the presence of thionins in dicots as well (Schrader-Fisher and Apel 1994; Stein et al. 1999). Thionins are named in the order of their discovery with descending letters of the Greek alphabets. On this basis, two well-characterized groups have been characterized viz. α/β-thionin (two α-helixes, double-stranded β-sheets and a C-terminal coil region) and γ-thionin (one α-helix and three anti-parallel β-sheets that form an amphipathic two-layer α/β sandwich) (Pelegrini and Franco 2005). Recently, γ-thionins have been redesignated as plant defensins.

6.5 TYPES OF THIONINS

α-/β-thionins have been divided into five classes each having somewhat conserved amino acid sequence (Stec 2006).

i. Type I thionins: Highly basic consisting of 45 amino acid residues, and 4 s–s bonds. These have been isolated from the endosperm of several members of the monocot family Poaceae (Table 6.4).
ii. Type II thionins: Composed of 46–47 amino acids and 4 s–s bonds, but are comparatively less basic than type I mentioned above. Type II ones are found in the foliage of barley and "buffalo nut" (Table 6.4).
iii. Type III thionins: Consist of 45/46 amino acids, 3–4 s–s bonds and similar basicity to type II thionins. These have been reported in different plant species, most of them belonging to order Santalales (Table 6.4).

TABLE 6.4
Different Types of Thionins and Their Sources

Thionin Type	Thionin	Plant	Family	References
Type I	α-hordothionin	*Hordeum vulgare*	Poaceae	Hernfindez-Lucas et al. (1986)
	β-hordothionin	*Hordeum vulgare*	Poaceae	Hernfindez-Lucas et al. (1986)
	ω-hordothionin	*Hordeum vulgare*	Poaceae	Mendez et al. (1996)
	α1-purothionin	*Triticum aestivum*	Poaceae	Jones and Mak (1976)
	β-purothionin	*Triticum aestivum*	Poaceae	Mak and Jones (1976)
	α-avenothionin	*Avena sativa*	Poaceae	Bekes and Lasztity (1981)
	β-avenothionin	*Avena sativa*	Poaceae	Bekes and Lasztity (1981)
Type II	Pyrularia thionin	*Pyrularia pubera*	Santalaceae	Vernon et al. (1985)
Type III	Denclatoxin B	*Dendrophthora clavata*	Santalaceae	Samuelsson and Pettersson (1977)
	Ligatoxin A	*Phoradendron liga*	Santalaceae	Thunberg and Samuelsson (1982)
	Phoratoxins A, B	*Phoradendron tomentosum*	Santalaceae	Mellstrand et al. (1974)
	Viscotoxin A2, A3, B	*Viscum album*	Santalaceae	Samuelsson et al. (1968); Samuelsson and Pettersson (1971); Olson and Samuelsson (1972)
Type IV	Crambin A, B	*Crambe abyssinica*	Brassicaceae	Teeter et al. (1981); Vermeulen et al. (1987)
	Hellothionin D	*Helleborus purpurascens*	*Ranunculaceae*	Milbradt et al. (2003)

iv. Type IV thionins: These are neutral and contain 46 amino acids and 3 s–s bonds.
v. Type V thionins: These are truncated forms of thionins foand Pectobacterium carotovorumund in some grains like wheat. Hellothionin D obtained from the underground parts of *Helleborus purpurascens, the* perennial flowering plant of family Ranunculaceae is sometimes categorized in this type.

6.6 STRUCTURE OF THIONINS

Thionins are rich in amino acids such as arginine, lysine and cysteine and have an end-to-end s–s bond linking the N- and C-termini due to which they are often considered as pseudocyclics. However, additional amino acids devoid of cysteines are present at both the N- and C-termini due to which they cannot be termed as true pseudocyclics. Thionins are highly conserved as compared to other plant AMPs with cysteine residues constituting about 12–17% of the amino acids and forming disulfide bridges (Schrader and Apel 1991). Moreover, the hydrophobic and hydrophilic residues distributed among the thionin molecules are highly conserved in nature.

Hendrickson and Teeter (1981) were the first to carry out the structural determination of crambin, a small hydrophobic seed storage protein obtained from the Abyssinian cabbage plant (*Crambe abyssinica*). This was followed by elucidation of structures of purothionins, viscotoxin A3 and β-hordothionin (Rao et al. 1995; Debreczeni et al. 2003; Johnson et al. 2005). The 3-D structures pointed out that types I, III and IV are rigid and amphipathic having a shape resembling the Greek letter Γ both in solution as well as in crystalline form (García-Olmedo et al. 1998). They share a conserved β1-α1-α2-β2-coil secondary structural motif, which forms a gamma (Γ) fold, a special turn consisting of three amino acid residues. The tertiary structure has revealed that the Ist and the IIIrd residue are connected with a hydrogen bond (Figure 6.1) with the hydrophobic residues present at the outer surface of the long arm of the Γ and the hydrophilic residues occurring at the inner surface and the periphery of the corner of the Γ (Hendrickson and Teeter 1981; Clore et al. 1986; Brünger et al. 1987).

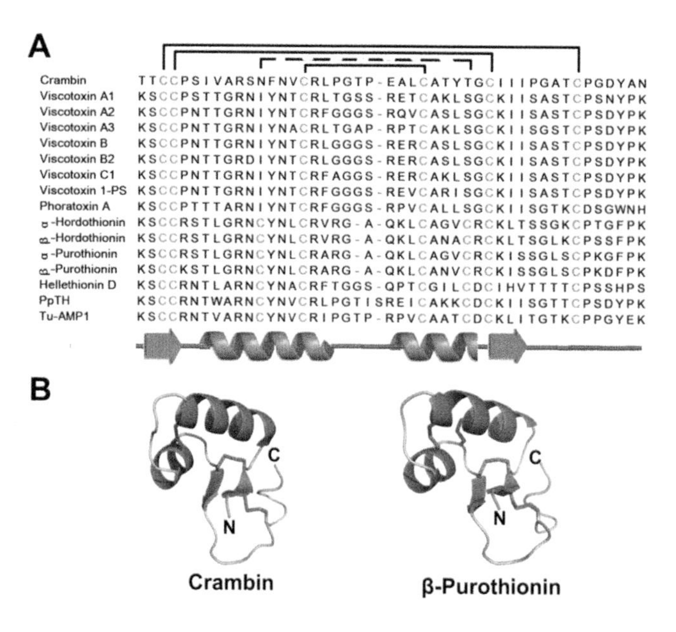

FIGURE 6.1 Sequences (A) and structures (B) of representative thionins. The secondary structure is represented by different colors: cyan-α helix; magenta-β strand; pink-random coil and yellow-disulfide bonds (Reprinted from Tam et al. 2015).

It is interesting to note that the α- and β-thionins equally share the same 3-D structure despite minor variation in length. The γ-thionins have low molecular weight, residual sequence similarity to the α- and β-thionins and are considered to have originated from a distant but common ancestor (Stec 2006). However, they are quite distinct from proteins sharing a significant sequence homology (~30%) and have evolved different forms by rearrangement of the s–s bridges and the β-sheet (Stec 2006).

6.7 BIOLOGICAL ACTIVITIES OF THIONINS

Thionins are toxic to several organisms, including yeasts, fungi, protozoans and insect larvae at low concentrations (Balls and Harris 1944; Fernandez de Caleya et al. 1972; Hernandez-Lucas et al. 1974; Molina et al. 1993; Melo et al. 2002). These are also toxic to nematodes (Almaghrabi et al. 2019) and cultured mammalian cells (Carrasco et al. 1981). Thionins display remarkable anticancer activity against different types of cancers which is quite evident in several published studies (Table 6.5). Besides this, legume AMPs have an essential role in rhizobium-legume symbiosis (Van de Velde et al. 2010; Maróti et al. 2015).

TABLE 6.5
The Cytotoxic and Anticancer Activity of Thionins

Name	Species	Activity Against	Cytotoxic Activity	Anticancer Activity	References
Pyrularia	*Pyrularia pubera*	B16, HeLa, rat hepatocytes, and lymphocytes	Yes	Yes	Evans et al. (1989)
Viscotoxin B2	*Viscum coloratum*	Rat sarcoma cells	Not tested	Yes	Kong et al. (2004)
Viscotoxins 1-PS, A1, A2, A3 and B	*Viscum album*	Human lymphocytes	Yes	Not tested	Büssing et al. (1999)
Viscotoxin C1	*Coloratum ohwi*	Rat sarcoma cells	Not tested	Yes	Romagnoli et al. (2003)
Ligatoxin B	*Phoradendron liga*	U-937-GTB ACHN	Not tested	Yes	Li et al. (2002)
Ligatoxin A	*Phoradendron liga*	Animal cells	Yes	Not tested	Thunberg and Samuelsson (1982)
Phoratoxins A and B	*Phoradendron tomentosum*	Mice	Yes	Not tested	Sauviat et al. (1985)
Phoratoxins C, D, E and F	*Phoradendron tomentosum*	10 cancer cell lines	Not tested	Yes	Johansson et al. (2003)
Thi2.1	*Arabidopsis thaliana*	HeLa, A549, MCF-7 and bovine mammary epithelial cells	Yes	Yes	Loeza-Ángeles et al. (2008)
β-Purothionin	*Tricum aestivum*	p388	Not tested	Yes	Hughes et al. (2000)

Source: Reprinted From Guzmán-Rodríguez et al. 2015.

6.8 THIONINS AS ANTIMICROBIAL AGENTS

Thionins display a wide array of effects on the viability and biochemical or physiological functions of microorganisms. Several *in vitro* studies have indicated that thionins have a pivotal role in plant defense against pathogenic organisms. These AMPs exhibit potent antibacterial (Bin Hafeez et al. 2021), antifungal (Garibotto et al. 2010; Li et al. 2016) and antiviral (VanCompernolle et al. 2005) properties that have been primarily attributed to increase in the membrane permeability and fluidity (Stec 2006). The antimicrobial activity of thionins is due to the generation of ion channels in cell membranes, resulting in leakage and dissipation of ion concentration gradients that are essential for the maintenance of cellular homeostasis (Hughes et al. 2000). The universal toxicity exhibited by thionins against diverse organisms suggests that the mortality induced by them does not rely on a specific cell surface receptor and the processes disrupted by thionins are quite common across most cells. With reference to the antifungal activity displayed by thionins, it has been observed that this feature is a result of direct protein–membrane interactions by electrostatic interaction of +vely charged thionins with the -vely charged phospholipids present in the fungal membranes, which results in pore formation or a specific interaction with a specific lipid domain (De Lucca et al. 2005).

Thionins from some plants display broad antimicrobial activity across a range of microorganisms. Molina et al. (1993) isolated and purified thionins of types I and II, from the endosperms of wheat and barley, and evaluated the biological activity against fungal and bacterial pathogens. Potent antibacterial activity at 2–3 × 10^{-7}M concentration was observed against *Clavibacter michiganensis* subsp. *sepedonicus* and *Pseudomonas solanacearum*. Among the fungal pathogens, *Rosellinia necatrix*, *Colletotrichum lagenarium* and *Fusarium solani* had EC-50 values in the range of 1–4 × 10^{-6}M. Another example of this feature is the thionins obtained from black seed (*Nigella sativa*), namely, NsW_1, NsW_2 and NsW_4 which inhibit viability of *Bacillus subtilis* and *Staphylococcus aureus*, as well as fungi like *Aspergillus flavus*, *A. fumigatus*, *A. oryzae* and *Candida albicans* (Vasilchenko et al. 2017; Barashkova et al. 2021).

Thionins exhibit substantial antibacterial activity against both Gram+ve and Gram-ve bacteria (Srivastava et al. 2021) (Table 6.6). However, some species of *Pseudomonas* and *Erwinia* appear to be unresponsive to these AMPs (Cammue et al. 1992; Pineiro et al. 1995). The susceptible bacterial genus includes *Agrobacterium*, *Burkholderia*, *Clavibacter*, *Corynebacterium*, *Escherichia*, *Pectobacterium*, *Pseudomonas*, *Staphylococcus* and *Xanthomonas*.

TABLE 6.6
Antibacterial Properties of Plant Thionins

Thionin	Plant Species	Family	Plant Part	Bacteria	References
CaThi	*Capsicum annuum*	Solanaceae	Fruit	*Escherichia coli*, *Pseudomonas aeruginosa*	Taveira et al. (2014)
Cp-Thionin II	*Vigna unguiculata*	Fabaceae	Seed	*Staphylococcus aureus*, *Pseudomonas syringae*	Franco et al. (2006)
NsW1, NsW2	*Nigella sativa*	Ranunculaceae	Seed	*Bacillus subtilis*, *Staphylococcus aureus*	Vasilchenko et al. (2017)
OsTHION15	*Oryza sativa*	Poaceae	Seed	*Xanthomonas oryzae* pv. *Oryzae*, *Pectobacterium carotovorum* pv. *atroseptica*	Boonpa et al. (2019)
PR-13/Thionin	*Nicotiana attenuata*	Solanaceae	nr	*Pseudomonas syringae* pv. *tomato*	Rayapuram et al. (2008)
Tu-AMP 1, Tu-AMP 2	*Tulipa gesneriana*	Liliaceae	Bulb	*Agrobacterium rhizogenes*, *A. radiobacter*, *Clavibacter michiganensis*	Fujimura et al. (2004)

TABLE 6.7
Antifungal Properties of Plant Thionins

Thionin	Plant	Family	Plant Part	Fungi/Yeast	References
AX1, AX2	*Beta vulgaris*	Amaranthaceae	Leaves	*Cercospora beticola*	Kragh et al. (1995)
CaThi	*Capsicum annuum*	Solanaceae	Fruit Fruit Fruit	*Candida buinensis, Candida parapsilosis* *Fusarium solani* *Candida albicans, Candida tropicalis*	Taveira et al. (2016) Taveira et al. (2017) Taveira et al. (2014)
Cp-thionin II	*Vigna unguiculata*	Fabaceae	Seed	*Fusarium culmorum*	Schmidt et al. (2019)
Hordothionin	*Hordeum vulgare*	Poaceae	Leaf	*Drechslera teres*	Bohlmann et al. (1988)
NsW1, NsW2, NsW4	*Nigella sativa*	*Ranunculaceae*	Seed	*Aspergillus flavus, A. fumigatus, A. oryzae*	Barashkova et al. (2021)
OsTHION15	*Oryza sativa*	Poaceae	Seed	*Fusarium oxysporum f. sp. cubense, Helminthosporium oryzae*	Boonpa et al. (2019)
PR protein-13	*Pennisetum glaucum*	Poaceae	Seed	*Sclerospora graminicola*	Chandrashekhara et al. (2010)
Purothionin	*Triticum aestivum*	Poaceae	Seed	*Rhizoctonia solani*	Oard et al. (2004)
Tu-AMP 1, Tu-AMP 2	*Tulipa gesneriana*	Liliaceae	Bulb	*Fusarium oxysporum, Geotrichum candidum*	Fujimura et al. (2004)
VtA3	*Viscum album*	Santalaceae	Leaves	*Fusarium solani*	Giudici et al. (2006)

The antifungal action of thionins has been reported against several plant pathogens such as *Sclerotinia sclerotiorum* (necrotrophic fungal pathogen), *Phytophtora infestans* (causal organism of late blight or potato blight), *Sclerospora graminicola* (causal organism of downy mildew of millets), *Cercospora beticola* (Cercospora leaf spot disease) and *Drechslera teres* (causal organism of Net blotch) (Table 6.7). Apart from this, thionins are also effective against *Neurospora crassa* (red bread mold), as well as several species of *Fusarium* (Giudici et al. 2006; Oard et al. 2004; Rayapuram et al. 2008; Chandrashekhara et al. 2010). *Ca*Thi, a thionin-like peptide isolated from the fruits of red pepper (*Capsicum annuum*), displays antifungal activity against several species of the genus *Candida* viz. *C. buinensis, C. parapsilosis, C. albicans* and *C. tropicalis*. In a recent study, Taveira et al. (2016) observed that *Ca*Thi induced membrane disruption in six species of *Candida*. However, nuclear localization and ROS production were exhibited only in *C. tropicalis*. In another study (Taveira et al. 2016), the mitochondrial membrane potential, cell surface pH, and extracellular H^+ fluxes were evaluated in *C. tropicalis* after exposure to *Ca*Thi. *Ca*Thi induced phosphatidylserine (PS) externalization in the outer leaflet of the cell membrane, activated caspases and dissipating mitochondrial membrane potential. The peptide also triggered apoptosis in *C. tropicalis* cells that involved a pH signaling mechanism.

6.9 BIOTECHNOLOGICAL INTERVENTIONS IN PLANT THIONINS

Biotechnology has an enormous role to play with respect to greater utilization of plant thionins as antimicrobial agents (Table 6.8). The overexpression of thionins is known to boost plant resistance against different plant pathogens, both bacterial and fungal, that is evident in several available

TABLE 6.8
Biotechnology and Thionins for Antimicrobial Action

Thionin	Source	Family	Transgenic Approach	Microbe Targeted	References
Thi2.1	*Arabidopsis thaliana*	Brassicaceae	Overexpression of Thi2.1 Development of transgenic tomato Expressed in BVE-E6E7 bovine endothelial cell line	*Fusarium oxysporum f sp matthiolae* *Ralstonia solanacearum* *Staphylococcus aureus, C. albicans*	Epple et al. (1997) Chan et al. (2005) Loeza-Angeles et al. (2008)
Asthi1	*Avena sativa*	Poaceae	Overexpression in rice	*Burkholderia glumae, B. plantarii*	Iwai et al. (2002)
			Ectopic expression in carnation	*Burkholderia caryophylli*	Shirasawa-Seo et al. (2002)
α-thionin	*Hordeum vulgare*	Poaceae	Overexpression in tobacco	*Pseudomonas syringae*	Carmona et al. (1993)
α-hordothionin	*Hordeum vulgare*	Poaceae	Expression in sweet potato	*Ceratocystis fimbriata*	Muramoto et al. (2012)
OsTHI7	*Oryza sativa*	Poaceae	Overexpression in *O. sativa*	*Pythium graminicola*	Ji et al. (2015)
OsTHION15	*Oryza sativa*	Poaceae	Production of recombinant peptide in *Escherichia coli*	*Xanthomonas oryzae, Pectobacterium carotovorum*	Boonpa et al. (2019)
Thi2.4	*Triticum spp.*	Poaceae	Using an anti-thionin 2.4 (Thi2.4) antibody	*F. graminearum*	Asano et al. (2013)

studies (Chan et al. 2005; Muramoto et al. 2012). Rice plants overexpressing *Avena sativa* thionin (Asthi1), which accumulates in the cell wall, have revealed pathogenic bacterial staining only on the surface of stomata. Thus, thionins have been proven to be highly effective in limiting bacteria at the target surface (Asano et al. 2013; Iwai et al. 2002).

Ji et al. (2015) carried out overexpression of *OsTHI7* gene in *O. sativa* cv. Nipponbare for assessing its protective effect against the root-knot nematode *Meloidogyne graminicola* and the root oomycete *Pythium graminicola*. The transgenic plant revealed low susceptibility to *P. graminicola* since the oomycete hyphae were largely restricted to the epidermis and less wilting of roots. The results suggested the protective role of rice thionins in controlling plant pathogens.

Hao et al. (2016) characterized endogenous citrus thionins and examined their expression in diverse plant tissues. The objective was to develop citrus plants having dual resistance to Huanglongbing (citrus greening disease) caused by *Candidatus* Liberibacter asiaticus and citrus canker caused by *Xanthomonas citri*. This was carried out by altering a sequence encoding the thionin and cloning it into pBinPlus/ARS, an improved binary plant transformation vector. This was followed by introducing the construct into *Agrobacterium* strain EHA105 for transformation which led to the generation of transgenic Carrizo plants. The transgenic Carrizo plants on exposure to different concentrations of *X. citri* 3213 showed substantial reduction in symptoms and a concomitant reduction in the bacterial growth. The transgenic plants when exposed to Huanglongbing through graft inoculation showed promise for engineering disease resistance against both Huanglongbing (HLB) and canker in citrus fruits.

Hammad et al. (2017) isolated two antifungal thionin genes viz. Thio60 and Thio63 from *Arabidopsis thaliana* and introduced them in two commercial cultivars (lady and spunta) of potato

(*Solanum tuberosum*) using Agrobacterium-mediated gene transfer. The transgenic potato cultivars were tested for efficacy by assessing the effect of spore suspension on potato organs and the inhibitory effect of thionin proteins on spore germination. The results showed that transgenic potato cultivars were highly resistant to pathogenic fungi like *Fusarium solani* and *F. oxysporum* with a decrease in the number of germinated spores as compared to control. The inhibition observed was around 52.49% for thio60 and 47.11% for thio63 against *F. solani*, 74.47% for thio60 and 84.08% for thio63 against *F. oxysporum*.

Boonpa et al. (2019) carried out in-silico expression analyses of 44 *Oryza sativa* thionins (OsTHIONs) under diverse conditions and selected OsTHION15 for further studies. The antimicrobial activity of OsTHION15 was assessed by producing a recombinant peptide in *Escherichia coli*. The recombinant displayed significant antibacterial activity against *Xanthomonas oryzae* pv. *Oryzae* (MIC- 112.6 $\mu g\ ml^{-1}$) and *Pectobacterium carotovorum* pv. *atroseptica* (MIC- 14.1 $\mu g\ ml^{-1}$). Substantial hyphal growth inhibition was also observed toward *F. oxysporum* f. sp. *cubense* and *Helminthosporium oryzae*. Additionally, in planta, antibacterial activity of the polypeptide was also exhibited in *Nicotiana benthamiana* (Family: Solanaceae) against *X. campestris* pv. *glycines*. The results indicated ample scope and possible utilization of OsTHION15 in plant disease control.

6.10 CONCLUSION

Thionins, displaying broad antibacterial, antifungal, antiparasitic and antiviral activities, might serve as alternative sources for the control of plant pests and pathogens leading to development of microbicide against phytopathogens. These amphipathic polypeptides, on account of their broad antimicrobial efficacy, also seem to be promising drug candidates for development of anti-infective drugs that are essential in combating emerging infections and diseases. Future applications of plant AMPs in general and thionins in particular need to address the following points:

a. Simplification and cost-reduction of screening, identification, and purification.
b. Production of active plant AMPs in large quantities.
c. Development of transgenic plants that efficiently express AMPs.

REFERENCES

Almaghrabi, B., Ali, M.A., Zahoor, A., Shah, K.H., Bohlmann, H., 2019. Arabidopsis thionin-like genes are involved in resistance against the beet-cyst nematode (*Heterodera schachtii*). *Plant Physiol. Biochem.* 140, 55–67.

Álvarez, C.A., Barriga, A., Albericio, F., Romero, M.S., Guzmán, F., 2018. Identification of peptides in flowers of *Sambucus nigra* with antimicrobial activity against aquaculture pathogens. *Molecules* 23, 1033.

Asano, T., Miwa, A., Maeda, K., Kimura, M., Nishiuchi, T., 2013. The secreted antifungal protein thionin 2.4 in *Arabidopsis thaliana* suppresses the toxicity of a fungal fruit body lectin from *Fusarium graminearum*. *PLoS Pathol.* 9, e1003581.

Arias, C.A., Murray, B.E., 2009. Antibiotic-resistant bugs in the 21st century — a clinical super-challenge. *N. Engl. J. Med.* 360(5), 439–443.

Ayukekbong, J.A., Ntemgwa, M., Atabe, A.N., 2017. The threat of antimicrobial resistance in developing countries: Causes and control strategies. *Antimicrob. Resist. Infect. Control* 6, 47.

Balls, A.K., Harris, T.H., 1944. The inhibitory effect of a protamine from wheat flour on the fermentation of wheat mashes. *Cereal Chem.* 21, 74–79.

Barashkova, A.S., Sadykova, V.S., Salo, V.A., Zavriev, S.K., Rogozhin, E.A., 2021. Nigellothionins from black cumin (*Nigella sativa* L.) seeds demonstrate strong antifungal and cytotoxic activity. *Antibiotics* 10, 166.

Bekes, F., Lasztity, R., 1981. Isolation and determination of amino acid sequence of avenothionin, a new purothionin analogue from oat. *Cereal Chem.* 58, 360–361.

Benko-Iseppon, A.M., Galdino, S.L., Calsa, T., Kido, E.A., Tossi, A., Belarmino, L.C., Crovella, S., 2010. Overview on plant antimicrobial peptides. *Curr. Protein Pept. Sci.* 11, 181–188.

Berrocal-Lobo, M., Segura, A., Moreno, M., López, G., García-Olmedo, F., Molina, A., 2002. Snakin-2, an antimicrobial peptide from potato whose gene is locally induced by wounding and responds to pathogen infection. *Plant Physiol.* 128, 951–961.

Bin Hafeez, A., Jiang, X., Bergen, P.J., Zhu, Y., 2021. Antimicrobial peptides: An update on classifications and databases. *Intern. J. Mol. Sci.* 22, 11691.

Bohlmann, H., Clausen, S., Behnke, S., Giese, H., Hiller, C., Reimann-Philipp, U., Schrader, G., Barkholt, V., Apel, K., 1988. Leaf-specific thionins of barley-a novel class of cell wall proteins toxic to plant-pathogenic fungi and possibly involved in the defence mechanism of plants. *EMBO J.* 7, 1559–1565.

Boonpa, K., Tantong, S., Weerawanich, K., Panpetch, P., Pringsulaka, O., Roytrakul, S., Sirikantaramas, S., 2019. In silico analyses of rice thionin genes and the antimicrobial activity of OsTHION15 against phytopathogens. *Phytopathol.* 109, 27–35.

Brahma, B., Patra, M.C., Karri, S., Chopra, M., Mishra, P., De, B.C., Kumar, S., Mahanty, S., Thakur, K., Poluri, K.M., Datta, T.K., De, S. 2015. Diversity, antimicrobial action and structure-activity relationship of buffalo cathelicidins. *PloS One* 10(12), e0144741.

Broekman, D.C., Zenz, A., Gudmundsdottir, B.K., Lohner, K., Maier, V.H., Gudmundsson, G.H., 2011. Functional characterization of codCath, the mature cathelicidin antimicrobial peptide from Atlantic cod (*Gadus morhua*). *Peptides* 32, 2044–2051.

Brünger, A.T., Campbell, R.L., Clore, G.M., Gronenborn, A.M., Karplus, M., Petsko, G.A., Teeter, M.M., 1987. Solution of a protein crystal structure with a model obtained from NMR interproton distance restraints. *Science* 235(4792), 1049–1053.

Büssing, A., Stein, G.M., Wagner, M., Wagner, B., Schaller, G., Pfüller, U., Schietzel, M., 1999. Accidental cell death and generation of reactive oxygen intermediates in human lymphocytes induced by thionins from *Viscum album* L. . *Eur. J. Biochem.* 262, 79–87.

Buynak, J.D., 2004. The discovery and development of modified penicillin- and cephalosporin-derived beta-lactamase inhibitors. *Curr. Med. Chem.* 11(14), 1951–1964.

Cammue, B.P., De Bolle, M.F., Terras, F.R., Proost, P., Van Damme, J., Rees, S.B., Broekaert, W.F., 1992. Isolation and characterization of a novel class of plant antimicrobial peptides form Mirabilis jalapa L. seeds. *J. Biol. Chem.* 267(4), 2228–2233.

Campos, M.L., de Souza, C.M., de Oliveira, K.B.S., Dias, S.C., Franco, O.L., 2018. The role of antimicrobial peptides in plant immunity *J. Exp. Bot.* 69(21), 4997–5011.

Carmona, M.J., Molina, A., Fernández, J.A., López-Fando, J.J., García-Olmedo, F., 1993. Expression of the alpha-thionin gene from barley in tobacco confers enhanced resistance to bacterial pathogens. *Plant J.* 3, 457–462.

Carrasco, L., Vazquez, D., Hernández-Lucas, C., Carbonero, P., Garcia-Olmedo, F., 1981. Thionins: Plant peptides that modify membrane permeability in cultured mammalian cells. *Eur. J. Biochem.* 116, 185–189.

Cascales, J.J.L., Zenak, S., de la Torre, J.G., Lezama, O.G., Garro, A., Enriz, R.D., 2018. Small cationic peptides: Influence of charge on their antimicrobial activity. *ACS Omega* 3, 5390–5398.

Castagnaro, A., Marana, C., Carbonero, P., Garcia-Olmedo, F., 1994. cDNA cloning and nucleotide sequences of alpha 1 and alpha 2 thionins from hexaploid wheat endosperm. *Plant Physiol.* 106, 1221–1222.

Chan, Y.L., Prasad, V., Chen, K.H., Liu, P.C., Chan, M.T., Cheng, C.P., 2005. Transgenic tomato plants expressing an *Arabidopsis* thionin (Thi2.1) driven by fruit-inactive promoter battle against phytopathogenic attack. *Planta* 221, 386–393.

Chandrashekhara, N.R.S., Deepak, S., Manjunath, G., Shetty, S.H., 2010. Thionins (PR protein 13) mediate pearl millet down mildew disease resistance. *Arch. Phytopathol Plant Protect.* 43, 1356–1366.

Chen, Y., Cai, S., Qiao, X., Wu, M., Guo, Z., Wang, R., Kuang, Y.Q., Yu, H., Wang, Y., 2017. As-CATH1-6, novel cathelicidins with potent antimicrobial and immunomodulatory properties from *Alligator sinensis*, play pivotal roles in host antimicrobial immune responses. *Biochem. J.* 474, 2861–2885.

Cheng, Y., Prickett, M.D., Gutowska, W., Kuo, R., Belov, K., Burt, D.W., 2015. Evolution of the avian β-defensin and cathelicidin genes. *BMC Evol. Biol.* 15, 188.

Cho, J.H., Sung, B.H., Kim, S.C., 2009. Buforins: Histone H2A-derived antimicrobial peptides from toad stomach. *Biochim. Biophys. Acta Biomembr.* 1788, 1564–1569.

Clore, G.M., Nilges, M., Sukumaran, D.K., Brünger, A.T., Karplus, M., Gronenborn, A.M., 1986. The three-dimensional structure of α1-purothionin in solution: Combined use of nuclear magnetic resonance, distance geometry and restrained molecular dynamics. *The EMBO J.* 5(10), 2729–2735.

Coates, A., Hu, Y., Bax, R., Page, C., 2002. The future challenges facing the development of new antimicrobial drugs. *Nat. Rev. Drug Discov.* 1(11), 895–910.

Colilla, F.J., Rocher, A., Mendez, E., 1990. Gamma-purothionins: Amino acid sequence of two polypeptides of a new family of thionins from wheat endosperm. *FEBS Lett.* 270, 191–194.

Daneshmand, F., Zare-Zardini, H., Ebrahimi, L., 2013. Investigation of the antimicrobial activities of snakin-z, a new cationic peptide derived from *Zizyphus jujuba* fruits. *Nat. Prod. Res.* 27, 2292–2296.

Debreczeni, J.E., Girmann, B., Zeeck, A., Krätzner, R., Sheldrick, G.M., 2003. Structure of viscotoxin A3: Disulfide location from weak SAD data. *Acta Crystallogr. Sec. D* 59(12), 2125–2132.

De Lucca, A.J., Cleveland, T.E., Wedge, D.E., 2005. Plant-derived antifungal proteins and peptides. *Can. J. Microbiol.* 51, 1001–1014.

Dubos, R.J., 1939a. Studies on a bactericidal agent extracted from a soil bacillus: I. Preparation of the agent. Its activity in vitro. *J. Exp. Med.* 70(1), 1–10.

Dubos, R.J., 1939b. Studies on a bactericidal agent extracted from a soil bacillus: II. Protective effect of the bactericidal agent against experimental Pneumococcus infections in mice. *J. Exp. Med.* 70(1), 11–17.

Duplantier, A.J., van Hoek, M.L., 2013. The human cathelicidin antimicrobial peptide LL-37 as a potential treatment for polymicrobial infected wounds. *Front. Immunol.* 4, 143.

Duquesne, S., Destoumieux-Garzón, D., Peduzzi, J., Rebuffat, S., 2007. Microcins, gene-encoded antibacterial peptides from enterobacteria. *Nat. Prod. Rep.* 24, 708–734.

Epand, R.M., Vogel, H.J., 1999. Diversity of antimicrobial peptides and their mechanisms of action. *Biochim. Biophys. Acta* 1462, 11–28.

Epple, P., Apel, K., Bohlmann, H., 1997. Overexpression of an endogenous thionin enhances resistance of *Arabidopsis* against *Fusarium oxysporum*. *Plant Cell* 9, 509–520.

Essig, A., Hofmann, D., Münch, D., Gayathri, S., Kunzler, M., Kallio, P.T., et al., 2014. Copsin, a novel peptide-based fungal antibiotic interfering with the peptidoglycan synthesis. *J. Biol. Chem.* 289, 34953–34964.

Evans, J., Wang, Y.D., Shaw, K.P., Vernon, L.P., 1989. Cellular responses to *Pyrularia* thionin are mediated by Ca^{2+} influx and phospholipase A_2 activation and are inhibited by thionin tyrosine iodination. *Proc. Natl. Acad. Sci. USA* 86, 5849–5853.

Fair, R.J., Tor, Y., 2014. Antibiotics and bacterial resistance in the 21st century. *Perspect. Med. Chem.* 6, 25–64.

Fernandez de Caleya, R., Gonzalez-Pascual, B., García-Olmedo, F., Carbonero, P., 1972. Susceptibility of phytopathogenic bacteria to wheat purothionins in vitro. *Appl. Microbiol.* 23, 998–1000.

Fischetti, V.A., 2008. Bacteriophage lysins as effective antibacterials. *Curr. Opin. Microbiol.* 11, 393–400.

Franco, O.L., Murad, A.M., Leite, J.R., Mendes, P.A., Prates, M.V., Bloch C., Jr., 2006. Identification of a cowpea gamma-thionin with bactericidal activity. *FEBS J.* 273, 3489–3497.

Fujimura, M., Ideguchi, M., Minami, Y., Watanabe, K., Tadera, K., 2004. Purification, characterization, and sequencing of novel antimicrobial peptides, Tu-AMP 1 and Tu-AMP 2, from bulbs of tulip (*Tulipa gesneriana* L.). *Biosci. Biotechnol. Biochem.* 68, 571–577.

García-Olmedo, F., Molina, A., Alamillo, J.M., Rodríguez-Palenzuéla, P., 1998. Plant defense peptides. *Biopolymers* 47(6), 479–491.

Garibotto, F.M., Garro, A.D., Masman, M.F., Rodríguez, A.M., Luiten, P.G.M., Raimondi, M., Zacchino, S.A., Somlai, C., Penke, B., Enriz, R.D., 2010. New small-size peptides possessing antifungal activity. *Bioorg. Med. Chem.* 18(1), 158–167.

Ghag, S.B., Shekhawat, U.K.S., Ganapathi, T.R., 2012. Petunia floral defensins with unique prodomains as novel candidates for development of *Fusarium* wilt resistance in transgenic banana plants. *PLoS ONE* 7, e39557.

Giudici, M., Poveda, J.A., Molina, M.L., de la Canal, L., González-Ros, J.M., Pfüller, K., Pfüller, U., Villalaín, J., 2006. Antifungal effects and mechanism of action of viscotoxin A3. *The FEBS J.* 273, 72–83.

Gour, S., Kumar, V., Singh, A., Gadhave, K., Goyal, P., Pandey, J., Giri, R., Yadav, J.K., 2019. Mammalian antimicrobial peptide protegrin-4 self assembles and forms amyloid-like aggregates: Assessment of its functional relevance. *J. Pept. Sci.* 25(3), e3151.

Guo, M., Wei, J., Huang, X., Huang, Y., Qin, Q., 2012. Antiviral effects of β-defensin derived from orange-spotted grouper (*Epinephelus coioides*). *Fish Shellfish Immunol.* 32, 828–838.

Guzmán-Rodríguez, J.J., Ochoa-Zarzosa, A., López-Gómez, R., López-Meza, J.E., 2015. Plant antimicrobial peptides as potential anticancer agents. *BioMed Res. Intern.* 2015, 735087.

Hamad, B., 2010. The antibiotics market. *Nat. Rev. Drug Discov.* 9(9), 675–676.

Hammad, I.A., Abdel-Razik, A.B., Soliman, E.R., Tawfik, E., 2017. Transgenic Potato (*Solanum tuberosum*) expressing two antifungal thionin genes confer resistance to *Fusarium* spp. *IOSR J. Pharmacy Biol. Sci.* 12, 69–79.

Hammami, R., Ben Hamida, J., Vergoten, G., Fliss, I., 2009. PhytAMP: A database dedicated to antimicrobial plant peptides. *Nucleic Acids Res.* 37, D963–D968.

Haney, E.F., Hancock, R.E., 2013. Peptide design for antimicrobial and immunomodulatory applications. *Biopolymers* 100(6), 572–583.

Haney, E.F., Mansour, S.C., Hancock, R.E., 2017. Antimicrobial peptides: An introduction. *Meth. Mol. Biol.* 1548, 3–22.

Hao, G., Stover, E., Gupta, G., 2016. Overexpression of a modified plant thionin enhances disease resistance to citrus canker and huanglongbing (HLB). *Front. Plant Sci.* 7, 1078.

Harris, F., Dennison, S.R., Phoenix, D.A., 2009. Anionic antimicrobial peptides from eukaryotic organisms. *Curr. Protein Pept. Sci.* 10, 585–606.

Hayes, B.M.E., Bleackley, M.R., Anderson, M.A., van der Weerden, N.L., 2018. The plant defensin NaD1 enters the cytoplasm of *Candida albicans* via endocytosis. *J Fungi (Basel)* 4, 20.

Hendrickson, W.A., Teeter, M.M., 1981. Structure of the hydrophobic protein crambin determined directly from the anomalous scattering of sulphur. *Nature* 290(5802), 107–113.

Herbel, V., Sieber-Frank, J., Wink, M., 2017. The antimicrobial peptide snakin-2 is upregulated in the defense response of tomatoes (*Solanum lycopersicum*) as part of the jasmonate-dependent signaling pathway. *J. Plant Physiol.* 208, 1–6.

Hernandez-Lucas, C., Fernandez De Caleya, R., Carbonero, P., 1974. Inhibition of Brewer's Yeasts by Wheat Purothionins. *Appl.Microbiol.*, 28, 165–168.

Hernfindez-Lucas, C., Royo, J., Paz-Ares, J., Ponz, F., Garcia-Olmedo, F., Carbonero, P., 1986. Polyadenylation site heterogeneity in mRNA encoding the precursor of the barley toxin α-hordothionin. *FEBS Lett.* 200, 103–105.

Höng, K., Austerlitz, T., Bohlmann, T., Bohlmann, H., 2021. The thionin family of antimicrobial peptides. *PLoS One* 16, e0254549.

Hu, Z., Zhang, H., Shi, K., 2018. Plant peptides in plant defense responses. *Plant Signal. Behav.* 13, e1475175.

Huang, P.H., Chen, J.Y., Kuo, C.M., 2007. Three different hepcidins from tilapia, *Oreochromis mossambicus*: Analysis of their expressions and biological functions. *Mol. Immunol.* 44, 1922–1934.

Huang, R.H., Xiang, Y., Liu, X.Z., Zhang, Y., Hu, Z., Wang, D.C., 2002. Two novel antifungal peptides distinct with a five-disulfide motif from the bark of *Eucommia ulmoides* oliv. *FEBS Lett.* 521, 87–90.

Hughes, P., Dennis, E., Whitecross, M., Llewellyn, D., Gage, P., 2000. The cytotoxic plant protein, beta-purothionin, forms ion channels in lipid membranes. *J. Biol. Chem.* 275, 823–827.

Iwai, T., Kaku, H., Honkura, R., Nakamura, S., Ochiai, H., Sasaki, T., Ohashi, Y., 2002. Enhanced resistance to seed-transmitted bacterial diseases in transgenic rice plants overproducing an oat cell-wall-bound thionin. *Mol. Plant-Microbe Inter.* 15, 515–521.

Jain, V., Mishra, P.K., Mishra, M., Prakash, V., 2022. Constitutive expression and discovery of antimicrobial peptides in *Zygogramma bicolorata* (Coleoptera: Chrysomelidae). *Proteins* 90, 465–475.

Jenssen, H., Hamill, P., Hancock, R.E.W., 2006. Peptide antimicrobial agents. *Clin. Microbiol. Rev.* 19(3), 491–511.

Ji, H., Gheysen, G., Ullah, C., Verbeek, R., Shang, C., De Vleesschauwer, D., Höfte, M., Kyndt, T., 2015. The role of thionins in rice defence against root pathogens. *Mol. Plant Pathol.* 16, 870–881.

Johansson, S., Gullbo, J., Lindholm, P., Ek, B., Thunberg, E., Samuelsson, G., Larsson, R., Bohlin, L., Claeson, P., 2003. Small, novel proteins from the mistletoe *Phoradendron tomentosum* exhibit highly selective cytotoxicity to human breast cancer cells. *Cell. Mol. Life Sci.* 60, 165–175.

Johnson, K.A., Kim, E., Teeter, M.M., Suh, S.W., Stec, B., 2005. Crystal structure of α-hordothionin at 1.9 Å resolution. *FEBS Lett.* 579(11), 2301–2306.

Jones, B.L., Mak, A.S., 1976. Amino acid sequences of the two α-purothionins of hexaploid wheat. *Cereal Chem.* 54, 511–523.

Kong, J.L., Du, X.B., Fan, C.X., Xu, J.F., Zheng, X.J., 2004. Determination of primary structure of a novel peptide from mistletoe and its antitumor activity. *Acta Pharm. Sin.* 39(10), 813–817.

Kovács, L., Virágh, M., Takó, M., Papp, T., Vágvölgyi, C., Galgóczy, L., 2011. Isolation and characterization of *Neosartorya fischeri* antifungal protein (NFAP). *Peptides* 32, 1724–1731.

Kragh, K.M., Nielsen, J.E., Nielsen, K.K., Dreboldt, S., Mikkelsen, J.D., 1995. Characterization and localization of new antifungal cysteine-rich proteins from *Beta vulgaris*. *Mol. Plant-Microbe Interac.* 8, 424–434.

Kuppusamy, S., Kakarla, D., Venkateswarlu, K., Megharaj, M., Yoon, Y.-E., Lee, Y.B., 2018. Veterinary antibiotics (VAs) contamination as a global agro-ecological issue: A critical view. *Agric. Ecosyst. Environ.* 257, 47–59.

Lay, F.T., Brugliera, F., Anderson, M.A., 2003. Isolation and properties of floral defensins from ornamental tobacco and petunia. *Plant Physiol.* 131, 1283–1293.

Li, S.-S., Gullbo, J., Lindholm, P., et al. 2002. Ligatoxin B, a new cytotoxic protein with a novel helix-turn-helix DNA-binding domain from the mistletoe Phoradendron liga. *Biochem. J.* 366, 405–413.

Li, L., Sun, J., Xia, S., Tian, X., Cheserek, M.J., Le, G., 2016. Mechanism of antifungal activity of antimicrobial peptide APP, a cell-penetrating peptide derivative, against *Candida albicans*: Intracellular DNA binding and cell cycle arrest. *Appl. Microbiol. Biotechnol.* 100, 3245–3253.

Li, W., Separovic, F., O'Brien-Simpson, N.M., Wade, J.D., 2021. Chemically modified and conjugated antimicrobial peptides against superbugs. *Chem. Soc. Rev.* 50(8), 4932–4973.

Liu, Y., Luo, J., Xu, C., Ren, F., Peng, C., Wu, G., Zhao, J., 2000. Purification, characterization, and molecular cloning of the gene of a seed-specific antimicrobial protein from pokeweed. *Plant Physiol.* 122, 1015–1024.

Lobo, D.S., Pereira, I.B., Fragel-Madeira, L., Medeiros, L.N., Cabral, L.M., Faria, J., Bellio, M., Campos, R.C., Linden, R., Kurtenbach, E., 2007. Antifungal Pisum Sativum Defensin 1 interacts with Neurospora crassa Cyclin F related to the cell cycle. *Biochemistry.* 46, 987–996.

Loeza-Angeles, H., Sagrero-Cisneros, E., Lara-Zárate, L., Villagómez-Gómez, E., López-Meza, J.E., Ochoa-Zarzosa, A., 2008. Thionin Thi2.1 from *Arabidopsis thaliana* expressed in endothelial cells shows antibacterial, antifungal and cytotoxic activity. *Biotechnol. Letters* 30, 1713–1719.

Mak, A.S., Jones, B.L., 1976. The amino acid sequence of wheat β-purothionin. *Can. J. Biochem.* 22, 835–842.

Manyi-Loh, C., Mamphweli, S., Meyer, E., Okoh, A., 2018. Antibiotic use in agriculture and its consequential resistance in environmental sources: Potential public health implications. *Molecules* 23, 795.

Maróti, G., Downie, J.A., Kondorosi, E., 2015. Plant cysteine-rich peptides that inhibit pathogen growth and control rhizobial differentiation in legume nodules. *Curr. Opin. Plant Biol.* 26, 57–63.

Masso-Silva, J.A., Diamond, G., 2014. Antimicrobial peptides from fish. *Pharmaceuticals* 7, 265–310.

Mellstrand, S.T., Samuelsson G., 1974. Phoratoxin, a toxic protein from the mistletoe *Phoradendron tomentosum* subsp. *macrophyllum* (Loranthaceae). The amino acid sequence. *Acta Pharm. Suec.* 11, 347–360.

Melo, F.R., Rigden, D.J., Franco, O.L., et al. 2002. Inhibition of trypsin by cowpea thionin: Characterization, molecular modeling, and docking. *Proteins.* 48, 311–319.

Mendez, E., Rocher, A., Calero, M., Girbes T., Citores L., Soriano F., 1996. Primary structure of omega-hordothionin, a member of a novel family of thionins from barley endosperm, and its inhibition of protein synthesis in eukaryotic and prokaryotic cell-free systems. *Eur. J. Biochem.* 239, 67–73.

Milbradt, A., Kerek, F., Moroder, L., Renner, C., 2003. Structural characterization of hellethionins from *Helleborus purpurascens*. *Biochem.* 42, 2404–2411.

Molina, A., Segura, A., García-Olmedo, F., 1993. Lipid transfer proteins (nsLTPs) from barley and maize leaves are potent inhibitors of bacterial and fungal plant pathogens. *FEBS Lett.* 316(2), 119–122.

Montero-Alejo, V., Acosta-Alba, J., Perdomo-Morales, R., Perera, E., Hernández-Rodríguez, E.W., Estrada, M.P., Porto-Verdecia, M., 2012. Defensin like peptide from *Panulirus argus* relates structurally with beta defensin from vertebrates. *Fish Shellfish Immunol.* 33, 872–879.

Mookherjee, N., Anderson, M.A., Haagsman, H.P., Davidson, D.J., 2020. Antimicrobial host defence peptides: functions and clinical potential. *Nat. Rev. Drug Discov.* 19, 311–332.

Morehead, M.S., Scarbrough, C., 2018. Emergence of global antibiotic resistance. *Prim. Care,* 45(3), 467–484.

Muramoto, N., Tanaka, T., Shimamura, T., Mitsukawa, N., Hori, E., Koda, K., Otani, M., Hirai, M., Nakamura, K., Imaeda, T., 2012. Transgenic sweet potato expressing thionin from barley gives resistance to black rot disease caused by *Ceratocystis fimbriata* in leaves and storage roots. *Plant Cell Rep.* 31, 987–997.

Mygind, P.H., Fischer, R.L., Schnorr, K.M., Hansen, M.T., Sönksen, C.P., Ludvigsen, S., Raventós, D., Buskov, S., Christensen, B., De Maria, L., 2005. Plectasin is a peptide antibiotic with therapeutic potential from a saprophytic fungus. *Nature* 437, 975.

Nawrot, R., Barylski, J., Nowicki, G., Broniarczyk, J., Buchwald, W., Goździcka-Józefiak, A., 2014. Plant antimicrobial peptides. *Folia Microbiol.* 59, 181–196.

Oard, S., Rush, M.C., Oard, J.H., 2004. Characterization of antimicrobial peptides against a US strain of the rice pathogen *Rhizoctonia solani*. *J. Appl. Microbiol.* 97, 169–180.

Ochiai, A., Ogawa, K., Fukuda, M., Ohori, M., Kanaoka, T., Tanaka, T., Taniguchi, M., Sagehashi, Y., 2018. Rice defensin OsAFP1 is a new drug candidate against human pathogenic fungi. *Sci. Rep.* 8, 11434.

Olson, T., Samuelsson, G., 1972. The amino acid sequence of viscotoxin A2 from the European mistletoe (*Viscum album* L., Loranthaceae). *Acta Chem. Scand.* 26, 585–595.

Parisien, A., Allain, B., Zhang, J., Mandeville, R., Lan, C.Q. 2008. Novel alternatives to antibiotics: Bacteriophages, bacterial cell wall hydrolases, and antimicrobial peptides. *J. Appl. Microbiol.* 104, 1–13.

Park, H.C., Kang, Y.H., Chun, H.J., Koo, J.C., Cheong, Y.H., Kim, C.Y., …Yoo, J.H., 2002. Characterization of a stamen-specific cdna encoding a novel plant defensin in Chinese cabbage. *Plant Mol. Biol.* 50, 59–69.

Pelegrini, P.B., Franco, O.L., 2005. Plant gamma-thionins: Novel insights on the mechanism of action of a multi-functional class of defense proteins. *Int. J. Biochem. Cell B* 37, 2239–2253.

Peng, S.Y., You, R.I., Lai, M.J., Lin, N.T., Chen, L.K., Chang, K.C., 2017. Highly potent antimicrobial modified peptides derived from the *Acinetobacter baumannii* phage endolysin LysAB2. *Sci. Rep.* 7, 11477.

Pfalzgraff, A., Brandenburg, K., Weindl, G., 2018. Antimicrobial peptides and their therapeutic potential for bacterial skin infections and wounds. *Front. Pharmacol.* 9, 281.

Pineiro, M., Diaz, I., Rodriguez-Palenzuela, P., Titarenko, E., Garcia-Olmedo, F., 1995. Selective disulphide linkage of plant thionins with other proteins. *FEBS Lett.* 369(2–3), 239–242.

Plotka, M., Kapusta, M., Dorawa, S., Kaczorowska, A.K., Kaczorowski, T., 2019. Ts2631 endolysin from the extremophilic *Thermus scotoductus* bacteriophage vB_Tsc2631 as an antimicrobial agent against gram-negative multidrug-resistant bacteria. *Viruses* 11, 657.

Qi, R.H., Chen, Y., Guo, Z.L., Zhang, F., Fang, Z., Huang, K., Yu, H.N., Wang, Y.P., 2019. Identification and characterization of two novel cathelicidins from the frog *Odorrana livida*. *Zool. Res*, 40, 94–101.

Rao, U., Stec, B., Teeter, M.M. 1995. Refinement of purothionins reveals solute particles important for lattice formation and toxicity. Part 1: Structure of β-purothionin at 1.7 Å resolution. *Acta Crystallogr.* 51(6), 904–913.

Rayapuram, C., Wu, J., Haas, C., Baldwin, I.T., 2008. PR-13/Thionin but not PR-1 mediates bacterial resistance in *Nicotiana attenuata* in nature, and neither influences herbivore resistance. *Mol. Plant-Microbe Interact.* 21(7), 988–1000.

Remuzgo, C., Oewel, T.S., Daffre, S., Lopes, T.R., Dyszy, F.H., Schreier, S., …Machini, M.T., 2014. Chemical synthesis, structure- activity relationship, and properties of shepherin I: A fungicidal peptide enriched in glycine-glycine histidine motifs. *Amino Acids* 46, 2573–2586.

Romagnoli, S., Fogolari, F., Catalano, M., et al. 2003. NMR solution structure of viscotoxin C1 from viscum album species *Coloratum ohwi*: Toward a structure-function analysis of viscotoxins. Biochem. 42, 12503–12510.

Roth, N., Käsbohrer, A., Mayrhofer, S., Zitz, U., Hofacre, C., Domig, K.J., 2019. The application of antibiotics in broiler production and the resulting antibiotic resistance in *Escherichia coli*: A global overview. *Poultry Sci.* 98, 1791–1804.

Roudi, R., Syn, N.L., Roudbary, M., 2017. Antimicrobial peptides as biologic and immunotherapeutic agents against cancer: a comprehensive overview. *Front Immunol.* 8, 1320.

Samuelsson, G., Pettersson, B.M., 1971. The amino acid sequence of viscotoxin B from the European mistletoe (*Viscum album* L. loranthaceae). *Eur. J. Biochem.* 21, 86–89.

Samuelsson, G., Pettersson, B.M., 1977. Toxic proteins from the mistletoe *Dendrophtora clavata*. II. The amino acid sequence of denclatoxin B. *Acta Pharm. Suecica* 14, 245–254.

Samuelsson, G., Seger, L., Olson, T., 1968. The amino acid sequence of oxidized viscotoxin A3 from the European mistletoe (Viscum album L, Loranthaceae). *Acta Chemica Scand.* 22(8), 2624–2642.

Sarkar, T., Chetia, M., Chatterjee, S., 2021. Antimicrobial peptides and proteins: From nature's reservoir to the laboratory and beyond. *Front. Chem.* 9, 691532.

Sauviat, M.P., Berton, J., Pater, C., 1985. Effect of phoratoxin B on electrical and mechanical activities of rat papillary muscle. *Acta Pharmacol. Sinica* 6, 91–93.

Schmidt, M., Arendt, E.K., Thery, T., 2019. Isolation and characterisation of the antifungal activity of the cowpea defensin Cp-thionin II. *Food Microbiol.* 82, 504–514.

Schrader, G., Apel, K., 1991. Isolation and characterization of cDNAs encoding viscotoxins of mistletoe (Viscum album). *Eur. J. Biochem.* 198(3), 549–553.

Schrader-Fisher, G., Apel, K., 1994. Organ specific expression of highly divergent thionin variants that are distinct from the seed-specific crambin in the crucifer *Crambe abyssinica*. *Mol. Gen. Genet.* 245, 380–389.

Selsted, M.E., Ouellette, A.J., 2005. Mammalian defensins in the antimicrobial immune response. *Nat. Immunol.* 6, 551–557.

Shirasawa-Seo, N., Nakamura, S., Ukai, N., Honkura, R., Iwai, T., Ohashi, Y., 2002. Ectopic expression of an oat thionin gene in carnation plants confers enhanced resistance to bacterial wilt disease. *Plant Biotechnol.* 19, 311–317.

Simons, A., Alhanout, K., Duval, R.E., 2020. Bacteriocins, antimicrobial peptides from bacterial origin: Overview of their biology and their impact against multidrug-resistant bacteria. *Microorganisms* 8, 639.

Sivakamavalli, J., Nirosha, R., Vaseeharan, B., 2015. Purification and characterization of a cysteine-Rich 14-kDa antibacterial peptide from the granular hemocytes of mangrove crab *Episesarma tetragonum* and its antibiofilm activity. *Appl. Biochem. Biotechnol.* 176, 1084–1101.

Slavokhotova, A.A., Shelenkov, A.A., Andreev, Y.A., Odintsova, T.I., 2017. Hevein-Like Antimicrobial Peptides of Plants. *Biochem.* 82, 1659–1674.

Bhargava, A., Srivastava, S., 2017. Bacterial integrons. In: Bhargava, A., Srivastava, S., (Eds.). *Biotechnology: Recent Trends and Emerging Dimensions.* CRC Press, Boca Raton, USA, pp. 57–74.

Srivastava, S., Dashora, K., Ameta, K.L., Singh, N.P., El-Enshasy, H.A., Pagano, M.C., Hesham, A.E., Sharma, G.D., Sharma, M., Bhargava, A., 2021. Cysteine-rich antimicrobial peptides from plants: The future of antimicrobial therapy. *Phytother. Res.* 35, 256–277.

Stec, B., 2006. Plant thionins--the structural perspective. *Cell. Mol. Life Sci.* 63, 1370–1385.

Stein, G.M., Schaller, G., Pfuller, U., Schietzel, M., Bussing, A., 1999. Thionins from *Viscum album* L: Influence of the viscotoxins on the activation of granulocytes. *Anticanc.Res.* 19, 1037–1042.

Subramaniam, G., Girish, M., 2020. Antibiotic resistance — a cause for reemergence of infections. *Indian J. Pediatr.* 87(11), 937–944.

Tam, J.P., Wang, S., Wong, K.H., Tan, W.L., 2015. Antimicrobial peptides from plants. *Pharmaceuticals* 8, 711–757.

Tassanakajon, A., Somboonwiwat, K., Amparyup, P., 2015. Sequence diversity and evolution of antimicrobial peptides in invertebrates. *Dev. Comp. Immunol.* 48, 324–341.

Taveira, G.B., Carvalho, A.O., Rodrigues, R., Trindade, F.G., Da Cunha, M., Gomes, V.M., 2016. Thionin-like peptide from *Capsicum annuum* fruits: Mechanism of action and synergism with fluconazole against *Candida* species. *BMC Microbiol.* 16, 12.

Taveira, G.B., Mathias, L.S., da Motta, O.V., Machado, O.L., Rodrigues, R., Carvalho, A.O., Teixeira-Ferreira, A., Perales, J., Vasconcelos, I.M., Gomes, V.M., 2014. Thionin-like peptides from *Capsicum annuum* fruits with high activity against human pathogenic bacteria and yeasts. *Biopolymers* 102, 30–39.

Taveira, G.B., Mello, E.O., Carvalho, A.O., Regente, M., Pinedo, M., de La Canal, L., Rodrigues, R., Gomes, V.M., 2017. Antimicrobial activity and mechanism of action of a thionin-like peptide from *Capsicum annuum* fruits and combinatorial treatment with fluconazole against *Fusarium solani. Peptide Sci.* 108, e23008.

Teeter, M.M., Mazer, J.A., L'Italien, J.J., 1981. Primary structure of the hydrophobic plant protein crambin. *Biochem.* 20, 5437–5443.

Ter Kuile, B.H., Kraupner, N., Brul, S., 2016. The risk of low concentrations of antibiotics in agriculture for resistance in human health care. *FEMS Microbiol. Lett.* 363, fnw210.

Thunberg, E., Samuelsson, G., 1982. Isolation and properties of ligatoxin A, a toxic protein from the mistletoe Phoradendron liga. *Acta Pharm. Suec.* 19, 285–292.

VanCompernolle, S.E., Taylor, R.J., Oswald-Richter, K., Jiang, J., Youree, B.E., Bowie, J.H., Tyler, M.J., Conlon, J.M., Wade, D., Aiken, C., Dermody, T.S., KewalRamani, V.N., Rollins-Smith, L.A., Unutmaz, D. 2005. Antimicrobial peptides from amphibian skin potently inhibit human immunodeficiency virus infection and transfer of virus from dendritic cells to T cells. *J. Virol.* 79(18), 11598–11606.

Van de Velde, W., Zehirov, G., Szatmari, A., Debreczeny, M., Ishihara, H., Kevei, Z., 2010. Plant peptides govern terminal differentiation of bacteria in symbiosis. *Science* 327, 1122–1126.

van Hoek, M.L., 2014. Antimicrobial peptides in reptiles. *Pharmaceuticals* 7, 723–753.

Vasilchenko, A.S., Smirnov, A.N., Zavriev, S.K., Grishin E.V., Vasilchenko, A.V., Rogozhin, E.A., 2017. Novel thionins from black seed (*Nigella sativa* L.) demonstrate antimicrobial activity. *Int. J. Pept. Res. Ther.* 23, 171–180.

Velkov, T., Thompson, P.E., Azad, M.A.K., Roberts, K.D., Bergen, P.J., 2019. History, chemistry and antibacterial spectrum. *Adv. Exp. Med. Biol.* 1145, 15–36.

Vermeulen, J.A.W.H., Lamerichs, R.M.J.N., Berliner, L.J., De Marco, A., Llinfis, M., Boelens, R., Alleman, J., Kaptein, R., 1987. ^{1}H NMR characterization of two crambin species. *FEBS Lett.* 219, 426–430.

Vernon, L.P., Evett, G.E., Zeikus, R.D., Gray, W.R., 1985. A toxic thionin from *Pyrularia pubera:* purification, properties, and amino acid sequence. *Arch. Biochem. Biophys.* 238, 18–29.

Wang, Y., Hong, J., Liu, X., Yang, H., Liu, R., Wu, J., Wang, A., Lin, D., Lai, R., 2008. Snake cathelicidin from *Bungarus fasciatus* is a potent peptide antibiotics. *PLoS ONE.* 3, e3217.

Wang, Y., Lu, Z., Feng, F., Zhu, W., Guang, H., Liu, J., He, W., Chi, L., Li, Z., Yu, H., 2011. Molecular cloning and characterization of novel cathelicidin-derived myeloid antimicrobial peptide from *Phasianus colchicus. Dev. Comp. Immunol.* 35, 314–322.

Wang, Y., Ouyang, J., Luo, X., Zhang, M., Jiang, Y., Zhang, F., Zhou, J., Wang, Y., 2021. Identification and characterization of novel bi-functional cathelicidins from the black-spotted frog (*Pelophylax nigromaculata*) with both anti-infective and antioxidant activities. *Dev. Comp. Immunol.* 116, 103928.

Wei, L., Gao, J., Zhang, S., Wu, S., Xie, Z., Ling, G., Kuang, Y.Q., Yang, Y., Yu, H., Wang, Y., 2015. Identification and characterization of the first cathelicidin from sea snakes with potent antimicrobial and anti-inflammatory activity and special mechanism. *J. Biol. Chem.* 290, 16633–16652.

Weiller, F., Moore, J.P., Young, P., Driouich, A., Vivier, M.A., 2016. The Brassicaceae species *Heliophila coronopifolia* produces root border-like cells that protect the root tip and secrete defensin peptides. *Ann. Bot.* 119, 803–813.

Willey, J.M., Van Der Donk, W.A., 2007. Lantibiotics: Peptides of diverse structure and function. *Annu. Rev. Microbiol.* 61, 477–501.

Williamson, D.A., Carter, G.P., Howden, B.P., 2017. Current and emerging topical antibacterials and antiseptics: agents, action, and resistance patterns. *Clin. Microbiol. Rev.* 30, 827–860.

Wisniewskia, M.E., Bassetta, C.L., Artlipa, T.S., Webba, R.P., Janisiewicza, W.J., Norellia, J.L., Goldwayb, M., 2003. Characterization of a defensin in bark and fruit tissues of peach and antimicrobial activity of a recombinant defensin in the yeast, *Pichia pastoris. Physiol. Plant.* 119, 563–572.

Wu, J., Gao, B., Zhu, S., 2014. The fungal defensin family enlarged. *Pharmaceuticals.* 7, 866–880.

Xu, X., Lai, R., 2015. The chemistry and biological activities of peptides from amphibian skin secretions. *Chem. Rev.* 115, 1760–1846.

Yacoub, H.A., Elazzazy, A.M., Mahmoud, M.M., Baeshen, M.N., Al-Maghrabi, O.A., Alkarim, S., Ahmed, E.S., Almehdar, H.A., Uversky, V.N., 2016. Chicken cathelicidins as potent intrinsically disordered biocides with antimicrobial activity against infectious pathogens. *Dev. Comp. Immunol.* 65, 8–24.

Zhang, Q.Y., Yan, Z.B., Meng, Y.M., Hong, X.Y., Shao, G., Ma, J.J., Cheng, X.R., Liu, J., Kang, J., Fu, C.Y., 2021. Antimicrobial peptides: Mechanism of action, activity and clinical potential. *Mil. Med. Res.* 8(1), 48.

Zhao, M., Ma, Y., Pan, Y.H., Zhang, C.H., Yuan, W.X., 2011. A hevein-like protein and a class I chitinase with antifungal activity from leaves of the paper mulberry. *Biomed. Chromatograph.* 25, 908–912.

Zimmer, M., Vukov, N., Scherer, S., Loessner, M.J., 2002. The murein hydrolase of the bacteriophage phi3626 dual lysis system is active against all tested *Clostridium perfringens strains. Appl. Environ. Microbiol.* 68, 5311–5317.

Section III

Nanobiotechnology

7 Curcumin Nanoemulsions

Recent Advances and Applications

T.P. Sari, Kiran Verma, and Prarabdh C. Badgujar
Department of Food Science and Technology, National Institute of Food Technology Entrepreneurship and Management, Haryana, India

Monika Sharma
Department of Botany, Sri Avadh Raj Singh Smarak Degree College, Vishunpur Bairya, Gonda , Uttar Pradesh, India

Benoît Moreau
Laboratoire de "Chimie verte et Produits Biobasés", Haute Ecole Provinciale de Hainaut-Condorcet, Département AgroBioscience et Chimie, 11, Rue de la Sucrerie, Belgium

Deborah Lanterbecq
Laboratoire de Biotechnologie et Biologie Appliquée, CARAH ASBL, Rue Paul Pastur, Belgium

Vinay B. Raghvendra
Department of Biotechnology (PG), Teresian College, Siddarthanagar, Mysore, India

Minaxi Sharma
Laboratoire de "Chimie verte et Produits Biobasés", Haute Ecole Provinciale de Hainaut-Condorcet, Département AgroBioscience et Chimie, 11, Rue de la Sucrerie, Belgium

Laboratoire de Biotechnologie et Biologie Appliquée, CARAH ASBL, Rue Paul Pastur, Belgium

CONTENTS

DOI: 10.1201/9781003324706-10

7.1 INTRODUCTION

Food nanotechnology is an emerging area which provides a range of options to food industry to enhance the quality, shelf life, safety and nutritional benefits to foods. The applications of nanotechnology in the food industry range from designing functional foods, improving bioavailability and water solubility, targeted delivery, protecting flavors and colors from processing conditions, biosensing in intelligent packaging, etc. A number of colloidal delivery approaches such as nanoliposomes, nanocomposites, nanoemulsions, microgels and solid-lipid nanoparticles are being used in food industry to resolve the low stability of many nutraceuticals in food systems. In particular, the food nanoemulsions have immense potential in improving the solubility and bioavailability of the encapsulated nutrient in foods. They are being used extensively in the delivery of nutraceuticals, plant polyphenols, enzymes, antimicrobial agents, etc. Nanoemulsion-based delivery has been found effective in improving the solubility and bioavailability of different polyphenols like beta carotene and curcumin which has poor metabolism and uptake in native form.

Various formulations of curcumin have been designed based on the purpose of use leaving the field of applications eternal. The Covid-19 pandemic had led to an intensification of research studies on curcumin and its formulations. The global curcumin market is estimated to reach USD 151.9 million by 2027 from USD 52.45 million in 2017 (Raduly et al. 2021). The excellent antioxidant and anti-inflammatory activities of curcumin had led to extensive studies especially to resolve its low solubility, stability and bioavailability issues. The versatility of curcumin made it possible to use its different formulations in different fields including dyeing agent in textiles, functional ingredient in food formulations, natural compound to enhance human health, ingredient in natural pesticides as well as a biosensor and chemosensor due to its fluorescent properties (Khorasani et al. 2019). In the present chapter, an attempt is made to review the advancement in the area of formulation, therapeutic benefits and food applications of curcumin-loaded nanoemulsions.

7.2 CURCUMIN

Curcumin, the polyphenolic compound extracted from the dried rhizomes of *Curcuma longa* (turmeric plant), is an important source of antioxidant in the balanced diet. Turmeric is a widely cultivated spice in the Asian continent especially India, Bangladesh and China (Tanvir et al. 2017). It is used on daily basis in spicy dishes and curries in Asian countries particularly in India. Curcumin, the bioactive compound of turmeric is responsible for the beneficiary properties of turmeric rhizome (1–6%) (Ariyarathna and Karunaratne 2016). Turmeric has been traditionally used in food and medicine in many Asian countries. The tubercles of turmeric are also used for flavor and color, and only little fraction (1–1.5 tons) is used for extraction of curcumin (Sousdaleff et al. 2013). The beneficial properties of curcumin include antioxidant, anticancer, anti-diabetic antimicrobial, anti-aging and other immunomodulatory properties (Karthikeyan et al. 2020). Curcumin has been extensively studied for its preventive and curative therapeutic benefits against a number of chronic ailments including neurogenerative diseases, (Ghosh et al. 2015) cardiovascular diseases (Sahebkar

2015), gastro-intestinal disorders (Jakubek et al. 2019) cancer (Liu and Ho 2018) chronic wounds (Kant et al. 2015) and an antiviral key for the treatment of SARS-CoV-2 (Dourado et al. 2021).

The numerous beneficial biological effects exerted by curcumin make it a functional ingredient in food with Generally Recognized As Safe (GRAS) status for use as a coloring and flavoring agent. Curcumin is a permitted food color in the EU (European Union), with levels from 20 to 500 mg/kg depending on food type. Based on No Observed Effect (NOE), JECFA (Joint FAO/WHO Expert Committee on Food Additives) has given an Acceptable Daily Intake (ADI) of 0–3 mg/kg body weight to curcumin.

7.2.1 Chemical Characterization of Curcumin

Turmeric, as major ingredient in Asian diet, considered to have approximately 235 compounds, primarily terpenoids and phenolics. Curcuminoids of turmeric show majority of bioactivities that have attracted the researchers to examine the depth of its various properties. Curcumin, the major curcuminoid, is a yellow-colored compound made of methoxyl and hydroxyl groups bonded to two aromatic rings linked with seven carbons in which two α,β unsaturated carbonyl groups are attached. As a result, curcumin exists in different isomeric form; the keto form at acidic/neutral conditions and enol form at alkaline conditions. The biological activities of curcumin are related to the different active sites present in it (Kotha and Luthria 2019) (Figure 7.1). The antioxidant properties of curcumin by scavenging reactive oxygen and nitrogen species are attributed to the phenolic OH- groups (Malik and Mukherjee 2014); the α,β unsaturated carbonyl groups are responsible for metal chelation properties there by reducing the oxidative stress and related diseases like Alzheimer's (Nelson et al. 2017); the β-diketo moiety provide anticancer properties by targeting thiol groups (Kotha and Luthria 2019). The major curcuminoids present in turmeric are shown in Figure 7.1.

The aqueous solubility of curcumin is reported to be negligible whereas it is highly soluble in polar solvents (Shin et al. 2016). The solubility can be enhanced by increasing the pH to alkaline conditions, but curcumin is more prone to degradation under higher pH (25% degradation at pH 8 at room temperature; Kurnik et al. 2020) and the degradation products are less active as compared to native curcumin (Wang et al. 2019). The solubility of curcumin in different oils depends on the conditions of preparation and oil type; much research studies found MCT (medium chain triglycerides) as a good carrier oil for curcumin (Sari et al. 2015; Tan et al. 2021) with least gravitational separation and droplet aggregation.

7.2.2 Key Challenges of Curcumin Administration

The potential therapeutic benefits of curcumin are restricted owing to its poor solubility in water, limited absorption and low stability in biological fluids. The metabolic degradation of curcumin in the intestinal pH conditions and the poor solubility in gastric/intestinal fluids is the major limiting factor contributing the low bioavailability of curcumin (Araiza-Calahorra et al. 2018). The very first report on the uptake and distribution of curcumin is done by Wahlstrom and Blennow (1978) showed that negligible amount of curcumin was detected from the blood plasma of rats orally administrated with 1 g/kg of curcumin. The hydrophobic nature of curcumin limits its food application and poor bioavailability limits its use in therapeutic applications. Taken together, to exert the required health benefit, the amount of native curcumin needed will be much higher than the safe dosage prescribed (up to 12 g/day) by many researchers (Minassi et al. 2013). Besides the good side of curcumin, at higher concentration it acts as pro-oxidant and can accelerate DNA damage, alter membrane permeation and cause deficiency of vital nutrients (Yoshino et al. 2004). To surmount these challenges, many researchers have introduced the use of nanoparticles of curcumin which outperform the native curcumin in terms of pharmacokinetic profile and functionality. Besides the improved functionality, the possibility of targeted delivery of curcumin in *in vivo* applications opens up a number of possibilities like greater control on dosage and bioavailability, co-delivery with other compounds and required lower concentrations for the desired benefit (Maradana et al. 2013).

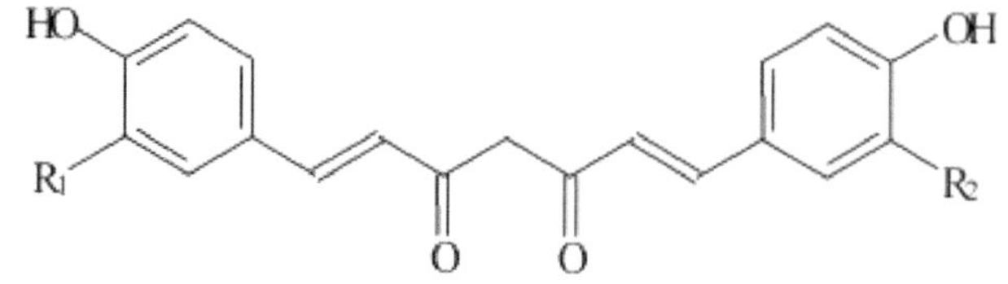

1) $R_1 = R_2 = OCH_3$

2) $R_1 = OCH_3$, $R_2 = H$

3) $R_1 = R_2 = H$

---

1) 1,7-Bis-(4-hydroxy-3-methoxyphenyl)-hepta-1,6-diene-3,5-dione
   (diferuloylmethane; curcumin)
   Chemical notation: $C_{21}H_{20}O_6$: C.A.S. number: 458-37-7, Molecular mass: 368 g/mol
2) 1-(4-Hydroxyphenyl)-7-(4-hydroxy-3-methoxyphenyl)-hepta-1,6-diene-3,5-dione
   p-hydroxycinnamoylferuloylmethane (Chemical notation: $C_{20}H_{18}O_5$: C.A.S. number: 33171-16-3, molecular mass: 338 g/mol)
3) 1,7-Bis-(4-hydroxyphenyl)-hepta-1,6-diene-3,5-dione
   p,p-dihidroxydicinnamoylmethane
   (Chemical notation: $C_{19}H_{16}O_4$: C.A.S. number: 33171-05-0, Molecular mass: 308 g/mol)

FIGURE 7.1 Different curcuminoids present in turmeric; 1) Curcumin (77%), Diferuloylmethane exist in both keto and enol form specific to pH. 2) demethoxycurcumin (17%) and 3) bisdemethoxycurcumin (3%). Goel et al. (2008); Slika and Patra (2020).

7.2.3 Nanostructures for Curcumin Delivery

Nanostructures of curcumin aim to improve solubility, reduce enzymatic and non-enzymatic degradation, and hydrolysis of curcumin. Different nanostructures used for encapsulation of curcumin include nanocomplexes, nanoliposomes, hydrogels, micelles and nanoemulsions with the choice depending on specific application and required thermodynamic stability. Nanoformulations of curcumin have proved to alter the pharmacokinetics of curcumin in a desired way. Among these formulations, nanoemulsions have been captivated by their ability to exert application in various fields including food, therapeutics and cosmetics (Ashaolu 2021). Compared to conventional emulsions, nanoemulsions provide greater stability toward droplet aggregation and separation (Choi and McClements 2020), over conventional emulsions as shown in Table 7.1.

7.3 CURCUMIN NANOEMULSIONS

In nanoemulsion delivery of curcumin, majorly two approaches are being used; curcumin-loaded nanoemulsion and excipient systems (Zou et al. 2015). In the former method, curcumin is encapsulated into a suitable delivery system, whereas in latter, turmeric-rich foods are consumed along with developed excipient formulations which allow the transfer of curcumin to these systems and improve bioavailability. Figure 7.2 is showing the basic representation of

TABLE 7.1
Ameliorative Potential of Nanoencapsulated Curcumin over Native Curcumin as a Functional Ingredient in Application Sector

Native Curcumin	Nanoencapsulated curcumin
	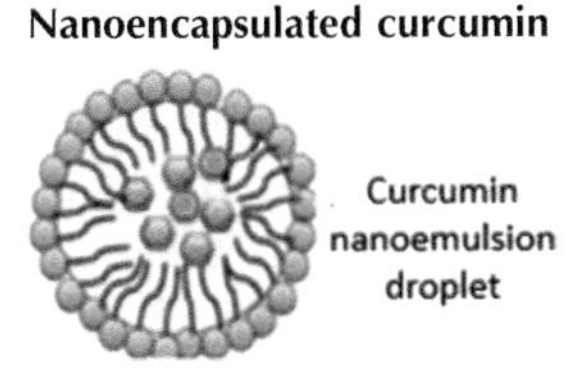
Poor water solubility	Improved water solubility
Poor bioavailability	Improved bioavailability
Rapid degradation into compounds with less activity	Increased stability and bioaccessibility
Targeted delivery is not possible	Targeted delivery to cells and organs
Activity specific to environmental conditions	Encapsulation protects the core curcumin
Prone to disintegration in native form, low levels in serum	Elevated concentration in body fluids

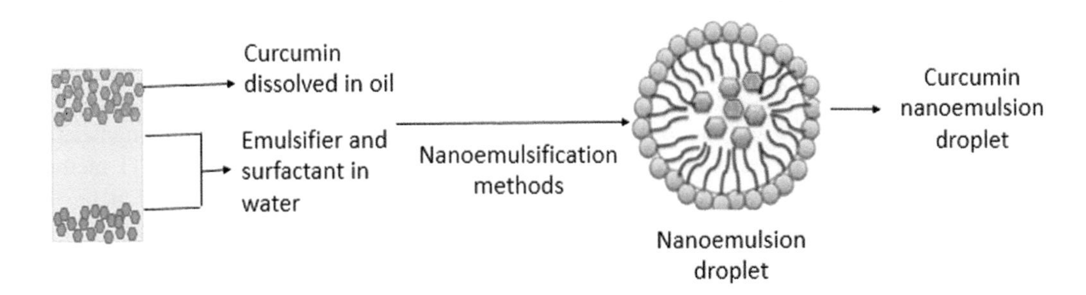

FIGURE 7.2 Schematic representation of preparation of curcumin nanoemulsion droplet.

preparation method of curcumin nanoemulsion, which is mostly preferred delivery system to encapsulated lipophilic bioactive compounds. Curcumin nanoemulsions are slightly opaque to transparent with a particle size range of 50–200 nm, prepared by either high-energy or low-energy methods (McClements 2015). The particle characteristics greatly depend on the parameters of the production methods like energy involved, time and temperature. High-energy methods have advantages over low-energy methods as it requires low surfactant/emulsifier concentration and provide higher thermodynamic stability of emulsion, at the same time it may be inappropriate for heat-sensitive formulations as the process generates much heat.

The choice of preparation of curcumin nanoemulsion greatly depends on appropriate applications, whether the entrapped curcumin has to be released slowly over a period of time or rapidly released at a particular site to get a higher local concentration. Curcumin has been encapsulated in a number of delivery agents by researchers, Ahmed et al. (2012) encapsulated curcumin in triacylglycerol where upon action of small intestinal lipases curcumin get released, whereas Sari et al. (2015), used whey protein concentrate to bypass the release of curcumin in small intestine and more than 90% of curcumin released in large intestine in *in vitro* trials. This approach can be used for targeted delivery of curcumin either in food applications or in drug delivery. Numerous studies, both *in vitro* and *in vivo* trials, proved the enhanced bioavailability and bioaccessibility of curcumin within nanoemulsions. The oral bioavailability was 33 times more for curcumin nanoemulsions as compared to native curcumin (Vecchione et al. 2016) whereas studies also report 100-fold increased bioavailability for nanocurcumin (Jamwal 2018).

7.3.1 Synthesis of Curcumin Nanoemulsion

Curcumin nanoemulsions can be fabricated using different methods; broadly categorized into high-energy or low-energy methods based on the underlying principle. The main components used in the production of curcumin nanoemulsions are curcumin with high purity, surfactant/emulsifier, oil and water. The production process involves two phases; emulsification phase and preparation phase. To incorporate curcumin into the colloidal systems, it must be dissolved in an oil phase of O/W emulsion, then the prepared emulsion is allowed to undergo appropriate methods to achieve desired particle size and stability. The loading capacity of curcumin into the oil depends on the type and polarity of fatty acid present in oil, method and conditions of preparation (time, temperature and energy involved). The commonly used emulsifiers in fabricating curcumin nanoemulsions are surfactants (Tween 80 and Tween 20), proteins (sodium caseinate), phospholipids, starches and polymers (Poly Ethylene Glycol; PEG).

7.3.1.1 Low-Energy Methods

7.3.1.1.1 Phase Inversion

This can be achieved by changing the net charge of surfactant to the complementary opposite charge by manipulating either temperature (phase inversion temperature) or composition (PIC) by keeping the other parameters of the emulsion constant (Ren et al. 2019). In phase inversion temperature (PIT) method, the hydrophobicity and hydrophilicity of the temperature dependant surfactant are allowed to vary with varying temperatures and phase of emulsion changes from O/W to W/O. The characteristics of the prepared emulsion greatly depend on the rate of cooling/heating which limits the viability of PIT in many applications. Several research studies used PIT method for formulating curcumin-encapsulated micro/nanoemulsions with detailed investigation on the particle characteristics (droplet size, ZP and perseverance) (Calligaris et al. 2016; Lahidjani et al. 2020). The transparent emulsion thus produced can have applications in food/beverage where the product transparency has to be maintained along with additional benefits of nanocurcumin. Lahidjani et al. (2020) reported that curcumin-loaded nanoemulsions prepared by PIT method resulted in emulsions with particle size as low as 10.34 nm. In phase inversion composition (PIC) method, the water/oil ratio of an emulsion is disturbed by adding either of the two at a constant temperature. The basic principle remains the same as in PIT method that is when water is being titrated into a system, the spontaneous curvature of surfactant changes and the system converts into O/W from W/O (Adena et al. 2021). Low-energy methods are energy-efficient alternatives and have many advantages over high energy as they require low input and less mechanical energy.

7.3.1.1.2 Spontaneous Emulsification

This is a low-energy method where O/W emulsions are prepared by adding water step-wise into a mixture of oil/surfactant containing the bioactive compound and stirring at a fixed temperature. It doesn't require any special equipment and can be conducted at room temperature. The spontaneity of the process depends on the characteristics (concentration, bulk density and structure) of the components and interfacial tension (Bouchemal et al. 2004).

7.3.1.1.3 Ionotropic Gelation

This technique is relatively simple and involves interaction of a cation/anion with an ionic edible polymer such as chitosan, alginate and carrageenan to generate a highly cross-linked structure. Due to the complexation between oppositely charged particles, the ionic polymer undergoes gelation and precipitates to form spherical particles. This method has been proven to improve the aqueous solubility and stability of curcumin (Guzman-Villanueva et al. 2013), and as a novel delivery approach for slow release of curcumin (Sharma et al. 2022).

7.3.1.2 High-Energy Methods

7.3.1.2.1 High-Pressure Homogenization

High-pressure homogenizers are generally used in dairy industries for preparation of homogenized milk with an intention to reduce the size of fat globules thereby reduce creaming. They can be used for preparation of nanoemulsions utilizing different forces like hydraulic shear, intense turbulence and cavitation thereby reducing the particle size. When a liquid with coarse emulsion containing large globules is allowed to pass through a tiny orifice at high velocity under high pressure (500–5000 psi) get converted into fine droplets with more uniform particle size, high surface area and stability toward aggregation (Badnjevic 2017). High-pressure homogenizers adapt well in small-scale and large-scale preparation of nanoemulsions and are also considered as highly efficient method for preparation of nanoemulsion, but utilizes large amount of energy and also increases the temperature of samples which might cause deterioration of heat-sensitive compounds (McClements 2015).

7.3.1.2.2 Ultrasonification

Ultrasonication is the best method for the production of nanoemulsions in laboratory/small-scale (Naseema et al. 2020). The sonicator probe is allowed to be in contact with the coarse emulsion and uses sound waves to create mechanical vibration and cavitation resulting in the size reduction (Leong et al. 2009). The particle characteristics depend on the frequency, time and power applied. The high energy input increases the temperature of samples, which has to be regulated using a cold-water bath. Ultrasonification has been used by many researchers to produce curcumin nanoemulsions using different surfactants as shown in Table 7.2. A schematic representation of ultrasonicator is presented in Figure 7.3.

7.3.1.2.3 Microfluidization

Microfluidization uses high-pressure displacement pump to produce fine nanoemulsions (Figure 7.3). The emulsions are allowed to collide from two opposite channels creating intense shear and high energy lead to the formation of fine emulsions. The particle size can be increased by increasing the operating pressure or by increasing the number of passes (Kentish et al. 2008). Recently, many studies have investigated the efficiency of microfluidization technique for the production of curcumin nanoemulsions (Song and McClements 2021; Verma et al. 2021). Páez-Hernández et al. (2019) evaluated the effect of different processing parameters and their effect on the characteristics of curcumin nanoemulsions. The particle size was found decreasing with increasing operating pressure and surfactant concentration.

7.3.2 Physico-chemical Characterization of Curcumin Nanoemulsion

In recent years, a strong upsurge in curcumin nanoemulsion fabrication can be seen in research studies using different carrier oils, emulsifiers, surfactants, different preparation methods and studying the physicochemical characteristics, bioavailability and bioaccessibility. The particle size, particle distribution measured in terms of polydispersity index (PDI) and zeta potential (ZP) influences the behavior of nanoemulsions in terms of stability, appearance and rheology (Faria et al. 2019). The droplet size and PDI can be determined by dynamic light scattering (DLS) based on their ability to scatter light.

Curcumin nanoemulsion droplets usually have a particle size between 50 and 200 nm diameter. The smaller the droplets, the higher the stability of the nanoemulsion. Smaller droplets can be prepared by optimizing the conditions of production which facilitate droplet disruption with minimum coalescence. The choice of emulsifier has a great impact on stabilizing the emulsion by rapidly absorbing at the interface. The surface charge of the particle measured as ZP also influences the stability and electrostatic interactions. Particles in a colloidal system are found to be stable when the

TABLE 7.2
Different Methods Used for Preparation of Curcumin-loaded Nanoemulsions

Method	Surfactant/Cmulsifier	Oil Phase	Production Conditions	Particle Characteristics	References
High-energy methods					
Ultrasonication	WPC (Whey protein concentrate)-70 and Tween 80	MCT	Magnetic stirring and sonification	Particle size 141.6 nm, zeta potential –6.9 mV and PDI 0.27	Sari et al. (2015)
	Hydroxylated soy lecithin	MCT, grapeseed oil or olive oil	High-speed mixer for 4 min) ultrasonication (130 W and 750 W for 30 min	Particle size ≤ 600 nm; PDI ≤ 0.35 and zeta potential ≤ –50	Páez-Hernández et al. (2019)
High-pressure homogenization	Tween 80	Corn oil	High shear mixing (2 min) followed by high-pressure homogenization (12000 psi, 5 cycles)	Droplet size- 270 nm; PDI ≤ 0.38; zeta potential –20.3 mV at pH 7 and +0.2 mV at pH 3	Zheng et al. (2017)
	Quillaja saponin	Corn oil	High-speed blending and HPH (12000 psi, 5 cycles	Particle size 170–200 nm; zeta potential –45 mV	Zheng et al. (2018)
	Whey protein isolate	MCT	Ultra –Turrax homogenizer (5000 rpm for 2 min), HPH (40 bar, 20 cycles)	Droplet size 184 nm, zeta potential –51.9 mV and PDI 0.124	Silva et al. (2019)
Microfluidization	Tween 80	Palm olein oil	High-speed homogenizer (7000 rev min for 1 hr) Microfluidizer (350–450 bar; 1–12 cycles)	Droplet size < 2170 nm, PDI ≤ 0.3 and zeta potential –43 to –31	Raviadaran et al. (2018)
	Tween 20, lecithin or sucrose mono palmitate	Corn oil	High shear homogenization (11000 rpm for 2 min) Microfluidization (150MPa; 5 cycles)	Particle size ≤ 400 nm, PDI ≤ 0.54 and zeta potential –37.	Artiga-Artigas et al. (2018)
Low-energy methods					
Phase inversion temperature method	Tween 80 (5, 10, 15, 20, 30 and 40 % w/w).	Extra virgin olive (EVO), castor oil, sunflower oil, peanut oil, tristearin and tripalmitinv	The aqueous phase (Tween 80 and NaCl) mixed with oil phase (with curcumin 5 mg/g) under magnetic stirring at 600 rpm /10 min. Emulsion heated for 30 min at 90°C and kept at 20°C up to 120 days	Highest solubility of curcumin obtained with castor oil. The PIT decreased with NaCl concentration. The minimum SOR for formation of transparent emulsions was 3.7.	Calligaris et al. (2016)
Spontaneous emulsification	Surfactant-Tween 80, co-surfactant- polyethylene glycol	Virgin coconut oil	Spontaneous mixing of oil phase (prepared by mixing at 400 rpm at 25°C for 30 min) with PEG loaded curcumin (prepared by sonication for 10 min at 45–50°C	Particle size- 15.92 mn; polydispersity index- 0.17	Md Saari et al. (2020)
	Surfactant- mixture of tween 80 and tween 85; Co-surfactant- ethanol	Soybean oil	Magnetic stirring at 100 rpm for 2 h	Mean particle size 215.66±16.8 nm and zeta potential –29.46±2.65 mV	Azami et al. (2018)

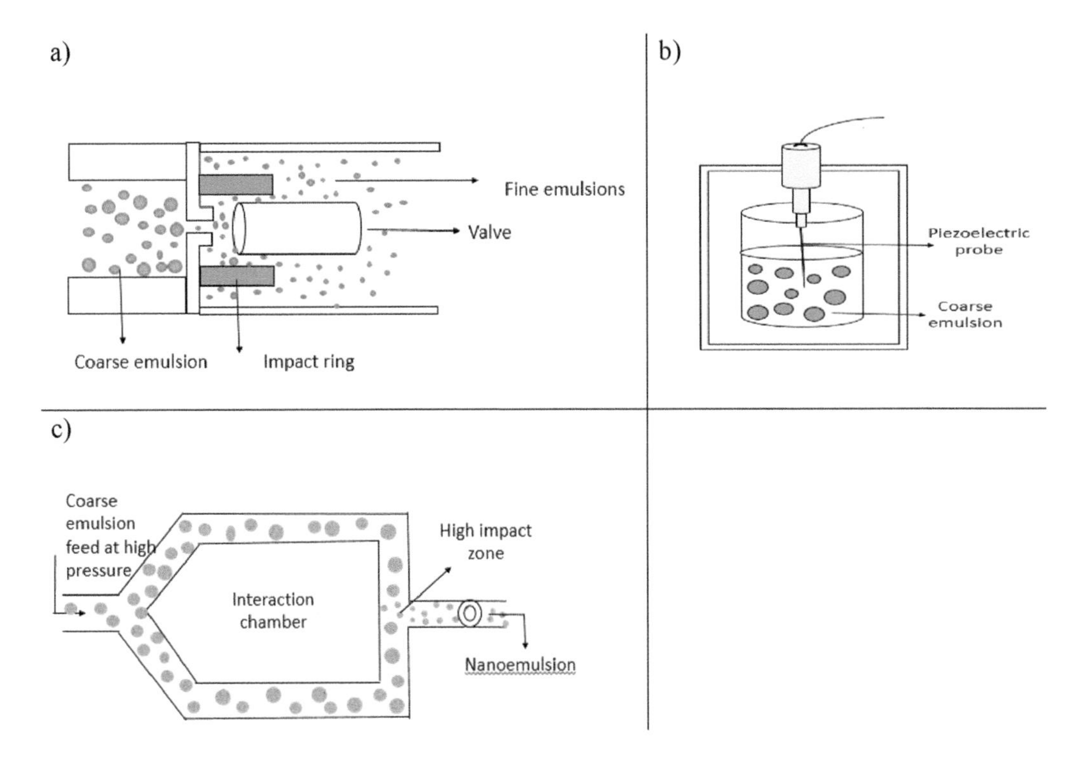

FIGURE 7.3 Schematic representation of different methods for the preparation of curcumin nanoemulsions, (a) High-pressure homogenizer, (b) ultrasonicator and (c) Microfluidizer. Redrawn from (McClements 2015).

ZP is in the range of +30 mV and −30 mV. The encapsulation efficiency of curcumin into the emulsion system depends on the properties of curcumin (purity, surface area, crystallinity and morphology), polarity of oil, nature of surfactant (ionic/non-ionic) and the energy applied for preparation (McClements 2012). Verma and co-workers (2021) report a decrease in particle size with increase in microfluidization pressure up to the third pass in different pressures ranging 50–200 MPa whereas after the fourth pass under all tested pressures, the particle size was found increasing. The team also reported an increase in encapsulation efficiency of microfluidized curcumin-loaded emulsion from 91.21 to 97.88% as compared to control sample (Verma et al. 2021).

7.3.3 Stability and Bioaccessibility of Curcumin Nanoemulsion

The stability of curcumin can be improved by encapsulating it in a nanoemulsion system. Various studies reported the increased stability of curcumin in nanoemulsion as compared to free curcumin (Xu et al. 2017; Artiga-Artigas et al. 2018). Xu and co-workers report an increased chemical stability of curcumin nanoemulsion prepared with milk protein and soybean soluble polysaccharide complex on storage under dark as compared to curcumin solutions. Only 25% curcumin was degraded after 40 days at 37°C whereas 94% curcumin degraded after 2 days of storage in un-encapsulated form (Xu et al. 2017). Bioaccesibility of curcumin generally depends on the pH of the system. According to Gupta et al. (2020), curcumin-loaded solid-lipid nanoparticles showed 20% of the curcumin release after 24 hours at an acidic pH of 1.2, as shown in Figure 7.4. According to their findings, free curcumin degraded by 77% in 6 hours and 89% in 24 hours at pH 6.8, but curcumin encapsulated in solid-lipid nanoparticles degraded by 10% in 6 hours and 30% in 24 hours. Similar degradation rates were seen for free curcumin at physiological pH 7.4, where 75 %

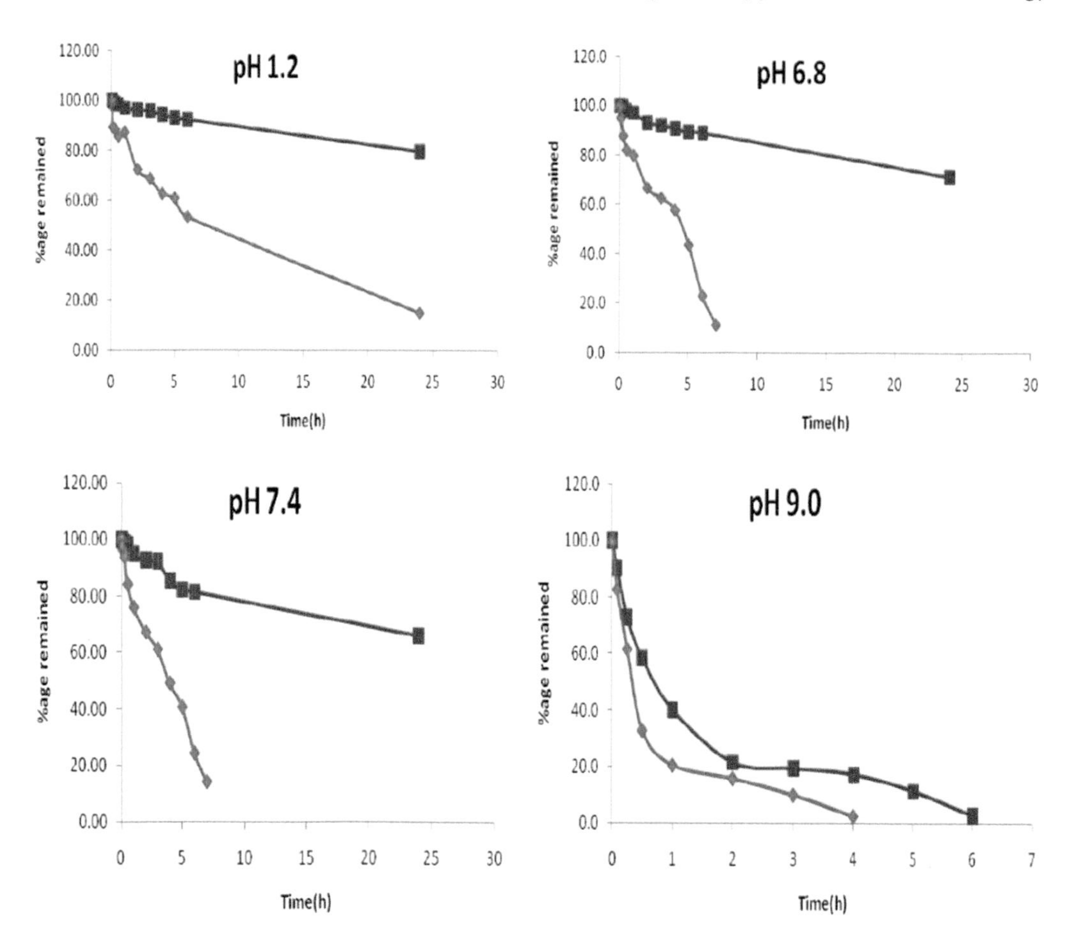

FIGURE 7.4 Percentage of remaining curcumin with time at different pH. Curcumin-encapsulated lipid nanoconstructs (red line) verses free curcumin (blue line). Adapted from Gupta et al. (2020), open access under Creative Commons Attribution License (CC BY).

of it was lost after 6 hours and 77% after 24 hours, and at extreme pH 9, where 98% of it was lost after 4 hours and the same amount after 6 hours.

Bioaccessibility of curcumin refers to the curcumin which solubilizes in the digestion and is available for absorption. Several research studies indicate the increased bioaccessibility of curcumin after encapsulation. Gonçalves et al. (2021) reported bioaccessibility of 53–71% for curcumin nanostructures. In a different study, Verma et al. (2021) reported 30% increase (from 65% for control cream curcumin emulsion to 84% after microfluidization) in bioaccessibility analysis. Jiang et al. (2020) report a fourfold increased bioaccessibility of curcumin in nanoemulsion form. Abbas and co-workers evaluated the influence of number of polymer layers on curcumin bioaccessibility and showed that with increase in polymer coating makes the emulsion less susceptible to lipid digestion and improves the bioaccessibility. The percentage bioaccessibility for curcumin-loaded single emulsion (NEI) was 15.25%, increased to 23.4% for curcumin-loaded double emulsion (NEII) and further increased to 29.2% with the addition of a third polymer layer for nanocapsules (Abbas et al. 2021) (Figure 7.5). The increased bioaccessibility of nanocurcumin could be due to the smaller particle size and greater surface area which makes the emulsion readily digested releasing the entrapped curcumin for absorption.

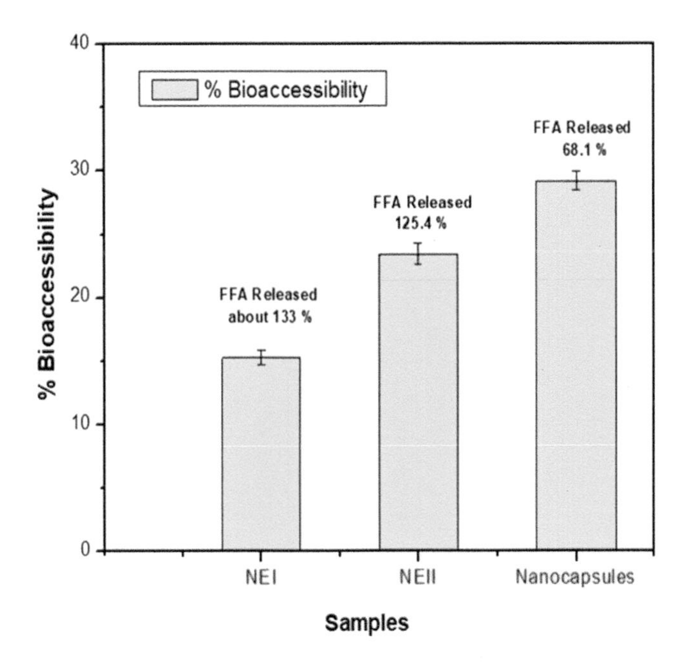

FIGURE 7.5 Bioaccessibility (*in vitro*) of NEI, NEII and nanocapsules after digestion for 2 h at 37 °C. Adapted from Abbas et al. (2021), Open Access article distributed under the terms of the Creative Commons Attribution License (http://creativecommons.org/licenses/by/4.0/).

7.4 APPLICATIONS OF CURCUMIN NANOEMULSIONS

7.4.1 Therapeutic Benefits of Curcumin Nanoemulsion

Medicinal uses of turmeric dated back to 4000 years in Vedic culture of Asia. For years, it is been considered that consumption of turmeric-rich food increases the longevity and decreases morbidities. Curcumin is believed to exhibit anti-inflammatory, antimicrobial, antioxidant, anticancer and antiviral activities (Páez-Hernández et al. 2019), as shown in Figure 7.6. The enhanced biological activities and bioaccessibility of curcumin nanoformulations have been confirmed via *in vivo* trials by many researchers. The phenolic OH- groups, the α, β unsaturated carbonyl groups and β-diketo moiety of curcumin (Figure 7.1) are attributed to different chemical/biochemical reactions including scavenging of reactive oxygen and nitrogen, metal chelation, antioxidant, anticancer and antimicrobial activity (Slika and Patra 2020). The nanoencapsulated curcumin has shown enhanced biological activities *in vivo* as compared to native curcumin. Some of the major therapeutic benefits of curcumin nanoemulsion are given below.

7.4.1.1 Antioxidant Activity

Curcumin has been reported to exhibit anti-oxidative properties including free radical and hydrogen peroxide scavenging and metal chelating which are responsible for inducing oxidative stress in the body contributing to various diseases. The complex biological reactions that take place in the body and exposure to environmental contaminants leads to the production of reactive oxygen species (ROS) and reactive nitrogen species (RNS), which can cause oxidative stress under various physiological conditions (Tanvir et al. 2017). ROS and RNS alter cellular protein, lipids, polysaccharides and even DNA. Antioxidants are consumed to reduce the adverse effect and potential hazards of oxidative damage caused by these reactive species.

The different methods such as micro and nano-encapsulation, liposomes, dispersions and nanoparticles used to protect curcumin from processing conditions not only serve as protecting agent

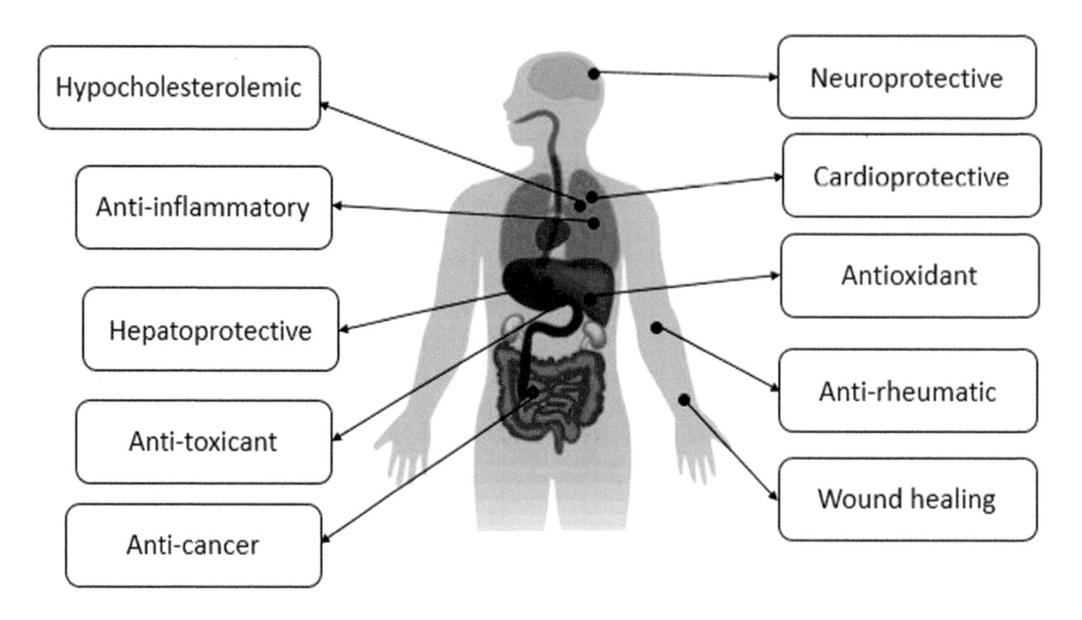

FIGURE 7.6 Therapeutic benefits/biological activities of curcumin nanoemulsion.

but also enhances the antioxidant activities (Verma et al. 2021). These methods produce an effective and stable emulsion/dispersion with smaller curcumin droplets/particles with larger surface area and consequently increase the potential antioxidant activity of the curcumin (Joung et al. 2016; Verma et al. 2021). The OH- groups of curcumin scavenge the free radicals contributing to the antioxidant activity. Few studies on antioxidant activity of curcumin in presented in Table 7.3.

7.4.1.2 Anticancer Activity

The anticancer activity of curcumin is one of the extensively investigated property against a number of uncontrolled growth of tumors such as breast, intestinal, oral cavity, blood, skin, lung, pancreas and human prostate cancer cell line 3 (PC3) (Adahoun et al. 2017). It has been studied and reported previously that curcumin has intense protective effect and therapeutic potential against several cancers. The prolonged release of curcumin from the drug formulations has shown increased solubility, controlled release and improved cytotoxicity with reduced side effects. Curcumin also exhibits potential to suppress proliferation, transformation and metastasis of tumors and tumor promotors (Bolat et al. 2020; Aggarwal et al. 2003). Curcumin affects cellular proliferation through modulation of signaling pathways that include transcription factors, protein kinases and growth factors (Liu et al. 2018). Suppression of anti-apoptotic proteins is another potential mechanism by which curcumin induces apoptosis in cells. The process of apoptosis is either mitochondria independent or mitochondria dependent. Studies have shown better cellular uptake and anticancer property of curcumin in nanoform than in native form (Jiang et al. 2020). The very first study on the anticancer potential of curcumin nanoformulation is done by Kuttan et al. (1985). The authors reported significantly higher anticancer activity of curcumin nanoliposomes as compared to free curcumin against Dalton's lymphoma cells. Thereafter a number of scientific research revealed the anticancer potential of curcumin against several cancers. Some of the work is shown in Table 7.3.

7.4.1.3 Anti-Inflammatory Activity

Turmeric has been traditionally used as an anti-inflammatory agent. Curcumin reduces the formation of pro-inflammatory cytokines which are responsible for upregulating inflammatory

TABLE 7.3
Therapeutic Application of Curcumin Nanoemulsion in Different Sectors

Study Design	Findings of Study	References
Antioxidant nanocurcumin		
Curcumin-milk cream emulsion was prepared by microfluidization	Microfluidized curcumin nanoemulsion showed 2-fold increase in DPPH scavenging activity and 25% increase in ferric-reducing antioxidant power	Verma et al. (2021)
PEGylated Nanoemulsions were prepared by high-speed homogenization	Curcumin-loaded nanoemulsions maintained antioxidant potential after 2 years of storage at room temperature.	Đoković et al. (2021)
Curcumin nanosuspensions were prepared using D-alpha-tocopheryl polyethylene glycol succinate by high-speed homogenization	Curcumin at concentrations 200 mg/kg or more exhibited excellent antioxidant properties in the lipid emulsion systems as assessed by the number of peroxides, conjugated dienes and aldehydes produced.	Kim et al. (2021)
Curcumin nanoemulsions were prepared with different surfactant concentrations using high-pressure homogenization	The antioxidant activity of curcumin nanoemulsions were influenced by concentrations of oil and surfactant.	Joung et al. (2016)
Liposome-encapsulated curcumin was prepared using homogenization and *in-vivo* plasma antioxidant activity was evaluated	The liposome-encapsulated curcumin showed significantly higher antioxidant activity in *in vivo* trial compared to free curcumin	Takahashi et al. (2009)
Binding of curcumin to β-lactoglobulin and its effect on antioxidant activity was evaluated	Antioxidant activity (ABTS scavenging) of complexes (curcumin and β-lactoglobulin) was higher than curcumin alone	Li et al. (2013)
Antimicrobial nanocurcumin		
Antimicrobial mechanism of curcumin was evaluated in Gram-negative (*Escherichia coli*) and Gram-positive (*Staphylococcus aureus*) bacteria isolates	Curcumin exerts antimicrobial activity by inducing oxidative stress and DNA damage.	Adeyemi et al. (2020)
Curcumin nanoparticles were prepared using ultrasonication	Curcumin in nanoform enhances the water solubility and antimicrobial activity.	Bhawana et al. (2011)
Curcumin nanoparticles were prepared using ultrasonication.	Antimicrobial activity improved by decreasing particle size to the nano-scale	Adahoun et al. (2017)
The antibacterial activity of curcumin against *H. pylori* was investigated	*H. pylori*-infected C57BL/6 mice showed reduced gastric damage due to antimicrobial activity of curcumin	De et al. (2009)
Curcumin incorporated polyurethane urea elastomers were studied	Curcumin polyurethane urea showed prominent antibacterial effect against *E. coli* and *S. aureus*	Shah et al. (2018)
Anticancer nanocurcumin		
Curcumin nanoparticles were prepared using ultrasonication	Nanocurcumin has potent anticancer activity	Adahoun et al. (2017)
Gum Arabic-curcumin conjugate micelles were investigated	The anticancer activity of the conjugate micelles is found to be higher in HepG2 cells than in MCF-7 cells	Sarika et al. (2015)
Curcumin-loaded gum Arabic-sodium alginate nanoparticles were prepared using the ionotropic gelation technique	Significant anticancer activity of curcumin-loaded gum Arabic-sodium alginate nanoparticles against HepG2 cells than HT29, A549 and MCF-7 cells.	Hassani et al. (2020)
Milk proteins (sodium caseinate, α-lactalbumin, β-lactoglobulin, whey protein concentrate and	sodium caseinate and whey protein concentrate as carrier for curcumin nanoparticles had the highest anticancer effect	Taha et al. (2020)

(Continued)

TABLE 7.3 (Continued)
Therapeutic Application of Curcumin Nanoemulsion in Different Sectors

Study Design	Findings of Study	References
whey protein isolate) as carrier for curcumin nanoparticles were evaluated		
Antiviral nanocurcumin		
Co-encapsulation of acyclovir and curcumin into microparticles was studied	Improved in vitro antiviral action of curcumin microparticles by co-encapsulation with acyclovir.	Reolon et al. (2019)
Interaction of curcumin and classical swine fever virus was investigated	Curcumin inhibited Classical swine fever virus replication by interfere lipid metabolism. curcumin promotes innate immune independent signaling pathway	Gao et al. (2021)
Curcumin bioconjugates bearing dipeptide, fatty acids and folic acid were studied	The molecules di-O-tryptophanyl phenylalanine curcumin and 3 di-O-decanoyl curcumin showed good results against Vesicular stomatitis virus (VSV) and Feline herpesvirus (FIPV/FHV)	Singh et al. (2010)
Antiviral property of curcumin was investigated	Curcumin inhibits Zika and chikungunya virus replication in human cells	Mounce et al. (2017)
Anti-aging nanocurcumin		
Evaluating anti-amyloid property of dietary curcumin and nanocurcumin in 5x-familial Alzheimer's disease mice	Solid-lipid curcumin particles (SLCPs) showed enhanced anti-amyloid, anti-inflammatory and neuroprotective effect.	Maiti et al. (2018)
PVA (poly vinyl alcohol)/Chitosan/curcumin patch was evaluated for wound healing properties in wistar rats against a commercial ointment having bactericidal properties.	Faster tissue regeneration was found in the wound area with PVA/Chitosan/curcumin patch; new skin produced appeared more normal as compared to commercial ointment	Niranjan et al. (2019)
Impact of Activatable curcumin polymer (ACP) is evaluated in monoiodoaetic acid-induced osteoarthritis mouse model	The ACP showed enhanced potential in reducing the levels of pro-inflammatory factors and hydrogen peroxide level in osteoarthritic joints as compared to control ones.	Kang et al. (2020)
Anti-inflammatory application of 5 different curcumin – chitosan nanogels (stabilized by Tween 80 and PEG 400) were evaluated for ex-vivo skin permeation and deposition.	The results showed a good ex-vivo skin permeation (75 – 107 μg/cm^3 /h for flux values) and skin deposition (17.46–46.07%) suggesting the potential use of curcumin nanoemulsions as anti-inflammatory agents compactable to human skin.	Thomas et al. (2017)

reactions in the body (Chen et al. 2017). Nanoencapsulated curcumin releases slowly from the encapsulating matrix enhance the retention time of curcumin and provide better anti-inflammatory activity in comparison to free curcumin. Apart from that the nanocurcumin has been studied for its ability to repair chronic wounds, which is usually compromised with agedness. (Shah and Amini-Nik 2017; Venkatasubbu and Anusuya 2017).

7.4.1.4 Antiviral Potential

Curcumin is an efficacious antiviral agent against viruses including Respiratory Syncytial Virus, influenza A and SARS-CoV (Thimmulappa et al. 2021). Curcumin interacts with proteins and intercedes with virus entry into lung cells (Pandey et al. 2021). Curcumin hinders hemagglutinin's binding (membrane glycoprotein) of influenza A virus with host cell receptors. Curcumin also attenuates the Chikungunya and Zika virus's infectivity by reforming the surface proteins of virus and preventing the establishment of an infection (Thimmulappa et al. 2021). It has also shown significant anti-SARS-CoV activity in recently reported studies. SARS-CoV-2S, a surface glycoprotein of 2019 novel coronavirus (Covid-19), helps the virus in entering the host cell and establish infection. Curcumin shows antiviral activity against enveloped virus by affecting its mechanism of multiplication by direct interaction with the virus envelop and proteins. Curcumin, as phytochemical, can be used for the treatment of Covid-19 in clinical trials owing to its safe and extensive antiviral activity against enveloped viruses (Dourado et al. 2021). Curcumin modifies the viral spike proteins, exerts immune-regulatory activity and reduces the oxidative stress in cells by increasing the antioxidant defences. Clinical trial showing the antiviral potency of nanocurcumin is shown in Table 7.3.

7.4.1.5 Antimicrobial Potential

The studies have demonstrated the improved antimicrobial activity of curcumin by reducing the particles into nanosize (Trigo-Gutierrez et al. 2021; Bhawana et al. 2011). In comparison to free curcumin, nanoencapsulated curcumin exhibited higher antimicrobial activities against bacteria both Gram-positive and Gram-negative. Additionally, curcumin exhibits detrimental effects against yeasts and molds. The enhanced activity may be attributed to the greater penetration of smaller particle into the cell leading to disruption of cell structure and organelles eventually leading to lysis of cell (Adahoun et al. 2017; Venkatasubbu and Anusuya 2017). However, studies majorly focused on the antimicrobial properties of nanoencapsulated curcumin rather than nanocurcumin enriched food matrices, so more studies are required to evaluate the influence of food components and processing conditions on the antimicrobial effect of curcumin in various food systems.

7.4.1.6 Anti-Aging Activity

Polyphenols have been studied extensively for their ability to modulate cellular senescence thereby providing beneficial effect against a number of aging-related diseases such as Alzheimer's disease, different types of cancers, type 2 diabetes, arthritis, kidney failure and Parkinson's disease. Different formulations of nanoencapsulated curcumin have been studied for their preventive and curative ability against aging-related health impairments. Studies also report ability of nanocurcumin to reduce joint inflammation associated with osteoarthritis (Kang et al. 2020), one of the most common causes of physical disability in elders. Some reports on the anti-aging activity of nanocurcumin are shown in Table 7.3.

7.4.2 Food Applications

Considering the consumer preferences for natural food additives, the food industry is encouraging the evolutions of novel techniques and functionality of natural bioactive compounds for enhanced activities and conservation (Adahoun et al. 2017). In spite of broad-spectrum activities of curcumin, the poor solubility and stability limit its use in many pharmaceutical and food applications. In order to defeat these limitations, numerous curcumin formulations have been developed by food researchers that are feasible to use in food systems (Table 7.4). Curcumin nanoformulations have been targeted to offer several additional benefits in comparison to conventional emulsions owing to their optical transparency, improved stability to particle aggregation and separation and rheological properties (Joung et al. 2016). Despite extensive research on optimizing and formulating nanoencapsulated curcumin using different carriers, their impact on incorporating into real food formulations is limited. Some of them are shown in Table 7.4.

TABLE 7.4
Potential Application of Nanocurcumin in Various Food Systems

Food System	Method of Preparation	Findings of the Study	References
Kulfi (Indian ice-cream)	Curcumin hydrogel beads by Ionotropic gelation	Increased overall acceptability and high sensory as compared to control sample with additional antimicrobial properties to the prepared kulfi.	Sharma et al. (2022)
Milk cream	Microfluidization	Microfluidization as a potential approach for generating stable and bio-accessible nanocurcumin emulsion.	Verma et al. (2021)
Yoghurt	Homogenization (10,000 rpm for 4 min) followed by ultrasonication (160 W, 20 kHz frequency and with 50% pulse)	The fortified yogurt with added caseinate-curcumin nanoparticles had higher antioxidant activity, improved dispersibility and textural properties.	Shehata and Soliman (2021)
Kefir	Homogenization at 7000 rpm for 7 min	Similar physicochemical properties of curcumin-fortified kefir samples as compared to control. Fortified kefir samples also reduced low-density lipoprotein (LDL), total cholesterol (TC), and triglycerides (TG) in rats as compared to standard diet.	Ershadi et al. (2021)
Salad dressing	Microfluidization at 82.7 mPa and 3 passes	The viscosity and mean particle diameter of the salad dressing was reduced and the color of the dressing turned to orange/yellow.	Song and McClements (2021)
Skim milk	Encapsulation of curcumin in skim milk by spray drying	Skim milk powder had yellowish color	Neves et al. (2019)
Milk	High-speed homogenization (5000 rpm for 10 min) followed by ultrasonication (15 min, 750 W and 40% amplitude).	The fortified flavored milk showed stability when stored for 21 days at 4°C, and the prepared emulsion can be used for targeted delivery of curcumin in intestine as > 90 % of curcumin retained simulated gastric phase.	Park et al. (2019)
Pineapple ice-cream	Emulsion inversion point method	Feasible alternative for artificial coloring agent in food systems as the incorporation of curcumin-loaded nanoemulsions in ice creams showed similar physicochemical, rheological properties and acceptable sensory properties as compared to control.	Borrin et al. (2018)
Ice-cream	Homogenization (200 kg cm^2); 2 passes	The curcumin nanoemulsion withstand the processing conditions without altering the sensory properties of ice-cream.	Kumar et al. (2016)
Milk	Microfluidization (1000 bar for 5 cycles)	The lipid oxidation in the unfortified milk was significantly higher as compared to curcumin-fortified milk	Joung et al. (2016)
Milk	Microfluidization (30,000 psi for three cycles)	The concentration of surfactant is critical in protecting the entrapped curcumin in nanoemulsion. The antioxidant activity of curcumin increased when added to milk.	Chuacharoen et al. (2019)
Canned ham	The spray-dried Turmeric nanoemulsion was incorporated in canned ham	Meat and meat products could not be used as a carrier for Turmeric nanoemulsion formulations as it exhibits unacceptable yellowing.	Kim et al. (2017)

TABLE 7.4 (Continued)
Potential Application of Nanocurcumin in Various Food Systems

Food System	Method of Preparation	Findings of the Study	References
Chicken fillets	Emulsion inversion point method	Retarded microbial spoilage and better sensory score were observed with curcumin-loaded emulsion as compared to control. curcumin-cinnamon essential oil nanoemulsion and curcumin-garlic essential oil nanoemulsion samples showed better water-holding capacity as compared to control and curcumin-sunflower oil nanoemulsion.	Abdou et al. (2018)

7.5 CONCLUSIONS AND FUTURE ASPECTS

Curcumin is a natural polyphenolic compound with wide range of therapeutic activities and has great potential for food application. Research conducted to date has shown the enhanced benefits of curcumin nanoemulsion over curcumin/turmeric. The present chapter was aimed to summarize the available preparation methods, physicochemical characteristics, therapeutic benefits and food applications of curcumin nanoemulsion. The available investigations on preparation and stability studies indicate the feasibility of incorporating these emulsions into the real food system, however, more focus is needed to formulate foods incorporated with curcumin nanoemulsions. Also, the feasibility of mimicking the laboratory conditions into large-scale commercial production of curcumin nanoemulsion needs attention.

In addition, the chapter outlines the therapeutic benefits of curcumin nanoemulsion *in vitro* and *in vivo*. The available reports indicate the enhanced pharmacological activities of nanocurcumin as compared to native curcumin. However, studies indicating the toxicological evaluations of the prepared nanoemulsions are fewer. To be used as a nano-additive, clinical trials and toxicological evaluations of these formulations have been conducted. Also, the undesirable outcomes of long-term consumption of these nanocurcumin are unknown. The possibility of co-delivery of other nutraceuticals along with curcumin can be a great innovation to the pharma sector. The food applications of curcumin nanoemulsions can open up a new range of functional foods in future. Further research in this area is needed to explore new techniques which are cost-effective and provide high entrapment efficiency without altering the natural characteristics of the food product in terms of sensory and stability. Simultaneously in future, it will be necessary to focus attention on medicinal applications of these nanoemulsions.

REFERENCES

Abbas, S., Chang, D., Riaz, N., Maan, A.A., Khan, M.K.I., Ahmad, I., et al., 2021. In-vitro stress stability, digestibility and bioaccessibility of curcumin-loaded polymeric nanocapsules. *J. Exp. Nanosci.* 16, 230–246.

Abdou, E.S., Galhoum, G.F., Mohamed, E.N., 2018. Curcumin loaded nanoemulsions/pectin coatings for refrigerated chicken fillets. *Food Hydrocol.* 83, 445–453.

Adahoun, M.A., Al-Akhras, M.H., Jaafar, M.S., Bououdina, M., 2017. Enhanced anti-cancer and anti-microbial activities of curcumin nanoparticles. *Artif. Cells, Nanomed. Biotechnol.* 45, 98–107.

Adena, S.K.R., Herneisey, M., Pierce, E., Hartmeier, P.R., Adlakha, S., Hosfeld, M.A.I., Drennen, J.K., Janjic, J.M., 2021. Quality by design methodology applied to process optimization and scale up of curcumin nanoemulsions produced by catastrophic phase inversion. *Pharmaceutics* 13, 880.

Adeyemi, O.S., Obeme-Imom, J.I., Akpor, B.O., Rotimi, D., Batiha, G.E., Owolabi, A., 2020. Altered redox status, DNA damage and modulation of L-tryptophan metabolism contribute to antimicrobial action of curcumin. *Heliyon* 6, e03495.

Aggarwal, B.B., Kumar, A., Bharti, A.C., 2003. Anticancer potential of curcumin: Preclinical and clinical studies. *Anticancer Res.* 23, 363–398.

Ahmed, K., Li, Y., McClements, D. J., & Xiao, H., 2012. Nanoemulsion-and emulsionbased delivery systems for curcumin: Encapsulation and release properties. *Food Chem.* 132, 799–807.

Araiza-Calahorra, A., Akhtar, M., Sarkar, A., 2018. Recent advances in emulsion-based delivery approaches for curcumin: From encapsulation to bioaccessibility. *Trends Food Sci. Technol.* 71, 155–169.

Ariyarathna, I.R., Karunaratne, D.N. 2016. Microencapsulation stabilizes curcumin for efficient delivery in food applications. *Food Pack. Shelf Life* 10, 79–86.

Artiga-Artigas, M., Lanjari-Pérez, Y., Martín-Belloso, O., 2018. Curcumin-loaded nanoemulsions stability as affected by the nature and concentration of surfactant. *Food Chem.* 266, 466–474.

Ashaolu, T.J., 2021. Nanoemulsions for health, food, and cosmetics: A review. *Environ. Chem. Lett.* 19, 3381–3395.

Azami, S.J., Teimouri, A., Keshavarz, H., Amani, A., Esmaeili, F., Hasanpour, H., Elikaee, S., Salehiniya, H., Shojaee, S., 2018. Curcumin nanoemulsion as a novel chemical for the treatment of acute and chronic toxoplasmosis in mice. *Int. J. Nanomed.* 13, 7363–7374.

Badnjevic, A., 2017. *CMBEBIH 2017. IFMBE Proceedings*. Springer, Singapore.

Bhawana, Basniwal, R.K., Buttar, H.S., Jain, V.K., Jain, N., 2011. Curcumin nanoparticles: Preparation, characterization, and antimicrobial study. *J. Agric. Food Chem.* 59, 2056–2061.

Bolat, Z.B., Islek, Z., Demir, B.N., Yilmaz, E.N., Sahin, F., Ucisik, M.H., 2020. Curcumin- and piperine-loaded emulsomes as combinational treatment approach enhance the anticancer activity of curcumin on HCT116 colorectal cancer model. *Front. Bioeng. Biotechnol.* 8, 50.

Borrin, T.R., Georges, E.L., Brito-Oliveira, T.C., Hadian, Z., Lorenzo, J.M., 2018. Technological and sensory evaluation of pineapple ice creams incorporating curcumin-loaded nanoemulsions obtained by the emulsion inversion point method. *Intern. J. Dairy Technol.* 71, 491–500.

Bouchemal, K., Briancon, S., Perrier, E., Fessi, H., 2004. Nanoemulsion formulation using spontaneous emulsification: solvent, oil and surfactant optimization. *Int. J. Pharm.* 280, 241–251

Calligaris, S., Valoppi, F., Barba, L., Pizzale, L., Anese, M., Conte, L., Nicoli, M.C., 2016. Development of transparent curcumin loaded microemulsions by phase inversion temperature (pit) method: Effect of lipid type and physical state on curcumin stability. *Food Biophys.* 12, 45–51.

Chen, Y.C., Shie, M.Y., Wu, Y.A., Lee, K.A., Wei, L.J., Shen, Y.F., 2017. Anti-inflammation performance of curcumin-loaded mesoporous calcium silicate cement. *J. Form. Med. Assoc.* 116, 679–688.

Choi, S.J., McClements, D.J., 2020. Nanoemulsions as delivery systems for lipophilic nutraceuticals: strategies for improving their formulation, stability, functionality and bioavailability. *Food Sci. Biotechnol.* 29, 149–168.

Chuacharoen, T., Prasongsuk, S., Sabliov, C.M., 2019. Effect of surfactant concentrations on physicochemical properties and functionality of curcumin nanoemulsions under conditions relevant to commercial utilization. *Molecules* 24, 2744.

De, R., Kundu, P., Swarnakar, S., Ramamurthy, T., Chowdhury, A., Nair, G.B., Mukhopadhyay, A.K., 2009. Antimicrobial activity of curcumin against *Helicobacter pylori* isolates from India and during infections in mice. *Antimicrob. Agents Chemother.* 53, 1592–1597.

Đoković, J.B., Savić, S.M., Mitrović, J.R., Nikolic, I., Marković, B.D., Randjelović, D.V., et al., 2021. Curcumin loaded PEGylated nanoemulsions designed for maintained antioxidant effects and improved bioavailability: A pilot study on rats. *Int. J. Mol. Sci.* 22, 7991.

Dourado, D., Freire, D.T., Pereira, D.T., Amaral-Machado, L., Alencar, E.N., de Barros, A.L.B., Egito, E.S.T., 2021. Will curcumin nanosystems be the next promising antiviral alternatives in COVID-19 treatment trials? *Biomed. Pharmacother.* 139, 111578.

Ershadi, A., Parastouei, K., Khaneghah, A.M., Hadian, Z., Lorenzo, J.M., 2021. Encapsulation of curcumin in Persian gum nanoparticles: An assessment of physicochemical, sensory, and nutritional properties. *Coatings* 11, 841.

Faria, M., Björnmalm, M., Thurecht, K.J., Kent, S.J., Parton, R.G., Kavallaris, M., et al., 2019. Europe PMC Funders Group minimum information reporting in bio-nano experimental literature. *Nat. Nanotechnol.* 13, 777–785.

Gao, Y., Hu, J.H., Liang, X.D., Chen, J., Liu, C.C., Liu, Y.Y., Cheng, Y., Go, Y.Y., Zhou, B. 2021. Curcumin inhibits classical swine fever virus replication by interfering with lipid metabolism. *Vetern. Microbiol.* 259, 09152.

Ghosh, S., Banerjee, S., Sil, P.C., 2015. The beneficial role of curcumin on inflammation, diabetes and neurodegenerative disease: A recent update. *Food Chem. Toxicol.* 83, 111–124.

Goel, A., Kunnumakkara, A.B., Aggarwal, B.B., 2008. Curcumin as "Curecumin": From kitchen to clinic. *Biochem. Pharmacol.* 75, 787–809.

Gonçalves, R.F.S., Martins, J.T., Abrunhosa, L., Baixinho, J., Matias, A.A., Vicente, A.A., Pinheiro, A.C., 2021. Lipid-based nanostructures as a strategy to enhance curcumin bioaccessibility: Behavior under digestion and cytotoxicity assessment. *Food Res. Intern.* 143, 110278.

Gupta, T., Singh, J., Kaur, S., Sandhu, S., Singh, G., Kaur, I.P., 2020. Enhancing bioavailability and stability of curcumin using solid lipid nanoparticles (CLEN): A covenant for its effectiveness. *Front. Bioeng. Biotechnol.* 8, 879.

Guzman-Villanueva, D., El-Sherbiny, I.M., Herrera-Ruiz, D., Smyth, H.D., 2013. Design and in vitro evaluation of a new nano-microparticulate system for enhanced aqueous-phase solubility of curcumin. *Biomed. Res. Int.* 2013, 724763.

Hassani, A., Mahmood, S., Enezei, H.H., Hussain, S.A., Hamad, H.A., Aldoghachi, A.F., Hagar, A., Doolaanea, A.A., Ibrahim, W.N., 2020. Formulation, characterization and biological activity screening of sodium alginate-gum Arabic nanoparticles loaded with curcumin. *Molecules* 25, 2244.

Jakubek, M., Kejík, Z., Kaplánek, R., Hromádka, R., Šandriková, V., Sýkora, D., Král, V., 2019. Strategy for improved therapeutic efficiency of curcumin in the treatment of gastric cancer. *Biomed. Pharmacother.* 118, 109278.

Jiang, T., Liao, W., Charcosset, C., 2020. Recent advances in encapsulation of curcumin in nanoemulsions: A review of encapsulation technologies, bioaccessibility and applications. *Food Res. Intern.* 132, 109035.

Jamwal, R. (2018). Bioavailable curcumin formulations: A review of pharmacokinetic studies in healthy volunteers. *J. Integr. Med.* 16, 367–374.

Joung, H.J., Choi, M.J., Kim, J.T., Park, S.H., Park, H.J., Shin, G.H., 2016. Development of food-grade curcumin nanoemulsion and its potential application to food beverage system: antioxidant property and in vitro digestion. *J. Food Sci.* 81, N745–N753.

Kang, C., Jung, E., Hyeon, H., Seon, S., Lee, D., 2020. Acid-activatable polymeric curcumin nanoparticles as therapeutic agents for osteoarthritis. *Nanomed. Nanotechnol. Biol. Med.* 23, 102104.

Kant, V., Gopal, A., Kumar, D., Pathak, N.N., Ram, M., Jangir, B.L., Kumar, D., 2015. Curcumin-induced angiogenesis hastens wound healing in diabetic rats. *J. Surg. Res.* 193, 978–988.

Karthikeyan, A., Senthil, N., Min, T., 2020. Nanocurcumin: A promising candidate for therapeutic applications. *Front Pharmacol.* 11, 487.

Kentish, S., Wooster, T., Ashokkumar, M., Balachandran, S., Mawson, R.L., Simons, L., 2008. The use of ultrasonics for nanoemulsion preparation. *Innov. Food Sci. Emerg. Technol.* 9, 170–175.

Khorasani, M.Y., Langari, H., Sany, S., Rezayi, M., Sahebkar, A., 2019. The role of curcumin and its derivatives in sensory applications. *Mater. Sci. Eng.* 103, 109792.

Kim, S.H., Lee, E.S., Lee, K.T., Hong, S.T., 2021. Stability properties and antioxidant activity of curcumin nanosuspensions in emulsion systems. *CyTA- J. Food* 19, 40–48.

Kim, S.W., Garcia, C.V., Lee, B.N., Kwon, H.J., Kim, J.T., 2017. Development of turmeric extract nanoemulsions and their incorporation into canned ham. *Korean J. Food Sci. Anim. Res.* 37, 889–897.

Kotha, R.R., Luthria, D.L., 2019. Curcumin: Biological, pharmaceutical, nutraceutical, and analytical aspects. *Molecules.* 24, 2930.

Kurnik, I.S., Noronha, M.A., Câmara, M.C., Mazzola, P.G., Vicente, A.A., Pereira, J.F., Lopes, A.M., 2020. Separation and purification of curcumin using novel aqueous two-phase micellar systems composed of amphiphilic copolymer and cholinium ionic liquids. *Sep. Purif. Technol.* 250, 117262.

Kumar, D. D., Mann, B., Pothuraju, R., Sharma, R., Bajaj, R., & Minaxi, M. (2016). Formulation and characterization of nanoencapsulated curcumin using sodium caseinate and its incorporation in ice cream. *Food Func.* 7, 417–424.

Kuttan, R., Bhanumathy, P., Nirmala, K., George, M.C., 1985. Potential anticancer activity of turmeric (*Curcuma longa*). *Cancer Lett.* 29, 197–202.

Lahidjani, L.K., Ahari, H., Sharifan, A., 2020. Influence of curcumin-loaded nanoemulsion fabricated through emulsion phase inversion on the shelf life of *Oncorhynchus mykiss* stored at 4°C. *J. Food Proc. Preserv.* 44, e14592.

Leong, T.S.H., Wooster, T.J., Kentish, S.E., Ashokkumar, M., 2009. Minimising oil droplet size using ultrasonic emulsification. *Ultrason. Sonochem.* 16, 721–727.

Li, M., Ma, Y., Ngadi, M.O., 2013. Binding of curcumin to β-lactoglobulin and its effect on antioxidant characteristics of curcumin. *Food Chem.* 141, 1504–1511.

Liu, H.-T., Ho, Y.-S., 2018. Anticancer effect of curcumin on breast cancer and stem cells. *Food Sci. Human Well.* 7, 134–137.

Maiti, P., Paladugu, L., & Dunbar, G. L. (2018). Solid lipid curcumin particles provide greater anti-amyloid, anti-inflammatory and neuroprotective effects than curcumin in the 5xFAD mouse model of Alzheimer's disease. *BMC Neurosci.* 19, 7.

Malik, P., Mukherjee, T.K., 2014. Structure-function elucidation of antioxidative and prooxidative activities of the polyphenolic compound curcumin. *Chinese J. Biol.* 2014, 396708.

Maradana, M.R., Thomas, R., & O'Sullivan, B J. (2013). Targeted delivery of curcumin for treating type 2 diabetes. *Mol. Nutr. Food Res.* 57, 1550–1556.

McClements, D.J., 2012. Crystals and crystallization in oil-in-water emulsions: Implications for emulsion-based delivery systems. *Adv. Colloid Interface Sci.* 174, 1–30.

McClements, D.J., 2015. *Food Emulsions: Principles, Practices, and Techniques.* CRC Press, Boca Raton, FL.

Md Saari, N.H., Chua, L.S., Hasham, R., Yuliati, L., 2020. Curcumin-loaded nanoemulsion for better cellular permeation. *Sci. Pharm.* 88, 44.

Minassi, A., Sanchez-Duffhues, G., Collado, J.A., Munoz, E., Appendino, G., 2013. Dissecting the pharmacophore of curcumin. Which structural element is critical for which action? *J. Nat. Prod.* 76, 1105–1112.

Mounce, B.C., Cesaro, T., Carrau, L., Vallet, T., Vignuzzi, M., 2017. Curcumin inhibits Zika and chikungunya virus infection by inhibiting cell binding. *Antiv. Res.* 142, 148–157.

Naseema, A., Kovooru, L., Behera, A.K., Kumar, K.P.P., Srivastava, P., 2020. A critical review of synthesis procedures, applications and future potential of nanoemulsions. *Adv. Colloid Interface Sci.* 287, 102318.

Nelson, K.M., Dahlin, J.L., Bisson, J., Graham, J., Pauli, G.F., Walters, M.A., 2017. The essential medicinal chemistry of curcumin. *J. Med. Chem.* 60, 1620–1637.

Neves, M.I.L., Desobry-Banon, S., Perrone, I.T., Desobry, S., Petit, J., 2019. Encapsulation of curcumin in milk powders by spray-drying: Physicochemistry, rehydration properties, and stability during storage. *Powder Technol.* 345, 601–607.

Niranjan, R., Kaushik, M., Prakash, J., Venkataprasanna, K.S., Arpana, C., Balashanmugam, P., Venkatasubbu, G.D., 2019. Enhanced wound healing by PVA/ chitosan/curcumin patches: In vitro and in vivo study. *Colloids Surf. B Biointerf.* 182, 110339.

Páez-Hernández, G., Mondragón-Cortez, P., & Espinosa-Andrews, H. (2019). Developing curcumin nanoemulsions by high-intensity methods: Impact of ultrasonication and microfluidization parameters. *LWT*, 111, 291–300.

Pandey, P., Rane, J.S., Chatterjee, A., Kumar, A., Khan, R., Prakash, A., Ray, S., 2021. Targeting SARS-CoV-2 spike protein of COVID-19 with naturally occurring phytochemicals: An *in silico* study for drug development. *J. Biomol. Struct. Dyn.* 39, 6306–6316.

Park, S.J., Hong, S.J., Garcia, C.B., Lee, S.B., Shin, G.H., Kim, J.T., 2019. Stability evaluation of turmeric extract nanoemulsion powder after application in milk as a food model. *J. Food Eng.* 259, 12–20.

Raduly, F.M., Raditoiu, V., Raditoiu, A., Purcar, V., 2021. Curcumin: Modern applications for a versatile additive. *Coatings* 11, 519.

Raviadaran, R., Chandran, D., Shin, L.H., Manickam, S., 2018. Optimization of palm oil in water nanoemulsion with curcumin using microfluidizer and response surface methodology. *LWT* 96, 58–65.

Ren, G., Sun, Z., Wang, Z., Zheng, X., Xu, Z., Sun, D. 2019. Nanoemulsion formation by the phase inversion temperature method using polyoxypropylene surfactants. *J. Colloid Interf. Sci.* 540, 177–184.

Reolon, J.B., Brustolin, M., Accarini, T., Viçozzi, G.P., Sari, M., Bender, E.A., Haas, S.E., Brum, M., Gündel, A., Colomé, L.M., 2019. Co-encapsulation of acyclovir and curcumin into microparticles improves the physicochemical characteristics and potentiates in vitro antiviral action: Influence of the polymeric composition. *Eur. J. Pharm. Sci.* 131, 167–176.

Sahebkar, A., 2015. Dual effect of curcumin in preventing atherosclerosis: The potential role of pro-oxidant–antioxidant mechanisms. *Nat. Prod. Res.* 29, 491–492.

Sarika, P.R., James, N.R., Kumar, P.R., Raj, D.K., Kumary, T.V., 2015. Gum arabic-curcumin conjugate micelles with enhanced loading for curcumin delivery to hepatocarcinoma cells. *Carbohyd. Polym.* 134, 167–174.

Shah, S.A.A., Imran, M., Lian, Q., Shehzad, F.K., Athir, N., Zhang, J., Cheng, J., 2018. Curcumin incorporated polyurethane urea elastomers with tunable thermo-mechanical properties. *React. Func. Polym.* 128, 97–103.

Shah, A., Amini-Nik, S., 2017. The role of phytochemicals in the inflammatory phase of wound healing. *Int. J. Mol. Sci.* 18, 1068.
Sharma, M., Inbaraj, B.S., Dikkala, P.K., Sridhar, K., Mude, A.N., Narsaiah, K., 2022. Preparation of curcumin hydrogel beads for the development of functional *Kulfi*: A tailoring delivery system. *Foods.* 11, 182.
Shin, G.H., Li, J., Cho, J.H., Kim, J.T., Park, H.J., 2016. Enhancement of curcumin solubility by phase change from crystalline to amorphous in Cur-TPGS nanosuspension. *J. Food Sci.* 81, N494–N501.
Shehata, S. H., & Nour Solim, Tarek (2021). Preparation and characterization of functional yoghurt using incorporated encapsulated curcumin by caseinate. *Int. J. Dairy Sci.* 16, 11–17.
Silva, H.D., Beldíková, E., Poejo, J., Abrunhosa, L., Serra, A.T., Duarte, C.M.M., et al., 2019. Evaluating the effect of chitosan layer on bioaccessibility and cellular uptake of curcumin nanoemulsions. *J. Food Eng.* 243, 89–100.
Singh, R.K., Rai, D., Yadav, D., Bhargava, A., Balzarini, J., De Clercq, E., 2010. Synthesis, antibacterial and antiviral properties of curcumin bioconjugates bearing dipeptide, fatty acids and folic acid. *Eur. J. Med. Chem.* 45, 1078–1086.
Slika, L., Patra, D., 2020. A short review on chemical properties, stability and nano-technological advances for curcumin delivery. *Expert Opin. Drug Deliv.* 17, 61–75.
Sousdaleff, M., Baesso, M.L., Medina Neto, A., Nogueira, A.C., Marcolino, V.A., Matioli, G., 2013. Microencapsulation by freeze-drying of potassium norbixinate and curcumin with maltodextrin: stability, solubility, and food application. *J. Agric. Food Chem.* 61, 955–965.
Sari, T.P., Mann, B., Kumar, R., Singh, R.R.B., Sharma, R., Bhardwaj, M., Athira, S., 2015. Preparation and characterization of nanoemulsion encapsulating curcumin. *Food Hydrocolloids.* 43, 540–546.
Song, H. Y., & McClements, D. J. (2021). Nano-enabled-fortification of salad dressings with curcumin: Impact of nanoemulsion-based delivery systems on physicochemical properties. *LWT*, 145, 111299.
Taha, S., El-Sherbiny, I., Enomoto, T., Salem, A., Nagai, E., Askar, A., Abady, G., Abdel-Hamid, M., 2020. Improving the functional activities of curcumin using milk proteins as nanocarriers. *Foods.* 9, 986.
Takahashi, M., Uechi, S., Takara, K., Asikin, Y., Wada, K., 2009. Evaluation of an oral carrier system in rats: Bioavailability and antioxidant properties of liposome-encapsulated curcumin. *J. Agric. Food Chem.* 57, 9141–9146.
Tan, K.X., Ng, L.E., Loo, S.C.J., 2021. Formulation development of a food-graded curcumin-loaded medium chain triglycerides-encapsulated kappa carrageenan (CUR-MCT-KC) gel bead based oral delivery formulation. *Materials* 14, 2783.
Tanvir, E.M., Hossen, M.S., Hossain, M.F., Afroz, R., Gan, S.H., Khalil, M.I., Karim, N., 2017. Antioxidant properties of popular turmeric (*Curcuma longa*) varieties from Bangladesh. *J. Food Qual.* 2017, 8471785.
Thimmulappa, R.K., Mudnakudu-Nagaraju, K.K., Shivamallu, C., Subramaniam, K., Radhakrishnan, A., Bhojraj, S., Kuppusamy, G., 2021. Antiviral and immunomodulatory activity of curcumin: A case for prophylactic therapy for COVID-19. *Heliyon.* 7, e06350.
Thomas, L., Zakir, F., Mirza, M.A., Anwer, M.K., Ahmad, F.J., Iqbal, Z., 2017. Development of Curcumin loaded chitosan polymer based nanoemulsion gel: In vitro, ex vivo evaluation and in vivo wound healing studies. *Intern. J. Biol. Macromol.* 101, 569–579.
Trigo-Gutierrez, J.K., Vega-Chacón, Y., Soares, A.B., Mima, E.G.d.O., 2021. Antimicrobial activity of curcumin in nanoformulations: A comprehensive review. *Int. J. Mol. Sci.* 22, 7130.
Vecchione, R., Quagliariello, V., Calabria, D., Calcagno, V., De Luca, E., Laffaioli, R.V., Netti, P.A., 2016. Curcumin bioavailability from oil in water nano-emulsions: In vitro and in vivo study on the dimensional, compositional and interactional dependence. *J. Control. Release* 233, 88–100.
Venkatasubbu, G.D., Anusuya, T., 2017. Investigation on Curcumin nanocomposite for wound dressing. *Int. J. Biol. Macromol.* 98, 366–378.
Verma, K., Tarafdar, A., Mishra, V., Dilbaghi, N., Kondepudi, K.K., Badgujar, P.C., 2021. Nanoencapsulated curcumin emulsion utilizing milk cream as a potential vehicle by microfluidization: Bioaccessibility, cytotoxicity and physico-functional properties. *Food Res. Intern.* 148, 110611.
Wahlstrom, B., Blennow, G.A., 1978. Study on the fate of curcumin in the rat. *Acta Pharmacol. Toxicol.* 43, 86–92.
Wang, N., Lu, N., Zhao, L., Qi, C., Zhang, W., Dong, J., Hou, X., 2019. Characterization of stress degradation products of curcumin and its two derivatives by UPLC–DAD–MS/MS. *Arab. J. Chem.* 12, 3998–4005.
Xu, G., Wang, C., Yao, P., 2017. Stable emulsion produced from casein and soy polysaccharide compacted complex for protection and oral delivery of curcumin. *Food Hydrocolloids.* 71, 108–117.

Yoshino, M., Haneda, M., Naruse, M., Htay, H.H., Tsubouchi, R., Qiao, S.L., Li, W.H., Murakami, K., Yokochi, T., 2004. Prooxidant activity of curcumin: copper-dependent formation of 8-hydroxy-2'-deoxyguanosine in DNA and induction of apoptotic cell death. *Toxicol. In Vitro* 18, 783–789.

Zheng, B., Peng, S., Zhang, X., McClements, D.J., 2018. Impact of delivery system type on curcumin bioaccessibility: comparison of curcumin-loaded nanoemulsions with commercial curcumin supplements. *J. Agric. Food Chem.* 66, 10816–10826.

Zheng, B., Zhang, Z., Chen, F., Luo, X., McClements, D.J., 2017. Impact of delivery system type on curcumin stability: Comparison of curcumin degradation in aqueous solutions, emulsions, and hydrogel beads. *Food Hydrocolloids.* 71, 187–197.

Zou, L., Liu, W., Liu, C., Xiao, H., McClements, D.J., 2015. Designing excipient emulsions to increase nutraceutical bioavailability: emulsifier type influences curcumin stability and bioaccessibility by altering gastrointestinal fate. *Food Func.*6, 2475–2486.

8 Mechanism and Method of Zinc Oxide Nanoparticles (ZnO NPs) Induced Toxicity in Biological Systems

Anurag Kumar Srivastav
Department of Clinical Immunology and Rheumatology, Sanjay Gandhi Postgraduate Institute of Medical Sciences, Lucknow, Uttar Pradesh, India

Supriya Karpathak
Department of Respiratory Medicine, King George's Medical University, Lucknow, Uttar Pradesh, India

Jyoti Prakash
Amity Institute of Biotechnology, Amity University Uttar Pradesh, Lucknow Campus, Lucknow, India

Mahadeo Kumar
Biochemistry Laboratory, Animal Facility, Regulatory Toxicology Group, CSIR-Indian Institute of Toxicology Research (CSIR-IITR), Lucknow, Uttar Pradesh, India

CONTENTS

DOI: 10.1201/9781003324706-11

8.1 INTRODUCTION

The desires of human always lift their imagination that often gives raise a new science and technology. Nanotechnology was born out of such types of dreams. Nanotechnology, the critical technology of the 21st century, is defined as the understanding and work on the matter at 1 and 100 nm scale range where their unique property facilitates novel applications (Roco 2000). Nanotechnology deals with the matter at the size range of 1 billionth of a meter (10^{-9} m = 1 nm; i.e., nanoscale), in which manipulating the matter at their atomic and molecular range. In principle, nanomaterials (NMs) are described as objects that must have one dimension in range of 1–100 nm as presented in Figure 8.1. In advanced nanotechnology, nanoparticles (NPs) have proven their usefulness in multidisciplinary sciences due to their unique physiochemical characteristics at the meeting point of physics, chemistry, biology, electronics, pharmaceutical, medicine and information technology.

8.2 HISTORY

The use of nano-biotechnology is dependent upon the knowledge which investigates the cell structure and their function as well as intra and inter-cellular processes. In 1902, Richard Zsigmondy and Henry Siedentopf had developed an ultramicroscope using ruby glasses and successfully detected the

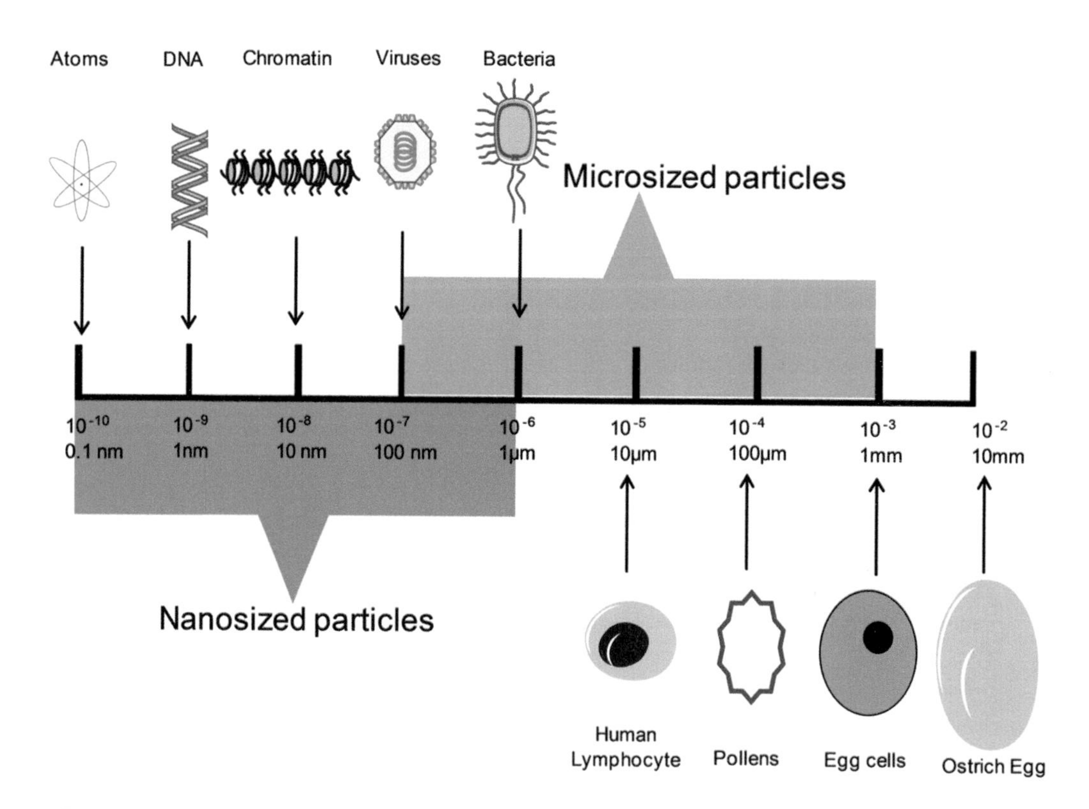

FIGURE 8.1 Representation of the nano and microscale object.

structure of nanometers range (Mappes et al. 2012) and further in 1912 he investigates the behavior of the colloidal solution. Since 1931, Max Knoll and Ernst Ruska were able to achieve better resolution with the transmission electron microscope (TEM) (Mappes et al. 2012).

Later, Erwin Müller in 1936 developed the field ion microscope (FIM), with which the studies of atoms and their arrangement on a surface were possible (Yamamoto et al. 2013). Gerd Binnig and Heinrich Rohrer (1982) developed the scanning tunneling microscope (STM), with the help of this microscope the graphical analysis of an atom was possible. Thus, the various methods of scanning probe microscopy were useful not only to demonstrate nanoscale structures but also were helpful to position and manipulate the particles in a controlled way.

The concept of a "nanometer" was first proposed by Richard Zsigmondy (Nobel Prize in chemistry in 1925). He coined the term nanometer and characterized the size of particles *such as* gold colloids by using a microscope. The Laureate Richard Feynman (Noble prize in Physics in 1965) presented a lecture titled, "There is Plenty of Room at the Bottom", in the American Physical Society meeting at Caltech in 1959, where they demonstrated the concept of manipulating matter at the atomic and molecular level (Feynman 1959).

This novel idea provides an advantage in nanotechnology and Feynman's hypotheses have since been proven correct. As a consequence, he is considered the father of modern nanotechnology. Fifteen years later, Norio Taniguchi was the first who use "nanotechnology" for demonstrating semiconductor processes based on the order of a nanometer. The nanotechnology consists of processing, separation, consolidation and deformation of materials by manipulating at a single atom or single molecule level. At coming till 1980, Kroto, Smalley and Curl discovered fullerenes (Hulla et al. 2015) and further in 1986, Eric Drexler, Massachusetts Institute of Technology (MIT), wrote a book titled, "*Engines of Creation: The Coming Era of Nanotechnology*" in which they addressed the ideas of Feynman and Taniguchi. Eric Drexler is the first who proposed the nanoscale assembler (Drexler 1986). Drexler's theory of nanotechnology is regularly called "molecular nanotechnology". Further a Japanese scientist, Iijima has developed carbon nanotubes, this approach has advanced the field of nanotechnology (Iijima 1991). Thus, the 1980s is known as the "Golden Era of Nanotechnology".

The beginning of the 21st century is increased attention in the emerging fields of nanoscience and nanotechnology. The United States was taken advances in the field of nanotechnology based on Feynman's concept, i.e., "manipulation of matter at the atomic level". President Bill Clinton favored and advocated for funding of research in this emerging technology during a speech at Caltech (January 21, 2000). Three years later, President George W. Bush signed into law the 21st century Nanotechnology Research and Development Act and created the National Technology Initiative (NNI) (Roco 2000). The NNI is the most wide-ranging R&D program in nanoscience and technology in the world. The focus of NNI is on the research and development of nanoscale science and technology for economic profit and national security. At the beginning of the new millennium, governments around the world created national nanotechnology programs in which spent billions of dollars and arranged the institution for the construction and reconfiguration of new research in the field of nanoscience (Roco Mihail C. 2003).

The government of India has implemented the 9th Five-Year Plan (1998–2002) in which first time taken concern for set up national facilities and core groups to promote research in frontier areas of S&T, included superconductivity, robotics, neurosciences and carbon, and NMs (Plan FFY 1998). The Government continuously initiated the Nanomaterials Science and Technology Mission (NSTM) in the 10th Five-Year Plan (2002–2007) for developments in the field of nanotechnology. Subsequently, the Department of Science and Technology of the Ministry of Science (DST) launched the National Nanoscience and Nanotechnology Initiative (NSTI) in 2001. After that, the 11th Five-Year Plan (2007–2012) projects have also been grouped to craft high impact on socio-economic delivery involving NM and nanodevices in health and disease (Plan EFY 2007). In the 12th Five-Year Plan (2012–2017) too, the government gave its approval for continuation of the Mission on Nano Science and Technology (Nano Mission) and from 2019, DST has started a scheme as Mission on Nano Science and Technology (Nano Mission). The Nano Mission, a new segment, would make a more excellent attempt to hold up application-oriented Research and Development.

8.3 NANOSTRUCTURE MATERIAL CLASSIFICATION

NPs are generally classified based on their dimensionality, morphology, composition, uniformity and agglomeration.

8.3.1 Dimensionality

According to Siegel (1994), nanostructured materials are classified into four groups based on their dimensionality (Siegel 1994) such as zero dimensional-0D NPs), one dimensional-1D (Graphene, thin film), two dimensional-2D (carbon nanotubes, nanowires) and three dimensional-3D (quantum dots or NPs and Fullerene) nanostructures as presented in Figure 8.2.

8.3.2 Nanoparticles Morphology

The NMs are classified based on morphologically by concerning the features: flatness, sphericity and aspect ratio. Commonly a classification of the NMs lies between high and low-aspect ratio. High aspect ratio NPs morphologies include nanotubes (helices, zigzags and belts) and nanowires (variant length). Small-aspect ratio NPs include oval, cubic, prism and helical types of particles.

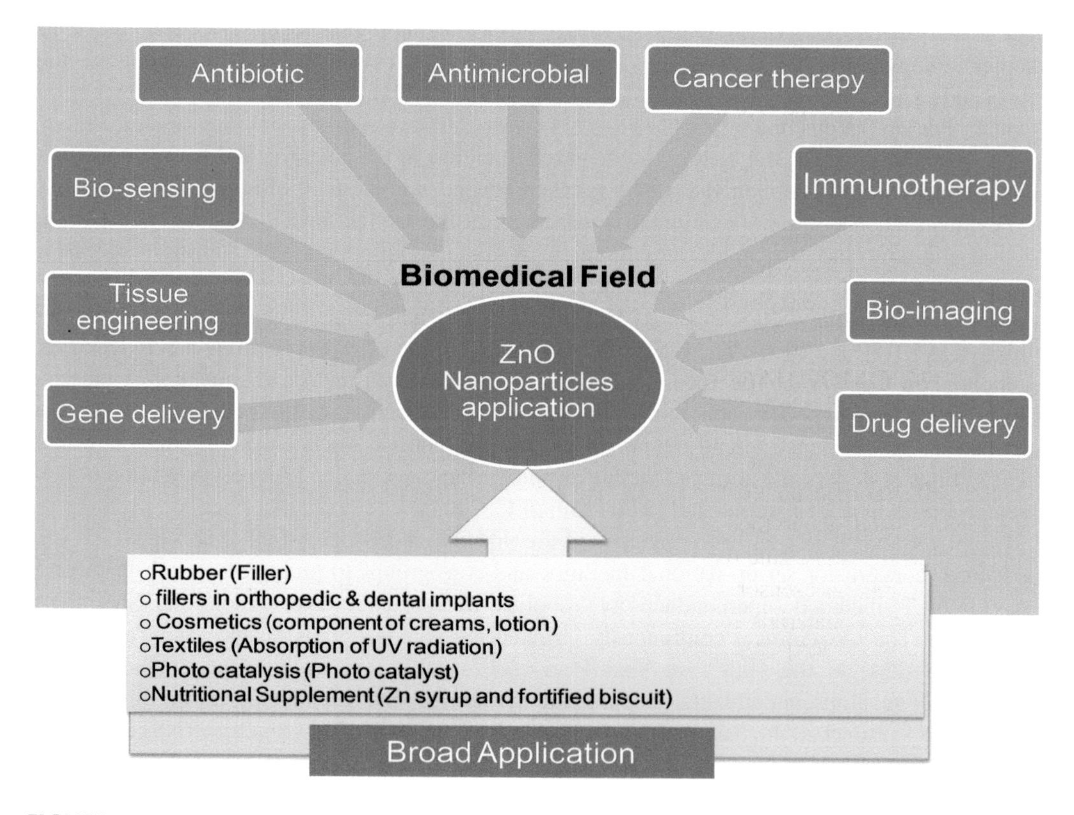

FIGURE 8.2 Representing use of ZnO nanoparticles in various fields.

8.3.3 Nanoparticles Composition

According to composition, the NPs may be prepared for a single constituent material or composite of several materials. In nature, the NPs are often found in the form of agglomerations of materials with various compositions, although the pure single type of composed NMs is synthesized in the laboratory by different methods (Jeevanandam et al. 2018).

8.3.3.1 Carbon-based Nanomaterials

These NMs contain a constituent of carbon and show definite morphologies such as hollow tubes, ellipsoids or spheres. Here, some well-known examples of carbon-based NMs are fullerenes (C-60), carbon nanotubes, carbon nanofibers, carbon black and graphene.

8.3.3.2 Inorganic-based Nanomaterials

These NMs consist of metal and metal oxide NPs. These NMs can be manufactured in the form of metals (Au or Ag NPs), or metal oxides (TiO_2, Fe_2O_3 and ZnO NPs), and also as semiconductors (silicon and ceramics).

8.3.3.3 Organic-based Nanomaterials

These NMs are made up of organic materials. The dendrimers, micelles, liposomes and polymeric NPs are examples of organic NMs.

8.3.3.4 Composite-based Nanomaterials

The composite NMs are formed by at least two parts, in which the first part must be NPs on the nanoscale dimension that combines with another type of NPs or with another type of bulk materials (such as a polymer and polysaccharides).

8.3.4 Nanomaterials Based on Their Origin

Based on the origin, the NMs can also be classified as natural or synthetic NMs.

i. Natural NMs are formed in nature either by biological species or due to anthropogenic activities.
ii. Synthetic or engineered NMs are produced unintentionally by the process of mechanical grinding, engine exhaust and smoking, or are synthesized based on the requirement by the following process physical, biological, chemical or hybrid methods.

8.4 ZINC OXIDE NANOPARTICLES (ZnO NPs): KEY FOCUS POINTS AND CHALLENGES IN THE FIELD OF NANOTOXICOLOGY

There are different types of NPs reported in the literature, e.g., metal NPs, metal oxide NPs and polymer NPs (composite NPs). Among these, metal oxide NPs are the most versatile materials and prominently used, due to their assorted properties and functionalities (Vaseem et al. 2010). Among all these metal oxides, zinc oxide (ZnO) NPs possess their importance due to their broad area of applications, e.g., gas sensor, chemical sensor, biosensor, cosmetics, storage, optical and electrical devices, window materials for displays, solar cells, biomedical and drug delivery (Sawai et al. 1996; Huang et al. 2001; Wang 2004; Baxter and Aydil 2005; Song et al. 2006; Zhang et al. 2013) as presented in Figure 8.2.

ZnO is excellent absorber of ultraviolet radiation of solar light (Özgür et al. 2005), hence highly used in optoelectronic applications, sunscreens and photocatalysts (Serpone et al. 2007; Wahab et al. 2011) due to their unique property, i.e., wide band gap 3.37 eV, excellent bond strength and large exciton binding energy (60 meV) at room temperature (Vaseem et al. 2010). Moreover, ZnO

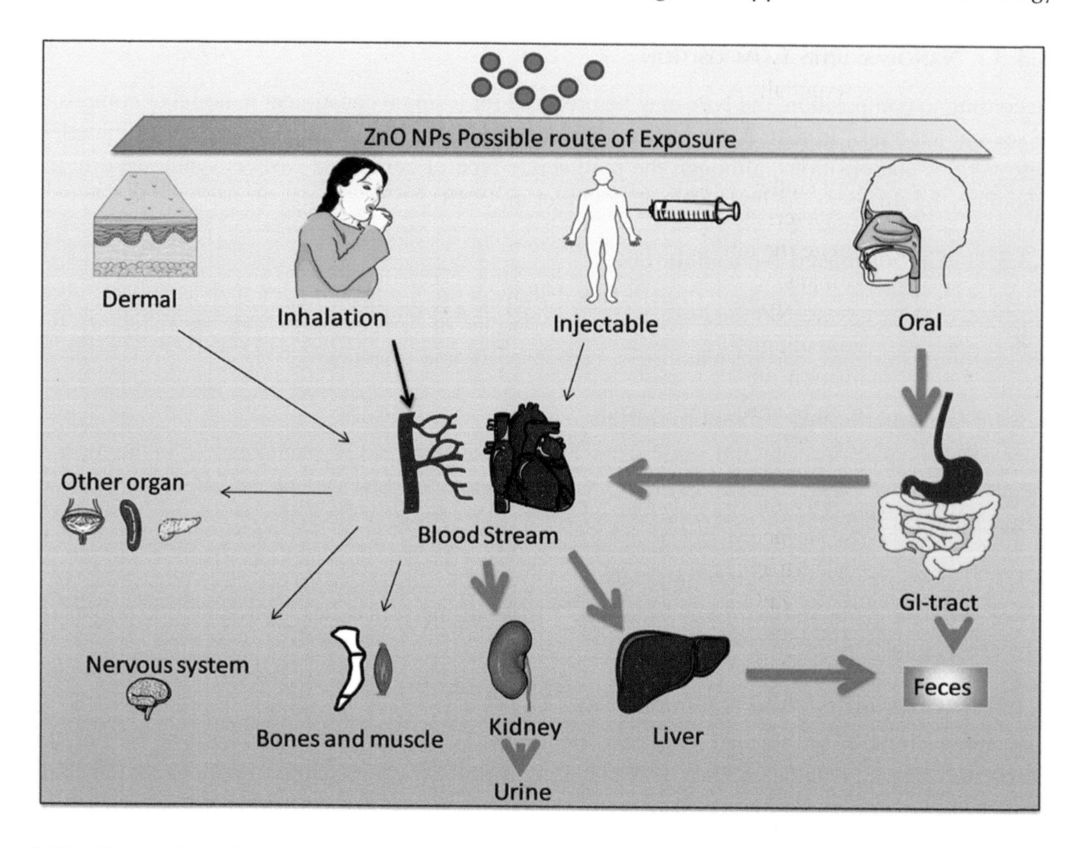

FIGURE 8.3 Possible route of exposure of ZnO NPs (Zinc Oxide Nanoparticles); the highlighted green arrow indicates the oral administration and their fate thereafter.

NPs can be applied as an antibacterial agent and anti-cancer drug because they can induce the oxidative stress in cells (Dwivedi et al. 2014; Wahab et al. 2014). ZnO NPs have been proven their usefulness in the biomedical application *such as* in the area of drug delivery, bioimaging, gene delivery and in biosensors (Nie et al. 2006; Kumar and Chen 2008; Cho N-H et al. 2011; Liu Y et al. 2011; Shinde et al. 2012).

The route of NPs exposure depends upon the risk factor of their surroundings and environment, such as in the context of nanotherapeutics, ingestion, dermal and oral are the prime route of exposure whereas in industrial level, inhalation and oral are the primary route of exposure (Figure 8.3). In the working environment, inhalation is a common exposure route than oral, but a part of the inhaled particles is removed from the lungs by the mucociliary defense system and then swallowed entering the gastrointestinal (GI) tract.

Despite the widespread use of ZnO NPs, the safety of this compound for humans is still unclear. Here, we want to summarize the current research on mammalian toxicity of ZnO NPs and discussed their relative cytotoxicity, cellular signaling, inflammation and genotoxicity, furthermore discuss the knowledge gap of study. Additionally, thrash out the *in vivo* consequences of ZnO NPs as induced *via* oral routes of exposure (Figure 8.3).

8.4.1 ZnO NPs Cytotoxicity Based on Cell Interaction

ZnO NPs induced a higher level of toxicity in cells as compared to another metal oxide (Saptarshi et al. 2015). ZnO NPs stimulated significant cytotoxicity in cancer cells (hepatocellular carcinoma cells (HepG2), lung epithelial cells (A549) and human bronchial epithelial cells (BEASB-2) but very least toxicity in primary rat astrocytes or hepatocytes, while mechanistic of ZnO NPs-induced cytotoxicity was preferentially not well outlined (Akhtar et al. 2012). However, dose-dependent cytotoxicity of ZnO NPs was also found in normal monocyte-derived dendritic cells and V79 lung cells (Andersson-Willman et al. 2012; Jain et al. 2019).

Usually, the toxicity studies of ZnO NPs have defined on the stable cell line, which represents only some level of characteristics of their parent cell lineage but not all the distinctiveness. Thus, it is challenging to evaluate the comparison of ZnO NP toxicity profile with possible adverse effects of NP exposure in humans.

Besides the cell type, NP physicochemical properties (size, shape, surface property and dissolution) and the surrounding environment may also affect the result of nanotoxicity. For instance, ZnO NPs after interaction with phosphates as present in the cell culture medium has shown to increase their cytotoxic potential (Everett et al. 2014). Additionally, the toxicity of NP has also been influenced based on the protein concentration of the culture medium (Yu et al. 2017). The adsorption of the protein around the NPs may change their surface property, and its circumstances can alter the bioreactivity of NP (Kurtz-Chalot et al. 2017), while relevant information on how protein adsorbs on the surface of ZnO NPs and their impact of nanotoxicological consequences is unavailable and needs further study. Similarly, surface coating or surface modification may also affect the cytotoxicity of ZnO NPs (Osmond-McLeod et al. 2013). On the other hand, size and shape are other essential factors that influenced the toxicity (Liu et al. 2017); ZnO NPs were shown to significantly affect on phagocytosis of THP-1 cells line as compared to 5 μm ZnO bulk particles (Sahu et al. 2014). Fascinatingly, spherical-shaped ZnO NPs (10–30 nm) have shown significantly higher cytotoxicity in Ana-1 cells in comparison to ZnO nano-rods (Song W et al. 2010). Furthermore, the dissolution of ZnO NPs in surrounding mediums to generate ions is natural property that can also concern the cytotoxicity by interfering with the chemical components/detection systems (Jiang et al. 2015). Although the mechanism of ZnO NPs causing cytotoxicity is not yet fully understood, ZnO NPs-induced ROS is considered to be one of the leading causes (Ng et al. 2017). Therefore, it is imperative to assess ZnO NPs cytotoxicity based on their physiochemical characteristics and correlates with NPs protein interaction (corona), to better interpret the mechanistic behind the induction of toxicity *in vivo* or human.

8.4.2 ZnO NPs-Mediated Toxicity Due to Dissolution

The dissolution has been considered to be one of the major underlying causes of ZnO NP cytotoxicity. Previously, it has been reported that the ZnO NPs release a high level of Zn^{2+} during cell exposure either intracellular (in acidic lysosomes) or extracellular environment consequently the reason for cytotoxicity (Saptarshi et al. 2015).

The dissolved ion in extracellular environment enters into the cell most likely through the receptor named zinc transporters (ZIP family) and accumulates in cells inside vesicles known as zincosomes (Eide 2006). Surplus accumulation of Zn^{2+} in a cell due to higher ZnO NP dissolution in media can lead for loss of cell membrane integrity, discharge of cytosolic LDH and apoptotic cell death (Akhtar et al. 2012). In contrast, Shen et al. demonstrated that ZnO NPs were capable of inducing cytotoxicity only *via* direct exposure to monocytes and not *via* indirect exposure (dissolved ZnO NPs) even at high NP doses used (Shen et al. 2013). Xu et al. pointed out that the amount of Zn^{2+} ions in cell culture media is very less and the total contribution of released Zn^{2+} ions those participated in cytotoxicity is only about 10% (Xu et al. 2013). Contradictorily, ZnO NP exposure upregulated the metallothionein genes, highlighting that Zn^{2+} ions are the primary source for ZnO NPs-induced cellular toxicity (Moos et al. 2011). Furthermore, it has been described that the ZnO NPs shows instability at slide acidic pH (4–6), thus the acidic microenvironment of cancer cells might be the reason for an extracellular dissolution of ZnO NPs (200 μM) (~100% after 24 h at pH 4–6), which can subsequently induce the reactive oxygen species (ROS) generation, damage the mitochondrial membrane and cell cycle arrest (Sasidharan et al. 2011). The finding supports the notion that ZnO NPs may induce higher cytotoxicity in cancer cell lines as compared to non-cancerous cells. The cumulative findings pointed out that the participation of ionic content in cytotoxicity may vary because it depends upon multiple factors such as cell types in which considering mechanisms of cellular uptake and number of NPs associated with the cells. Furthermore, the finding of the *in vitro* result is failed to explain the exact mechanistic for *in vivo* toxicity. On the other hand, when ZnO NPs were administered in *in vivo*, then the dissolution property and their internalization percentage were changed based on their route of exposure basically during oral administration. Previously, the dissolution property was examined in gastric fluid, intestinal fluid and serum to interpret the exact mode of action during *in vivo* exposure. Some studies contended that the dissolved Zn^{+2} from ZnO NPs were preferentially absorbed into the systemic circulation. Baeks et al. reported that orally administrated ZnO NPs first dissolved in their ion in GI tract before absorption (Baek et al. 2012). Cho et al. also attributed the higher absorption of ZnO NPs in rats that were in the form of ion in gastric juice (Cho et al. 2013). Umrani et al. and Hong et al. also confirmed that ZnO NPs primarily dissolved into ions that would induce the distribution and relative toxicity (Hong et al. 2013; Umrani and Paknikar 2014). Contradictorily, Yu et al. examined the solubility of ZnO NPs in different biological fluids and found the dissolutions of ~24%, 0.2% and 2.8% in simulated gastric fluid, intestinal fluid and plasma, respectively. The finding pointed out that ZnO NPs are absorbed primarily in particulate forms, and only extent part, i.e., 25% dissolved in the GI fluid after oral administration (Yu et al. 2017). Similarly, Du et al., also reported that only up to 12 % of ZnO NPs dissolve in the GI tract rest were in the particulate form before uptake (Du et al. 2018). Thus, it has been confirmed that ZnO NPs after oral administration were absorbed in both form particulate as well as some extent as ion, and the dissolution occurred due to acidic gastric fluid whereas in serum only 2% fraction of ZnO NPs were dissolved. To decipher this dissimilarity further studies are required to explore the contribution of ionic and non-ionic content in toxicity induction. Additional studies are required to find out the tissue accumulation and distribution percentage of ionic and non-ionic content of ZnO NPs after exposure.

8.4.3 ZnO NPs-Mediated Toxicity Due to ROS Generation

ROS or free radicals are the most reactive molecule made up of anionic oxygen atoms, e.g., hydroxyl radicals, superoxide molecules and hydrogen peroxide are ubiquitous examples among these. Generation of ROS and induction of oxidative stress has been defined to be the significant contributing factor in ZnO NPs mediated toxicity (Senapati et al. 2015). It has been explored that ZnO NPs generate ROS in different cell types in size-dependent manner. The level of Zn^{2+} in the cells upon exposure to ZnO NPs causes alteration in zinc-dependent protein activity, which may lead to inactivation of cellular redox maintaining systems and escorting to cell death (Shen et al. 2013). A recent study pointed out that ZnO NPs induce toxicity by activating PINK1/Parkin-mediated mitophagy in CAL 27 oral cancer cell lines (Wang et al. 2018). In the same way, it has been attributed that ZnO NPs enhanced ROS and ROS-mediated mitochondrial dysfunction and apoptotic cell death (Ryu et al. 2014; Jain et al. 2019). Similarly, *in vivo* study also defined that oxidative stress and generation of ROS is the primary reason for ZnO NPs toxicity (Sharma et al. 2011; Jain et al. 2019). Likewise, treatment of ZnO NPs was induced toxicity in liver and kidney tissues of rats in ROS-dependent pathways (Tang et al. 2015). Yang et al. also defined that generation of ROS-governed ER stress and autophagy in the liver after subchronic oral exposure of ZnO NPs (Yang et al. 2016).

Thus, the finding explored that the ZnO NPs exposure may disrupt the cellular zinc homeostasis and generates the level of ROS that further leads to toxicity on the cellular level. Accordingly, the impairment of the cellular redox machinery due to the exposure of ZnO NPs is a probable responsible cause for induction of toxicity (Saptarshi et al. 2015).

Although, ZnO NPs-induced ROS-mediated acute and chronic neurotoxic, genotoxic effects were investigated in time and concentration-dependent manner using in *in-vivo* models, but their biocompatibility and intrinsic mechanistic and effect of biological environment on the toxicity of ZnO NPs is emerging.

8.4.4 ZnO NPs-Induced Immunomodulatory Responses

NPs can cause immunotoxicity by stimulating the immune cells to release cytokines that consequently lead to inflammation. A dose-dependent increase in expression of TNF-α and upregulation of markers CD80 and CD86 was reported in RAW264.7 macrophages, and mouse primary dendritic cells, respectively, after the treatment of spherical/sheet-shaped ZnO NPs (Heng et al. 2011). Furthermore, Setapati et al. have also been found a significant increase in expression of inflammatory cytokine (TNF-α and IL-1β) in ZnO NPs exposed THP-1 monocyte cell (Senapati et al. 2015). On the contrary, ZnO NPs exposed to THP-1 monocyte cells failed to up-regulate the proinflammatory markers (HLA DE or CD14) (Prach et al. 2013). There are several mechanisms that have been defined to explore the role of ZnO NPs in the stimulation of cytokines production. A recent study has been reported that ZnO NPs enhanced ROS-ERK-Egr-1 mediated inflammation by inducing the expression of cytokine TNF-α in keratinocytes (Jeong et al. 2013). Another report suggested that ZnO NPs also involved in transcriptional regulation of NF-κB or C/EBPβ and posttranscriptional regulation of IL-8 expression in human bronchial cells (Yan et al. 2014). Chen et al. explored the involvement of ZnO NPs with the TLR4 receptor on the surface of bronchial epithelial cells, the leading cause of an increase in the expression of IL-8. Furthermore, the MAPK signaling mediated by TLR6 and upregulation of IL-1β, IL-6, TNF-α and COX-2 was also shown in macrophages when treated with ZnO NPs (Roy et al. 2014). On the other hand, it has also been defined that the upregulation of proinflammatory markers also depends upon the metal metabolic proteins chaperonin and protein folding (Moos et al. 2011). The previous study also explored that ZnO NPs-induced ROS, oxidative DNA damage along with proinflammatory cytokines in mice.

Similarly, Senapati et al. reported immunotoxicity in BALB/c mice after subacute exposure by a change in the CD4 and CD8 cells, increasing cytokine release (IL-6, IFN-γ and TNF-α) and oxidative stress (Senapati et al. 2017). Abass et al. evaluate immunotoxic response in rats after oral administration (Abass et al. 2017). They concluded that ZnO NPs were highly upregulated the expression of proinflammatory (IL-6, IFN-γ and TNF-α), immunomodulatory (CD3, CD11b, heme oxygenase (HO-1)) and the inflammatory (toll-like receptors 4 and 6 (TLR4 and TLR6) genes in the spleen. The published literature defines the ZnO NPs immunomodulatory effect in the *in vitro* and *in vivo* model, but the induction of immune response and their mechanistic explanation in the *in-vivo* model system may change due to the interaction of NPs with a biological component (Monopoli et al. 2012), another effect of NP-mediated stimulation of inflammation is possible due to the adsorption of undesirable proteins onto their surface. Interestingly, for better correlation of NPs effect *in vivo* model, need to evaluate the inflammatory condition/immune responses of NPs in context to the role of corona genesis; as only minimal information is available for ZnO NPs.

8.4.5 ZnO NPs-Induced Apoptotic Cell Death and Autophagy

Another mechanism of toxicity that has been associated with various types of NPs is the modulation of autophagy and apoptosis. Mitochondrial strength is directly responsible factor for apoptotic cell death. Along with being the energy-generating apparatus of the cell, mitochondria also act in response to DNA damage, necrosis and apoptosis by activating several signaling pathways (Wang and Youle 2009). Depolarization of the mitochondrial membrane by NMs can cause cell death as reported in the case of lung adenocarcinoma (H1355) cells exposed to ZnO NPs (Kao et al. 2011). The exposure of ZnO NPs to the cells induced ROS-mediated mitochondrial dysfunctions that prominently carry on the apoptotic pathway (Forman and Torres 2002). The loss of mitochondrial integrity initiates the release of apoptogenic factors out from the intermembrane space. Release of cytochrome c commence the formation of apoptosome (Apaf-1 and the initiator caspase-9) and ultimately activate the caspase-9 (Shi 2002). The activity of proapoptotic genes caspase-3, bax and tumor-suppressor gene p53 along with fragmentation of DNA was reported after ZnO NPs treatment (Meyer et al. 2011; Senapati et al. 2015; Jain et al. 2019). Cleavage of their specific substrates leads to the characteristic apoptotic cell death.

In *in vivo*, subacute oral exposure to ZnO NPs (330 mg/kg) in mice leads to oxidative stress-mediated DNA damage and apoptosis in liver (Sharma et al. 2011). ZnO NPs-induced neurotoxicity after oral exposure in the form of ROS-mediated DNA damage that was governed to apoptosis as the increased expression level of caspase-3 and Fas (Attia et al. 2017). Stress in the endoplasmic reticulum (ER stress) is also a prominent reason for apoptotic cell death. The ZnO NPs or their ionic part induced ROS-mediated ER stress and mitochondrial damage resultant altered the expression of interlinked signaling factors such as caspase-12, NrF2 and NOX-2 pathways, which are responsible for apoptotic cell death as pointed out in various in vitro studies (Sharma et al. 2012; Chen et al. 2014; Senapati et al. 2015; Ng et al. 2017).

Furthermore, oxidative stresses (formation of ROS) also effectuate autophagy, with these means; to eliminate ROS-mediated degradation of cellular organelles, typically mitochondria (Peynshaert et al. 2014). In recent years, it has also been shown that exposure to NPs may induce ROS and ROS may lead to the accumulation of unfolded protein in the ER, a condition called as ER stress as a possible mechanism for NP-induced toxicity to endothelial cells (Cao et al. 2017; Hoseinzadeh et al. 2017). Thus the ZnO NPs induce ROS would be the reason for initiation of unfolded protein response pathway in ER-associated with autophagy (Ali et al. 2017). Kuang et al. explained size-dependent ZnO NP-induced-ER stress; they explained 30 nm ZnO NPs induced a higher expression of ER stress genes compared with that of bulk and 90 nm NPs (Peynshaert et al. 2014). Yang et al. conducted a subchronic study in the mouse model and explored that the oral exposure of ZnO NPs-induced-ER stress and ROS-mediated hepatotoxicity (Yang et al. 2016). They found the increased in the expression of ER stress-associated genes (grp78, grp94, pdi-3 and xbp-1) and increased phosphorylation of protein kinase-RNA like ER kinase (PERK) and eukaryotic initiation factor 2a (eIF2a) that correlated with stress in ER-mediated autophagy (Yang et al. 2016). Furthermore, they reported an increase in expression of apoptotic caspase-3, caspase-9, caspase-12, CHOP/GADD153 and proapoptotic genes (chop and bax) in liver tissue. Another *in vivo* study also defined the size-dependent hepatotoxicity, associated with elevated expression of PERK, eIF2α, ATF4, Chop, JNK, caspase-12, caspase-9 and GRP94 (ER stress markers) (Kuang et al. 2017).

8.4.6 ZnO NPs-Induced Genotoxicity

Due to the small size and increased surface area coupled with physiochemical characteristics such as charged surfaces, these NPs may exhibit unpredictable genotoxic properties by direct interaction with the genetic material or by indirect DNA damage induced through ROS (Kisin et al. 2007; Barnes et al. 2008). Genotoxicity is defined as the damage to the cellular DNA that may take place either directly due to the interaction of NPs with the nuclear material or indirectly due to the generation of ROS, other reactive ions or mechanical injury (Ghosh et al. 2016). ZnO NPs were shown significant DNA damage as demonstrated by comet assay in a human epidermal cell line (A431) and human lymphoblastoid cell line (TK6) in a concentration-dependent manner (Sharma et al. 2009; Demir et al. 2014). Furthermore, concentration-dependent oxidative DNA damage at the characteristics of induction of nuclear condensation, DNA fragmentation, the formation of hypodiploid DNA nuclei and apoptotic bodies were found due to the treatment of ZnO NPs in macrophages (Wilhelmi et al. 2013). Nuclear distribution of the NPs is also a possible explanation for genotoxicity (Jain et al. 2019). In contrast, THP-1 cells exposed to nano or bulk ZnO particles did not show any genotoxic effect (Sahu et al. 2014).

Although *in vivo* model system, Sharma et al. demonstrated ZnO NPs-induced oxidative stress and defined its role in DNA damage and apoptosis in mouse liver after subacute oral exposure

(Sharma et al. 2011). Ghosh et al. also defined the *in vivo* genotoxicity of ZnO on bone marrow cells, they observed a reduction of mitochondrial membrane potential, high ROS generation and G0/G1 cell cycle arrest along with chromosome aberrations and micronuclei development (Ghosh et al. 2016). However, Kwon et al. recently reported no genotoxic potential of capped ZnO NPs up to doses of 2000 mg/kg in mice based on the micronuclei, comet assays and ames test (Kwon et al. 2014).

Thus, for revealing the clarity, there is a need to study the genotoxic potential of ZnO NPs and unveil their probable correlation with ROS formation in an animal model. Overall, there is still a need to fully understand the mechanism in which ZnO NPs inflict cellular DNA damage.

8.5 ZnO NPs TOXICITY ON ANIMAL MODELS: CURRENT SCENARIO

Zinc is an essential trace element, and ZnO NPs are often used as supplements and food additives; therefore, it is imperative to assess their potential toxicological *in vivo*. Biologically systems are significantly more complex, consisting of several protein-rich environments and cellular diversity that may simultaneously impact the bioreactive nature of NMs. The *in vivo* animal models thus provide an essential platform to assess nanotoxicity because it involves significantly in particles' biological conditioning, distribution pattern and macrophage processing than a cellular model system. Here we discussed the recent literature (within ten years) on the effect of ZnO NPs in animal models and their subsequent mechanistic explanation by concern the various route of exposure.

Mostly the mechanisms underlying ZnO NP toxicity *in vivo* and *in vitro* are similar as reported and include generation of ROS and subsequent DNA damage. Previously, it has been reported that subacute oral exposure of ZnO NPs-induced hepatotoxicity and nephrotoxicity followed by oxidative stress-mediated DNA damage and apoptosis (Sharma et al. 2011). Yousef et al. also defined the ROS-mediated genotoxicity of ZnO NPs in rat (Yousef and Mohamed 2015). Wang et al. observed that zinc oxide exhibited toxicity at 2 g/kg body weight and found lesions in the liver, pancreas, heart and stomach (Wang B et al. 2008). Another pathway has also been defined that ZnO NPs after internalization accumulated in the primary and secondary target organ and induced the toxicity by Apoptotic cell death or autophagy cell death, these circumstances might be initiated by direct generation of ROS or mitochondrial damage along with ER stress (Peynshaert et al. 2014; Saptarshi et al. 2015; Cao et al. 2017; Kuang et al. 2017). The NPs biodistribution in the different organs of the body after administration depends upon their unique characters, i.e., their size. Baek et al. described the tissue distribution pattern and excretion profile of zinc in a single oral dose (2000 mg/kg) of ZnO NPs in rats and defined those smaller NPs having higher bioavailability and higher accumulation rather than large-size particles (Baek et al. 2012). The higher accumulation of NPs in organ might be a probable reason for higher toxicity. Some studies defined the size variant toxicity of ZnO NPs in Swiss mice after acute oral administration. They found significantly higher toxicity as compared to their bulk particles by concerning their hemato-biochemical parameters (Wang et al. 2008; Baek et al. 2012; Almansour et al. 2017; Kuang et al. 2017). The ZnO NPs toxicity is influence not only size but also other physicochemical properties such as dissolution, surface property, aggregation and concentration; these are also the responsible factor of NPs toxicity.

As decreasing size results in increasing NPs specific surface area to volume ratio, which promotes not only the accumulation of NPs but also increases reactivity and enhanced interaction between NPs and biomolecules of the target animal (Yu et al. 2017).

Thus it has been confirmed that ZnO NPs-induced ROS-mediated cellular toxicity and genotoxicity although few studies have shown that excessive Zn due to the exposure of ZnO NPs impaired the homeostasis of other metals (Cu and Iron) (Willis et al. 2005; Manuel et al. 2006). Hemolytic anemia was caused in a dog due to the exposure of zinc oxide. Chronic exposure of ZnO (640 mg/kg body weight/day) decreases the iron level in rats (Clancey and Murphy 2006). ZnO NPs-induced toxicity on various target organs through internalization of different routes such as dermal, intravenous and intratracheal, intraperitoneal and oral (Ilves et al. 2014; Fujihara et al.

2015; Jacobsen et al. 2015; Saptarshi et al. 2015; Du et al. 2018). Although, the defined toxicity mechanism of NPs may be influenced due to the interaction of the new biological environment in the *in vivo* study was not well defined. Thus, the study for evaluating the interactions between ZnO and biomatrices is necessitatable that probably affects the oral absorption, distribution and toxicity and may be influenced the overall biological consequences. Thus, the approach to identifying the NPs-biological interaction may provide a precise mechanism of NPs toxicity in the *in vivo* model system.

8.6 NPs-PROTEIN CORONA

Recent studies showed that physicochemical properties of NPs could be altered by different surrounding conditions, such as different buffers as well as blood and cell culture media (Maiorano et al. 2010). Particularly, NPs after entering into the biological fluid immediately coated by biomolecules most likely with proteins, which is called as "Protein Corona" (Sopotnik et al. 2015). Further formation of corona around NPs is modulated by the entry of NPs in the body. The corona formation mainly influenced by two factors, first one i) is the characteristics of NPs *viz.* size, surface charge and surface morphology and second ii) is the adsorption of the different categories of molecules such as proteins, enzymes, lipids and carbohydrates but protein majorly participate in Corona (Pareek et al. 2018). The dynamics of NPs corona genesis and their intricate affinity are very challenging to study due to the inherent complexity and rendering extremely discrete experimental results (Monopoli et al. 2012). During corona genesis these biomolecules competitively interact with the surface of the NPs; popularly it has been found that proteins compete with other molecules to surround NPs (Sopotnik et al. 2015).

8.7 DRAWBACK OF CORONA FORMATION

The protein corona configuration on to NPs may have special significance in biological explanation to an equilibrium state that has been unknown for NMs in blood. Previous studies have shown that biological responses of NPs mostly depend upon the surface area rather than relative mass (Brown et al. 2001; Schmid and Stoeger 2016). Because the nanoscale materials have greater surface-to-volume ratios that may facilitate for more proteins binding around a NP than a particle of larger size (Kurtz-Chalot et al. 2017). The adsorption of protein may change their confirmation and probably influences the biodistribution of NPs during *in vivo* administration (Gessner et al. 2002; Dutta et al. 2007). Keselowsky et al. reported that the attachment of various proteins on the surface of NPs (with varying properties), alters its conformation/structure that would finally affect its cell adhesion capability (Keselowsky et al. 2003). There are various types of proteins that may participate in corona formation; among 3700 plasma proteins approximately 125–500 proteins are adsorbed on diverse NMs in different amounts after incubated in plasma. The adsorbed proteins may involve in various cellular processes, specifically, complement activation, pathogen recognition and blood coagulation according to their binding. Thus the binding of proteins on to NPs may lead to change in the unique property of the NPs. Commonly, Immunoglobulin G, serum albumin, fibrinogen, clusterin and apolipoproteins are present in the PC of most analyzed NPs but it may vary from NPs to NPs. Thus, the changes in the confirmation of the corona proteins can create new epitopes on the NPs surface that can exert aberrant cellular signaling, immunogenicity and ultimately induced toxicity (Nel et al. 2009; Prapainop et al. 2013; Pareek et al. 2018).

8.8 CONCLUSION

ZnO NPs are one of the most widely used NMs. NPs are currently used in chemical, cosmetic, pharmaceutical and electronic products resulting in increased human exposure to ZnO. Following exposure, these NPs can accumulate in various organs and induce the toxicity due to their physiochemical characteristics. Nevertheless, limited safety information is available, especially in terms of their interactions with various binding proteins, leading to potential toxic effects. Principally, particles administered orally encounter diverse biological matrices, such as GI fluids and blood, and these interactions lead to the formation of particle–biomatrix corona of different affinity. This can alter their physicochemical property, biological interaction and biological fate. The NPs after interacting with blood formed protein corona around the NPs enriched with opsonin protein that may be recognized by macrophages and immune effectors cells, resulting rapid clearance with induced toxicity. Hence, the potential of ZnO NPs toxicity pathway also depends on proteins on to surface of NPs, called protein corona, in order to understand their potential mechanisms in vivo and explored their protective effect by surface modification.

REFERENCES

Abass, M.A., Selim, S.A., Selim, A.O., El-Shal, A.S., Gouda, Z.A., 2017. Effect of orally administered zinc oxide nanoparticles on albino rat thymus and spleen. *IUBMB Life* 69, 528–539.

Akhtar, M.J., Ahamed, M., Kumar, S., Khan, M.A.M., Ahmad, J., Alrokayan, S.A., 2012. Zinc oxide nanoparticles selectively induce apoptosis in human cancer cells through reactive oxygen species. *Int. J. Nanomed.* 7, 845.

Ali, I., Shah, S.Z.A., Jin, Y., Li, Z.-S., Ullah, O., Fang, N-Z., 2017. Reactive oxygen species-mediated unfolded protein response pathways in preimplantation embryos. *J. Vet. Sci.* 18, 1–9.

Almansour, M.I., Alferah, M.A., Shraideh, Z.A., Jarrar, B.M., 2017. Zinc oxide nanoparticles hepatotoxicity: Histological and histochemical study. *Environ. Toxicol. Pharmacol.* 51, 124–130.

Andersson-Willman, B., Gehrmann, U., Cansu, Z., Buerki-Thurnherr, T., Krug, H.F., Gabrielsson, S., Scheynius, A., 2012. Effects of subtoxic concentrations of TiO_2 and ZnO nanoparticles on human lymphocytes, dendritic cells and exosome production. *Toxicol. Appl. Pharmacol.* 264, 94–103.

Attia, H., Nounou, H., Shalaby, M., 2017. Zinc oxide nanoparticles induced oxidative DNA damage, inflammation and apoptosis in rat's brain after oral exposure. *Toxics* 6, 29.

Baek, M., Chung, H.-E., Yu, J., Lee, J.-A., Kim, T.-H., Oh, J.-M., Lee, W.-J., Paek, S.-M., Lee, J.K., Jeong, J., 2012. Pharmacokinetics, tissue distribution, and excretion of zinc oxide nanoparticles. *Int. J. Nanomed.* 7, 3081.

Barnes, C.A., Elsaesser, A., Arkusz, J., Smok, A., Palus, J., Lesniak, A., et al., 2008. Reproducible comet assay of amorphous silica nanoparticles detects no genotoxicity. *Nano Lett.* 8, 3069–3074.

Baxter, J.B., Aydil, E.S., 2005. Nanowire-based dye-sensitized solar cells. *Appl. Phys. Lett.* 86, 053114.

Binnig, G. & Rohrer, H. (1982). Scanning tunneling microscopy. *Helvetica Physica Acta.* 55, 726–735.

Brown, D.M., Wilson, M.R., MacNee, W., Stone, V., Donaldson, K., 2001. Size-dependent proinflammatory effects of ultrafine polystyrene particles: A role for surface area and oxidative stress in the enhanced activity of ultrafines. *Toxicol. Appl. Pharmacol.* 175, 191–199.

Cao, Y., Long, J., Liu, L., He, T., Jiang, L., Zhao, C., Li, Z., 2017. A review of endoplasmic reticulum (ER) stress and nanoparticle (NP) exposure. *Life Sci.* 186, 33–42.

Chen, R., Huo, L., Shi, X., Bai, R., Zhang, Z., Zhao, Y., Chang, Y., Chen, C., 2014. Endoplasmic reticulum stress induced by zinc oxide nanoparticles is an earlier biomarker for nanotoxicological evaluation. *ACS Nano* 8, 2562–2574.

Cho, N.-H., Cheong, T.-C., Min, J.H., Wu, J.H., Lee, S.J., Kim, D., et al., 2011. A multifunctional core-shell nanoparticle for dendritic cell-based cancer immunotherapy. *Nature Nanotechnol.* 6, 675.

Cho, W.-S., Kang, B.-C., Lee, J.K., Jeong, J., Che, J.-H., Seok, S.H., 2013. Comparative absorption, distribution, and excretion of titanium dioxide and zinc oxide nanoparticles after repeated oral administration. *Part. Fibre Toxicol.* 10, 9.

Clancey, N.P., Murphy, M.C., 2006. Zinc-induced hemolytic anemia in a dog caused by ingestion of a game-playing die. *Can. Vet. J.* 53, 383.

Demir, E., Creus, A., Marcos, R., 2014. Genotoxicity and DNA repair processes of zinc oxide nanoparticles. *J. Toxicol. Environ. Health Part A* 77, 1292–1303.

Drexler K.E., 1986. Engines of creation: The coming era of nanotechnology. Anchor Press/Doubleday: New York.

Du, L.-J., Xiang, K., Liu, J.-H., Song, Z.-M., Liu, Y., Cao, A., Wang, H., 2018. Intestinal injury alters tissue distribution and toxicity of ZnO nanoparticles in mice. *Toxicol. Lett.* 295, 74–85.

Dutta, D., Sundaram, S.K., Teeguarden, J.G., Riley, B.J., Fifield, L.S., Jacobs, J.M., Addleman, S.R., Kaysen, G.A., Moudgil, B.M., Weber, T.J., 2007. Adsorbed proteins influence the biological activity and molecular targeting of nanomaterials. *Toxicol. Sci.* 100, 303–315.

Dwivedi, S., Wahab, R., Khan, F., Mishra, Y.K., Musarrat, J., Al-Khedhairy, A.A., 2014. Reactive oxygen species mediated bacterial biofilm inhibition via zinc oxide nanoparticles and their statistical determination. *PLoS One* 9, e111289.

Eide, D.J., 2006. Zinc transporters and the cellular trafficking of zinc. *Biochim. Biophys. Acta* 1763, 711–722.

Everett, W.N., Chern, C., Sun, D., McMahon, R.E., Zhang, X., Chen, W.-J.A., Hahn, M.S., Sue, H.J., 2014. Phosphate-enhanced cytotoxicity of zinc oxide nanoparticles and agglomerates. *Toxicol. Lett.* 225, 177–184.

Feynman, R.P., 1959. There's plenty of room at the bottom: An invitation to enter a new field of physics. *Handbook of Nanoscience, Engineering, and Technology*. CRC Press.

Forman, H.J., Torres, M., 2002. Reactive oxygen species and cell signaling: Respiratory burst in macrophage signaling. *Am. J. Resp. Crit. Care Med.* 166, S4–S8.

Fujihara, J., Tongu, M., Hashimoto, H., Yamada, T., Kimura-Kataoka, K., Yasuda, T., Fujita, Y., Takeshita, H., 2015. Distribution and toxicity evaluation of ZnO dispersion nanoparticles in single intravenously exposed mice. *J. Med. Invest.* 62, 45–50.

Gessner, A., Lieske, A., Paulke, B.R., Müller, R.H., 2002. Influence of surface charge density on protein adsorption on polymeric nanoparticles: Analysis by two-dimensional electrophoresis. *Eur. J. Pharm. Biopharm.* 54, 165–170.

Ghosh, M., Sinha, S., Jothiramajayam, M., Jana, A., Nag, A., Mukherjee, A., 2016. Cyto-genotoxicity and oxidative stress induced by zinc oxide nanoparticle in human lymphocyte cells in vitro and Swiss albino male mice in vivo. *Food Chem. Toxicol.* 97, 286–296.

Heng, B.C., Zhao, X., Tan, E.C., Khamis, N., Assodani, A., Xiong, S., Ruedl, C., Ng, K.W., Loo, J.S.-C., 2011. Evaluation of the cytotoxic and inflammatory potential of differentially shaped zinc oxide nanoparticles. *Arch. Toxicol.* 85, 1517–1528.

Hong, T.-K., Tripathy, N., Son, H.-J., Ha, K.-T., Jeong, H.-S., Hahn, Y.-B., 2013. A comprehensive in vitro and in vivo study of ZnO nanoparticles toxicity. *J. Mater. Chem. B* 1, 2985–2992.

Hoseinzadeh, E., Makhdoumi, P., Taha, P., Hossini, H., Stelling, J., Amjad Kamal, M., 2017. A review on nano-antimicrobials: Metal nanoparticles, methods and mechanisms. *Curr. Drug Metab.* 18, 120–128.

Huang, M.H., Mao, S., Feick, H., Yan, H., Wu, Y., Kind, H., Weber, E., Russo, R., Yang, P., 2001. Room-temperature ultraviolet nanowire nanolasers. *Science* 292, 1897–1899.

Hulla, J.E., Sahu, S.C., Hayes, A.W., 2015. Nanotechnology: History and future. *Human Exp. Toxicol.* 34, 1318–1321.

Iijima, S., 1991. Helical microtubules of graphitic carbon. *Nature* 354, 56.

Ilves, M., Palomäki, J., Vippola, M., Lehto, M., Savolainen, K., Savinko, T., Alenius, H., 2014. Topically applied ZnO nanoparticles suppress allergen induced skin inflammation but induce vigorous IgE production in the atopic dermatitis mouse model. *Part. Fibre Toxicol.* 11, 38.

Jacobsen, N.R., Stoeger, T., van den Brule, S., Saber, A.T., Beyerle, A., Vietti, G., et al., 2015. Acute and subacute pulmonary toxicity and mortality in mice after intratracheal instillation of ZnO nanoparticles in three laboratories. *Food Chem. Toxicol.* 85, 84–95.

Jain, A.K., Singh, D., Dubey, K., Maurya, R., Pandey, A.K., 2019. Zinc oxide nanoparticles induced gene mutation at the HGPRT locus and cell cycle arrest associated with apoptosis in V-79 cells. *J. Appl. Toxicol.* 39, 735–750.

Jeevanandam, J., Barhoum, A., Chan, Y.S., Dufresne, A., Danquah, M.K., 2018. Review on nanoparticles and nanostructured materials: history, sources, toxicity and regulations. *Beilstein J. Nanotech.* 9, 1050–1074.

Jeong, S.H., Kim, H.J., Ryu, H.J., Ryu, W.I., Park, Y.-H., Bae, H.C., Jang, Y.S., Son, S.W., 2013. ZnO nanoparticles induce TNF-Î± expression via ROS-ERK-Egr-1 pathway in human keratinocytes. *J. Dermatol. Sci.* 72, 263–273.

Jiang, X., Miclăuş, T., Wang, L., Foldbjerg, R., Sutherland, D.S., Autrup, H., Chen, C., Beer, C., 2015. Fast intracellular dissolution and persistent cellular uptake of silver nanoparticles in CHO-K1 cells: implication for cytotoxicity. *Nanotoxicology* 9, 181–189.

Kao, Y.-Y., Chen, Y.-C., Cheng, T.-J., Chiung, Y.-M., Liu, P.-S., 2011. Zinc oxide nanoparticles interfere with zinc ion homeostasis to cause cytotoxicity. *Toxicol. Sci.* 125, 462–472.

Keselowsky, B.G., Collard, D.M., García, A.J., 2003. Surface chemistry modulates fibronectin conformation and directs integrin binding and specificity to control cell adhesion. *J. Biomed. Mater. Res. A* 66, 247–259.

Kisin, E.R., Murray, A.R., Keane, M.J., Shi, X.-C., Schwegler-Berry, D., Gorelik, O., Arepalli, S., Castranova, V., Wallace, W.E., Kagan, V.E., 2007. Single-walled carbon nanotubes: Geno-and cytotoxic effects in lung fibroblast V79 cells. *J. Toxicol. Env. Health A* 70, 2071–2079.

Kuang, H., Yang, P., Yang, L., Aguilar, Z.P., Xu, H., 2017. Size dependent effect of ZnO nanoparticles on endoplasmic reticulum stress signaling pathway in murine liver. *J. Hazard. Mater.* 317, 119–126.

Kumar, S.A., Chen, S.M., 2008. Nanostructured zinc oxide particles in chemically modified electrodes for biosensor applications. *Anal. Lett.* 41, 141–158.

Kurtz-Chalot, A., Villiers, C., Pourchez, J., Boudard, D., Martini, M., Marche, P.N., Cottier, M., Forest, V., 2017. Impact of silica nanoparticle surface chemistry on protein corona formation and consequential interactions with biological cells. *Mater. Sci. Eng C* 75,16–24.

Kwon, J.Y., Lee, S.Y., Koedrith, P., Lee, J.Y., Kim, K.-M, Oh, J.-M., Yang, S.I., Kim, M.-K., Lee, J.K., Jeong, J., 2014. Lack of genotoxic potential of ZnO nanoparticles in in vitro and in vivo tests. *Mut. Res.* 761, 1–9.

Liu, J., Kang, Y., Yin, S., Song, B., Wei, L., Chen, L., Shao, L., 2017. Zinc oxide nanoparticles induce toxic responses in human neuroblastoma SHSY5Y cells in a size-dependent manner. *Int. J. Nanomed.* 12, 8085.

Liu, Y., Ai, K., Yuan, Q., Lu, L., 2011. Fluorescence-enhanced gadolinium-doped zinc oxide quantum dots for magnetic resonance and fluorescence imaging. *Biomaterials* 32, 1185–1192.

Maiorano, G., Sabella, S., Sorce, B., Brunetti, V., Malvindi, M.A., Cingolani, R., Pompa, P.P., 2010. Effects of cell culture media on the dynamic formation of protein-nanoparticle complexes and influence on the cellular response. *ACS Nano* 4, 7481–7491.

Manuel, O., Fernando, P., Manuel, R., 2006. Inhibition of iron absorption by zinc: Effect of physiological and pharmacological doses: TL015. *Pediatr. Res.* 60, 636.

Mappes, T., Jahr, N., Csaki, A., Vogler, N., Popp, J.R., Fritzsche, W., 2012. The invention of immersion ultramicroscopy in 1912- the birth of nanotechnology? *Angew. Chem. Int. Ed.* 51, 11208–11212.

Meyer, K., Rajanahalli, P., Ahamed, M., Rowe, J.J., Hong, Y., 2011. ZnO nanoparticles induce apoptosis in human dermal fibroblasts via p53 and p38 pathways. *Toxicol. In Vitro* 25, 1721–1726.

Monopoli, M.P., Aberg, C., Salvati, A., Dawson, K.A., 2012. Biomolecular coronas provide the biological identity of nanosized materials. *Nat. Nanotechnol.* 7, 779.

Moos, P.J., Olszewski, K., Honeggar, M., Cassidy, P., Leachman, S., Woessner, D., Cutler, N.S., Veranth, J.M., 2011. Responses of human cells to ZnO nanoparticles: A gene transcription study. *Metallomics* 3, 1199–1211.

Nel, A.E., Mädler, L., Velegol, D., Xia, T., Hoek, E.M., Somasundaran, P., Klaessig, F., Castranova, V., Thompson, M., 2009. Understanding biophysicochemical interactions at the nano-bio interface. *Nat. Mater.* 8, 543.

Ng, C.T., Yong, L.Q., Hande, M.P., Ong, C.N., Yu, L.E., Bay, B.H., Baeg, G.H., 2017. Zinc oxide nanoparticles exhibit cytotoxicity and genotoxicity through oxidative stress responses in human lung fibroblasts and *Drosophila melanogaster*. *Int. J. Nanomed.* 12, 1621.

Nie, L., Gao, L., Feng, P., Zhang, J., Fu, X., Liu, Y., Yan, X., Wang, T., 2006. Three-dimensional functionalized tetrapod-like ZnO nanostructures for plasmid dna delivery. *Small* 2, 621–625.

Osmond-McLeod, M.J., Osmond, R.I.W., Oytam, Y., McCall, M.J., Feltis, B., Mackay-Sim, A., Wood, S.A., Cook, A.L., 2013. Surface coatings of ZnO nanoparticles mitigate differentially a host of transcriptional, protein and signalling responses in primary human olfactory cells. *Part. Fibre Toxicol.* 10, 54.

Özgür, Ü., Alivov, Y.I., Liu, C., Teke, A., Reshchikov, M., DoÄŸan, S., Avrutin, V., Cho, S.J., Morkoc, H., 2005. A comprehensive review of ZnO materials and devices. *J. Appl. Phys.* 98, 11.

Pareek, V., Bhargava, A., Bhanot, V., Gupta, R., Jain, N., Panwar, J., 2018. Formation and characterization of protein corona around nanoparticles: A review. *J. Nanosci. Nanotechnol.* 18, 6653–6670.

Peynshaert, K., Manshian, B.B., Joris, F., Braeckmans, K., De Smedt, S.C., Demeester, J., Soenen, S.J., 2014. Exploiting intrinsic nanoparticle toxicity: the pros and cons of nanoparticle-induced autophagy in biomedical research. *Chem. Rev.* 114, 7581–7609.

Plan, F.F.Y., 1998. The Fifth Five Year Plan. *Planning Commission, Government of India.*

Plan, E.F.Y., 2007. The Eleventh Five Year Plan. *Planning Commission, Government of India.*

Prach, M., Stone, V., Proudfoot, L., 2013. Zinc oxide nanoparticles and monocytes: Impact of size, charge and solubility on activation status. *Toxicol. Appl. Pharm.* 266, 19–26.

Prapainop, K., Witter, D.P., Wentworth, Jr. P., 2013. A chemical approach for cell-specific targeting of nanomaterials: Small-molecule-initiated misfolding of nanoparticle corona proteins. *J. Am. Chem. Soc.* 134, 4100–4103.

Roco, M.C., 2000. *National Nanotechnology Initiative (NNI).* National Science Foundation.

Roco, M.C., 2003. Broader societal issues of nanotechnology. *J. Nanopart. Res.* 5, 181–189.
Roy, R., Singh, S.K., Das, M., Tripathi, A., Dwivedi, P.D., 2014. Toll-like receptor 6 mediated inflammatory and functional responses of zinc oxide nanoparticles primed macrophages. *Immunol.* 142, 453–464.
Ryu, W.-I., Park, Y.-H., Bae, H.C., Kim, J.H., Jeong, S.H., Lee, H., Son, S.W. 2014. ZnO nanoparticle induces apoptosis by ROS triggered mitochondrial pathway in human keratinocytes. *Mol. Cell. Toxicol.* 10, 387–391.
Sahu, D., Kannan, G.M., Vijayaraghavan, R., 2014. Size-dependent effect of zinc oxide on toxicity and inflammatory potential of human monocytes. *J. Toxicol. Env. Health Part A* 77, 177–191.
Saptarshi, S.R., Duschl, A., Lopata, A.L., 2015. Biological reactivity of zinc oxide nanoparticles with mammalian test systems: An overview. *Nanomed.* 10, 2075–2092.
Sasidharan, A., Chandran, P., Menon, D., Raman, S., Nair, S., Koyakutty, M., 2011. Rapid dissolution of ZnO nanocrystals in acidic cancer microenvironment leading to preferential apoptosis. *Nanoscale* 3, 3657–3669.
Sawai, J., Kojima, H., Igarashi, H., Hashimoto, A., Shoji, S., Sawaki, T., et al., 1996. Antibacterial characteristics of magnesium oxide powder. *J. Chem. Eng. Japan* 29, 556.
Schmid, O., Stoeger, T., 2016. Surface area is the biologically most effective dose metric for acute nanoparticle toxicity in the lung. *J. Aerosol Sci.* 99, 133–143.
Senapati, V.A., Kumar, A., Gupta, G.S., Pandey, A.K., Dhawan, A., 2015. ZnO nanoparticles induced inflammatory response and genotoxicity in human blood cells: A mechanistic approach. *Food Chem. Toxicol.* 85, 61–70.
Senapati, V.A., Gupta, G.S., Pandey, A.K., Shanker, R., Dhawan, A., Kumar, A., 2017. Zinc oxide nanoparticle induced age dependent immunotoxicity in BALB/c mice. *Toxicol. Res.* 6, 342–352.
Serpone, N., Dondi, D., Albini, A., 2007. Inorganic and organic UV filters: Their role and efficacy in sunscreens and suncare products. *Inorg. Chim. Acta.* 360, 794–802.
Sharma, V., Anderson, D., Dhawan, A., 2012. Zinc oxide nanoparticles induce oxidative DNA damage and ROS-triggered mitochondria mediated apoptosis in human liver cells (HepG2). *Apoptosis* 17, 852–870.
Sharma, V., Shukla, R.K., Saxena, N., Parmar, D., Das, M., Dhawan, A., 2009. DNA damaging potential of zinc oxide nanoparticles in human epidermal cells. *Toxicol. Lett.* 185, 211–218.
Sharma, V., Singh, P., Pandey, A.K., Dhawan, A., 2011. Induction of oxidative stress, DNA damage and apoptosis in mouse liver after sub-acute oral exposure to zinc oxide nanoparticles. *Mut. Res.* 745, 84–91.
Shen, C., James, S.A., de Jonge, M.D., Turney, T.W., Wright, P.F.A., Feltis, B.N., 2013. Relating cytotoxicity, zinc ions, and reactive oxygen in ZnO nanoparticle exposed human immune cells. *Toxic. Sci.* 136, 120–130.
Shi, Y., 2002. Apoptosome: The cellular engine for the activation of caspase-9. *Structure.* 10, 285–288.
Shinde, N.C., Keskar, N.J., Argade, P.D., 2012. Nanoparticles: Advances in drug delivery systems. *Res. J. Pharm. Biol. Chem. Sci.* 3, 922–929.
Siegel, R.W., 1994. Nanophase materials: Synthesis, structure, and properties. In: Fujita, F.E. (Ed.) *Physics of New Materials. Springer Series in Materials Science*, vol 27. Springer, Berlin, Heidelberg. pp. 65–105.
Song, J., Zhou, J., Wang, Z.L., 2006. Piezoelectric and semiconducting coupled power generating process of a single ZnO belt/wire. A technology for harvesting electricity from the environment. *Nano Lett.* 6, 1656–1662.
Song, W., Zhang, J., Guo, J., Zhang, J., Ding, F., Li, L., Sun, Z., 2010. Role of the dissolved zinc ion and reactive oxygen species in cytotoxicity of ZnO nanoparticles. *Toxicol. Lett.* 199, 389–397.
Sopotnik, M., Leonardi, A., Križaj, I., Dušak, P., Makovec, D., Mesarič, T., Ulrih, NaP., Junkar, I., Sepčić, K., Drobne, D., 2015. Comparative study of serum protein binding to three different carbon-based nanomaterials. *Carbon* 95, 560–572.
Tang, H.-Q., Xu, M., Rong, Q., Jin, R.-W., Liu, Q.-J., Li, Y.-L., 2015. The effect of ZnO nanoparticles on liver function in rats. *Int. J. Nanomed.* 11, 4275.
Umrani, R.D., Paknikar, K.M., 2014. Zinc oxide nanoparticles show antidiabetic activity in streptozotocin-induced Type 1 and 2 diabetic rats. *Nanomed.* 9, 89–104.
Vaseem, M., Umar, A., Hahn, Y.-B., 2010. ZnO nanoparticles: growth, properties, and applications. In: Umar, A., Hahn, Y.B., (Eds.) *Metal Oxide Nanostructures and Their Applications*. American Sci. Publ., USA. pp. 1–36.
Wahab, R., Hwang, I.H., Kim, Y.-S., Musarrat, J., Siddiqui, M.A., Seo, H.-K., Tripathy, S.K., Shin, H,-S., 2011. Non-hydrolytic synthesis and photo-catalytic studies of ZnO nanoparticles. *Chem. Eng. J.* 175, 450–457.

Wahab, R., Siddiqui, M.A., Saquib, Q., Dwivedi, S., Ahmad, J., Musarrat, J., Al-Khedhairy, A.A., Shin, H.-S., 2014. ZnO nanoparticles induced oxidative stress and apoptosis in HepG2 and MCF-7 cancer cells and their antibacterial activity. *Coll. Surf. B: Biointer.* 117, 267–276.

Wang, B., Feng, W., Wang, M., Wang, T., Gu, Y., Zhu, M., Ouyang, H., Shi, J., Zhang, F., Zhao, Y., 2008. Acute toxicological impact of nano-and submicro-scaled zinc oxide powder on healthy adult mice. *J. Nano. Res.* 10, 263–276.

Wang, C., Youle, R.J., 2009. The role of mitochondria in apoptosis. *Ann. Rev. Genet.* 43, 95–118.

Wang, J., Gao, S., Wang, S., Xu, Z., Wei, L., 2018. Zinc oxide nanoparticles induce toxicity in cal 27 oral cancer cell lines by activating PINK1/Parkin-mediated mitophagy. *Int. J. Nanomed.* 13, 3441.

Wang, Z.L., 2004. Functional oxide nanobelts: Materials, properties and potential applications in nanosystems and biotechnology. *Annu. Rev. Phys. Chem.* 55, 159–196.

Wilhelmi, V., Fischer, U., Weighardt, H., Schulze-Osthoff, K., Nickel, C., Stahlmecke, B., et al., 2013. Zinc oxide nanoparticles induce necrosis and apoptosis in macrophages in a p47phox-and Nrf2-independent manner. *PLoS One* 8, e65704.

Willis, M.S., Monaghan, S.A., Miller, M.L., McKenna, R.W., Perkins, W.D., Levinson. B.S., Bhushan. V., Kroft. S.H., 2005. Zinc-induced copper deficiency: a report of three cases initially recognized on bone marrow examination. *Am. J. Clin. Pathol.* 123, 125–131.

Xu, M., Li, J., Hanagata, N., Su, H., Chen, H., Fujita, D., 2013. Challenge to assess the toxic contribution of metal cation released from nanomaterials for nanotoxicology – the case of ZnO nanoparticles. *Nanoscale* 5, 4763–4769.

Yamamoto, V., Suffredini, G., Nikzad, S., Hoenk, M.E., Boer, M.S., Teo, C., Heiss, J.D., Kateb, B., 2013. From Nanotechnology to Nanoneuroscience/Nanoneurosurgery and Nanobioelectronics: A Historical Review of Milestones. *The Textbook of Nanoneuroscience and Nanoneurosurgery*. CRC Press, Boca Raton, USA.

Yan, Z., Xu, L., Han, J., Wu, Y.-J, Wang, W., Yao, W., Wu, W., 2014. Transcriptional and posttranscriptional regulation and endocytosis were involved in zinc oxide nanoparticle-induced interleukin-8 over-expression in human bronchial epithelial cells. *Cell Biol. Toxicol.* 30, 79–88.

Yang, X., Shao, H., Liu, W., Gu, W., Shu, X., Mo, Y., Chen, X., Zhang, Q., Jiang, M., 2016. Endoplasmic reticulum stress and oxidative stress are involved in ZnO nanoparticle-induced hepatotoxicity. *Toxicol. Lett.* 234, 40–49.

Yousef, J.M., Mohamed, A.M., 2015. Prophylactic role of B vitamins against bulk and zinc oxide nanoparticles toxicity induced oxidative DNA damage and apoptosis in rat livers. *Pak. J. Pharm. Sci.* 28, 175–184.

Yu, J., Kim, H.-J., Go, M.-R., Bae, S.-H, Choi, S.-J., 2017. ZnO interactions with biomatrices: Effect of particle size on ZnO-protein corona. *Nanomaterials* 7, 377.

Zhang, Y.R., Nayak, T., Hong, H., Cai, W., 2013. Biomedical applications of zinc oxide nanomaterials. *Curr. Mol. Med.* 13, 1633–1645.

Section IV

Environmental Biotechnology

9 Waste Management by Thermophilic Bacteria

Nevadita Sharma
B.S. Anangpuria Institute of Technology and Management, Faridabad, Haryana, India

The Public Health Research Institute at the International Centre for Public Health (ICPH), Newark, NJ, USA

Nishant Sharma
Translational Health Science and Technology Institute, Faridabad, Haryana, India

The Public Health Research Institute at the International Centre for Public Health (ICPH), Newark, NJ, USA

CONTENTS

9.1 INTRODUCTION

Microorganisms have been divided into three main groups based on their optimal growth temperatures viz. psychrophiles (below 20°C), mesophiles (moderate temperatures) and thermophiles (high temperatures, above 55°C) (Gleeson et al. 2013). Thermophiles have been isolated from geothermal regions and hot environments and have been widely reported from volcanoes, tectonically active faults, hot springs, geothermal soils, coal refuse, hot-water tanks, sunlit soil and manmade environments like compost piles (Panda et al. 2019). Most of them are microaerophilic and exhibit a chemolithoautotrophic mode of nutrition. Microorganisms and thermostable enzymes have been topics for much research during the last two decades. The pioneering work in the 1960s by Brock and his colleagues shows how thermophiles and their proteins are able to function at elevated temperatures (Brock and Freeze 1969; Brock 1986). Table 9.1 provides information on the growth parameters of thermophilic organisms. Only a few eukaryotes are known to grow above this temperature, but some fungi grow in the temperature range 50–55°C (Maheshwari et al. 2000). Several years ago, Kristjansson and Stetter (1992) suggested a further division of thermophiles and created a hyperthermophile boundary (growth at and above 80°C) that has today reached general acceptance (Kristjansson and Stetter 1992; Stetter 1996, 1999). Most thermophilic bacteria characterized today grow below the hyperthermophilic boundary (with some exceptions, such as *Thermotoga* and *Aquifex*), while hyperthermophilic species are dominated by the Archaea

DOI: 10.1201/9781003324706-13

TABLE 9.1
Thermophilic Diversity in Nature

Sl. No.	Organism	Optimum Temperature	Growth Physiology	References
1	*Anaerocellum thermophilum*	75°C	Strictly anaerobic	Bolshakova et al. (1994)
2	*Acidobacterium capsulatum*	65°C	Facultative anaerobic	Inagaki et al. (1998)
3	*Alicyclobacillus acidocaldarius*	60°C	Strictly anaerobic	Groenewald et al. (2009)
4	*Anaerocellum thermophilum*	74°C	Strictly anaerobic	Yang et al. (2010)
5	*Anoxybacillus flavithermus*	60°C	Facultative anaerobic	Kambourova et al. (2007)
6	*Anoxybacillus kamchatkensis NASTPD13*	65°C	Facultative anaerobic	Yadav et al. (2018)
7	*Bacillus licheniformis*	70°C	Facultative aerobic	Lu et al. (2013)
8	*Bacillus licheniformis*	70°C	Facultative aerobic	Azlina and Norazila (2013)
9	*Bacillus pumilus*	70°C	Endospore forming aerobic	Reiss et al. (2011)
10	*Bacillus stearothermophilus*	65°C	Facultative aerobic	Abedi and Arjunan (2014)
11	*Bacillus thermoleovorans ID-1*	74°C	Facultative aerobic	Lee et al. (1999)
12	*Bacillus vallismortis RG 07*	65°C	Aerobic	Gaur et al. (2015)
13	*Burkholderia ubonensis SL-4*	65°C	Obligate aerobic	Yang et al. (2016)
14	*Caldicellulosiruptor saccharolyticus*	70°C	Anaerobic	Rainey et al. (1994)
15	*Caldanaerobacter subterraneus*	75°C	Anaerobic	Royter et al. (2009)
16	*Caldibacillus cellulovorans*	80°C	Aerobic or Facultative anaerobic	Huang and Monk (2004)
17	*Caldicellulo siruptorbescii*	70°C	Anaerobic	An et al. (2015)
18	*Caldicellulosiruptor saccharolyticus*	69°C	Extremely thermophilic anaerobic	Bergquist et al. (1999)
19	*Cellulosimicrobium cellulans CKMX1*	60°C	Xylanolytic anaerobic	Walia et al. (2015)
20	*Clostridium thermocellum*	(55–68)°C	Anaerobic	Akinosho et al. (2014)
21	*Clostridium thermosulfurogenes*	60°C	Strictly anaerobic	Swamy et al. (1996)
22	*Dehalococcus restrictus*	63°C	Strictly anaerobic	Kengen et al. (1999)
23	*Desulfitobacterium frappieri TCE1*	65°C	Strictly anaerobic	Drzyzga et al. (2001)
24	*Fervidobacterium islandicum AW-1*	70°C	Extremely thermophilic anaerobic	Nam et al. (2002)
25	*Fervidobacterium pennavorans Ven5*	80°C	Extremely thermophilic anaerobic	Bertoldo et al. (1999)
26	*Geobacilluss tearothermophilus*	70°C	Aerobic or Facultative anaerobic	Guagliardi et al. (1996)
27	*Geobacilluss tearothermophilus*	(60–80)°C	Aerobic or Facultative anaerobic	Karaguler et al. (2007)
28	*Geobacillus thermocatenulatus MS5*	(60–65)°C	Aerobic or Facultative anaerobic	Verma and Shirkot (2014)
29	*Meiothermus taiwanensis WR-220*	65°C	Aerobic	Wu et al. (2017)
30	*Methanopyrus kandleri*	110°C	Obligate anaerobic	Kurr et al. (1991)
31	*Methanothermobacter thermoautotrophicus*	(60–70)°C	Obligate anaerobic	Kosaka et al. (2013)

TABLE 9.1 (Continued)
Thermophilic Diversity in Nature

Sl. No.	Organism	Optimum Temperature	Growth Physiology	References
32	*Nocardiopsissp B2*	70°C	Strictly aerobic	Stamford et al. (2001)
33	*Pyrococcus furiosus*	100°C	Hyperthermophilic anaerobic	Fiala and Stetter (1986)
34	*Pyrococcus furiosus*	>72°C	Thermophilic anaerobic	Cline et al. (1996)
35	*Pyrococcus abyssi*	96°C	Hyperthermophilic anaerobic	Cohen et al. (2003)
36	*Pyrococcus horikoshii*	98°C	Hyperthermophilic anaerobic	Ando et al. (2002)
37	*Rhodothermus marinus*	70°C	Obligate aerobic	Hamed et al. (2017)
38	*Rhodothermus marinus*	80°C	Obligate aerobic	Gomes et al. (2003)
39	*Staphylococcus aureus ALA1*	60°C	Facultative anaerobic	Bacha et al. (2018)
40	*Streptomyces thermocerradoensis I3*	75°C	Anaerobic	Brito-Cunha et al. (2015)
41	*Sulfolobus solfataricus*	85°C	Obligate aerobic	Cannio et al. (2004)
42	*Thermoanaerobacter ethanolicus*	60°C	Anaerobic	Shao et al. (2016)
43	*Thermoanaerobacter subterraneus*	65°C	Anaerobic	Fardeau et al. (2000)
44	*Thermoanaerobacter tengcongensis*	75°C	Anaerobic	Wang et al. (2012)
45	*Thermoanaerobacter thermohydrosulfuricus*	(60–70)°C	Obligate anaerobic	Verbeke et al. (2013)
46	*Thermoanaerobacterium saccharolyticum NTOU1*	63°C	Anaerobic	Hung et al. (2011)
47	*Thermococcus kodakaraensis*	85°C	Obligate anaerobic	Fukui et al. (2005)
48	*Thermococcus litoralis*	>72°C	Anaerobic	Cariello et al. (1991)
49	*Thermomonosporafusca*	75°C	Aerobic or Facultative anaerobic	Irwin et al. (1994)
50	*Thermoplasma acidophilum*	60°C	Facultative anaerobic	Jung and Lee (2006)
51	*Thermosyntropha lipolytica*	75°C	Anaerobic	Gumerov et al. (2012)
52	*Thermotaga maritima*	80°C	Anaerobic	Van Fossen et al. (2008)
53	*Thermotaga neapolitana*	80°C	Anaerobic	Ooteghem et al. (2004)
54	*Thermotoga naphthophila*	80°C	Anaerobic	Takahata et al. (2001)
55	*Thermotoga maritima*	76–82°C	Anaerobic	Huber et al. (1986)
56	*Thermotoga lettingae*	65°C	Anaerobic	Balk et al. (2002)
57	*Thermotoga petrophila*	80°C	Anaerobic	Takahata et al. (2001)
58	*Thermus aquaticus*	>75°C	Aerobic	Farazmandfar et al. (2013)
59	*Thermus caldophilus*	(70–77)°C	Obligate aerobic	Taguchi (2017)
60	*Thermus caldophilus*	75°C	Aerobic	Kim et al. (1996)
61	*Thermus thermophila*	>75°C	Aerobic	Aye et al. (2018)

Source: Reproduced with Permission from Ghosh et al. 2020.

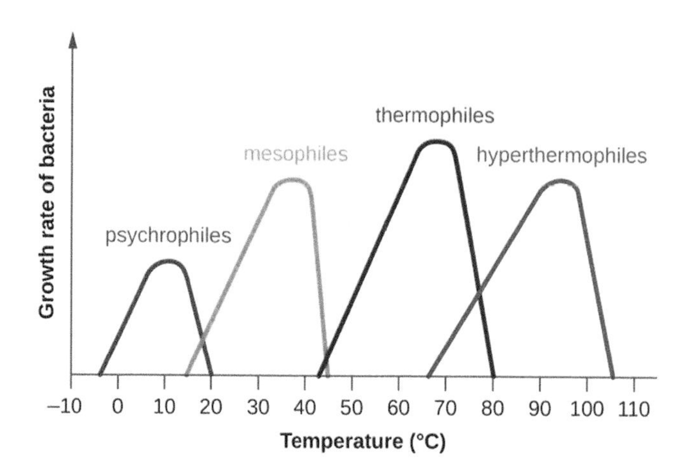

FIGURE 9.1 Different types of microbial communities (Reproduced from Alisawi 2020).

(Figure 9.1). This polyextremophilic group is found in all three domains of life and is further classified into thermoalkaliphiles and thermoacidophiles since they thrive at pH values of 0–12.

During the 1990s, thermostable enzyme work got dramatically increased due to intense biochemical studies on thermophilic enzyme purification and characterization followed by overexpression of the encoding genes in prokaryotic hosts which facilitated higher level of production (Ghosh et al. 2020). Various techniques stimulated isolation of a number of thermophilic microbes in order to access thermostable enzymes which can increase the window for enzymatic bioprocess operations. The first thermostable enzyme *Taq*-polymerase, isolated from the thermophilic eubacterium *Thermus aquaticus*, has been one of the most successful and commercialized thermostable enzymes used in polymerase chain reactions (PCR) for amplification of DNA (Terpe 2013). A number of other DNA polymerases from thermophilic sources have since then been isolated and some of these commercialized in this area (Satyanarayana et al. 2005; Podar and Reysenbach 2006). These include Tfl from *Thermus flavus*, Tth from *Thermus thermophilis* HB8, Tbr from *Thermus brockianus*, Tca from *Thermophilus caldophilus*, Tma from *Thermotoga maritima*, KOD from *Thermococcus kodakarensis*, Tli from *Thermococcus litoralis* and Pfu from *Pyrococcus furiosus* (Kaledin et al. 1980; Rüttimann et al. 1985; Carballeira et al. 1990; Lundberg et al. 1991; Park et al. 1993; Harrell and Hart 1994). Another area of interest has been the prospecting for industrial enzymes for use in technical products and processes, often on a very large scale. These enzymes can be advantageous as industrial catalysts as they rarely require toxic metal ions for functionality, hence creating the possibility to use more environmentally friendly processing (Comfort et al. 2004). Thermostable enzymes offer robust catalyst alternatives since they are able to withstand the often relatively harsh conditions of industrial processing.

Energy utilization was a topic of concern about 30 years ago with respect to the conversion of biomass into sugars. Renewed interest in biocatalytic conversions has recently emerged, with growing concern on the instability and possible depletion of fossil oil resources as well as environmental concerns, and focus is again put on biorefining, and the biorefinery concept. In biorefining, renewable resources such as agricultural crops or wood are utilized for extraction of intermediates or for direct bioconversion into chemicals, commodities and fuels (Fernando et al. 2006; Kamm and Kamm 2004). Thermostable enzymes have an obvious advantage as catalysts in these processes since high temperatures often promote better enzyme penetration and cell-wall disorganization of the raw materials (Paes and O'Donohue 2006). Table 9.2 enlists some of the industrially important thermostable enzymes obtained from thermophilic organisms. By the parallel development in molecular biology, novel and developed stable enzymes also have a good chance to be produced at suitable levels. So, there are potential possibilities of thermostable

TABLE 9.2
Thermophilic Enzymes Obtained from Heat-tolerant Microbes

Enzyme Class/Enzyme	Obtained From	References
Amylases		
α-amylase (BLA)	*Bacillus licheniformis*	Hwang et al. (1997)
α-amylase	*Paecilomyces variotti*	Michelin et al. (2010)
Dextrinogenic and saccharogenic amylases	*Rhizomucor pusillus*	Silva et al. (2005)
Proteases		
Stetterlysin	*Thermococcus stetteri*	Voorhorst et al. (1997)
Pyrolysin	*Pyrococcus furious*	Gao et al. (2016)
Ak.1 protease	*Bacillus* sp.	Toogood et al. (2000)
Tk-SP, Tk-1689, and Tk subtilisins	*Thermococcus kodakaraensis*	Foophow et al. (2010)
Aqualysin I	*Thermus aquaticus*	Sakaguchi et al. (2014)
Proteolysin	*Coprothermobacter proteolyticus*	Toplak et al. (2013)
EI, EII and EIII	*Pseudomonas aeruginosa*	Flores-Fernandez et al. (2019)
Cellulases		
CelA	*Caldicellulosiruptor bescii*	Brunecky et al. (2018)
Endo-1,4-β-d-glucanase	*Sulfolobus shibatae*	Boyce and Walsh (2018)
Glycoside hydrolase	*Thielavia terrestris*	Gao et al. (2017)
GH45 endoglucanase	*Chaetomium thermophilum*	Zhou et al. (2017)
GH45 endoglucanase	*Sclerotinia sclerotiorum*	Chahed et al. (2018)
Xylanases		
Xylanase T-6	*Bacillus stearothermophilus*	Khasin et al. (1993)
Xylanase I	*Thermomyces lanuginosus*	Leggio et al. (1999)
XynA	*Thermomyces lanuginosus*	Schlacher et al. (1996)
Lipases		
TA lipase	*Geobacillus thermoleovorans*	Fotouh et al. (2016)
HydS14	*Actinomadura sp.*	Sriyapai et al. (2015)

enzymes, developed or isolated from thermophiles. This chapter shows the fundamental and applied aspects of thermophilic bacteria with an industrial perspective. There are different advantages of thermozymes such as reduced risk of microbial contamination, increased mass transfer, lower viscosity and improved susceptibility of some proteins to enzyme molecules. This chapter covers thermostabilizing strategies and some application of thermostable enzymes in waste management, food processing, agriculture and other industries.

9.2 IMPORTANCE OF THERMOPHILIC MICROBES OVER MESOPHILIC MICROBES

The environmental and rising economic costs of petroleum-based products have regained interest in the biological production of commodity fuels. Most research in this area has focused on microbes growing in the mesophilic temperature range (25–37°C). But most chemical refineries work in high-temperature fermentations, so there is immense scope for use of thermophilic ($T_{opt} \geq 70°C$) microbial hosts, which offer potential advantages over mesophilic biorefineries. High range of thermophilic enzymes is also of considerable interest in biotechnology ever since the development of the PCR (Bartlett and Stirling 2003). The thermostable enzymes are more useful and powerful tool for industrial catalysis as well (Zamost et al. 1991; Vieille and Zeikus 2001; Atomi et al. 2011). It is also becoming increasingly possible to improve the thermostability of mesophilic enzymes, either through protein engineering or techniques such as enzyme immobilization (Lehmann and Wyss 2001; Harris et al. 2010; Steiner and Schwab 2012; Singh et al. 2013), so that thermally based bioprocessing can be considered. There are number of advantages of high-temperature bioprocessing, including reduced risk of contamination as compared to mesophilic hosts, improved solubility of substrates such as lignocellulosic biomass, lowered chances of phage infection, continuous recovery of volatile chemical products directly from fermentation broth, and reduced cooling costs due to the greater temperature differential between the fermenter and the ambient air, which is the ultimate heat acceptor (Frock and Kelly 2012; Keller et al. 2014). Hydrolysis of large biopolymers into smaller components by cellulases, chitinases, proteinases, and other carbohydrate-degrading enzymes is accomplished by using *in vitro* single-step reactions of thermostable enzymes (Vieille and Zeikus 2001). However, more complex multistep chemical conversions require an intact cellular host (Ladkau et al. 2014). Therefore, many interesting and potentially industrially relevant pathways require multiple enzyme steps, regeneration of co-factors, energy conservation via coupling to a transmembrane gradient or input of additional chemical energy. The potential for extreme thermophiles to serve as intact platforms for metabolic engineering and whole-cell biocatalysts is now increasingly being considered (Taylor et al. 2011; Frock and Kelly 2012).

Besides these, there are many more application of thermostable enzyme in the industrial setting where higher temperature favors good production. The following are some striking examples:

i. Industrial production of fructose from corn syrup via glucose isomerase, where higher temperatures favor the fructose side of the reaction, creating a final product with better sweetening power (Bhosale et al. 1996);
ii. Hydrogen production becomes more favorable at high temperatures, leading to increased hydrogen productivities in thermophiles (Verhaart et al. 2010);
iii. Bioleaching of highly refractory ores such as chalcopyrite is more favorable under thermophilic conditions (Du Plessis et al. 2007); and
iv. In methane production, the use of thermophilic methanogens improves methane production since more methane must be generated to provide the same amount of cellular energy (Ahring et al. 1999; Amend and Shock 2001).

9.3 WASTE MANAGEMENT

The continuous increase in worldwide industrialization has made researchers find economical ways to fulfill the growing demand. Industries such as automobiles, textiles, animal feed, detergent, paper, health care needs, food processing, manure, wine making and waste management have shown a gradual increase in their demand (Bhardwaj et al. 2021). Hence, it is necessary to fulfill these requirements without affecting the economy or any harsh effects on the environment. All industries require various parameters to be considered, such as the use of enzyme-based catalysis, low risk of environmental effects and biodegradable and cost-effective raw materials (Ding et al. 2008). Many different types of industrial waste managements are used on large scale which briefly mention in Figure 9.2 (Bhardwaj et al. 2021).

9.3.1 Food Waste Management

Food waste impacts the environment as discarded food waste would rot in the landfill and discharge methane gas, a very potent gas that causes the greenhouse effect (Lights 2019). Organic food waste composting minimizes the impact on many sectors. For instance, reduction of methane gas production from landfills directly minimizes the greenhouse effect, and production of compost material reduces the dependability on pesticides and synthetic fertilizers resulting in maintaining the natural pH level of the soil. Various types of organic food waste composting systems

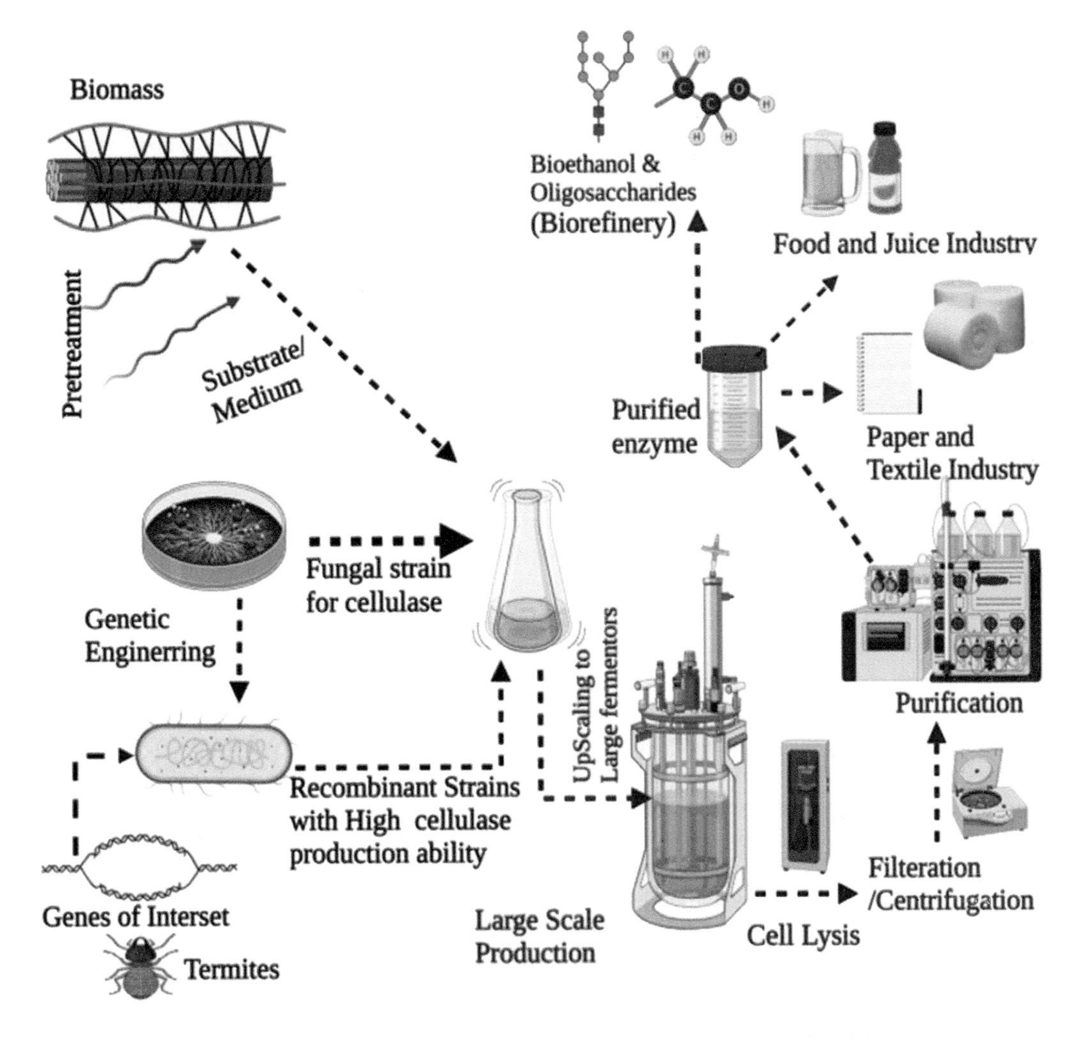

FIGURE 9.2 Large scale management system (Reproduced from Bhardwaj et al. 2021).

implemented worldwide with techniques such as vermi-composting, windrow composting, aerated static pile composting and in-vessel composting are of considerable interest. Organic food waste composting is only possible with the involvement of function-specific microbial community (Rastogi et al. 2020). Microbial community comprises of a variety of microbial species ranging from culturable and non-culturable strains that can interact with each other in the environment they co-exist in (Boers et al. 2019).

Microorganisms such as aerobic and anaerobic can consume and reduce the amounts of food wastes which contain various organic compounds (by breaking the complex molecules into simpler ones). In recent years, with the advancement in the field and with increasing knowledge of these bacteria, there has been extensive research on food waste treatments with the help of microorganisms including thermophiles (Choi et al. 1996; Shin et al. 2004; Yi et al. 2006; Sudharhsan et al. 2007; Kiran et al. 2014; Kwon et al. 2014). Among the various microorganisms, bacteria are widely considered useful for food waste treatment. Several mesophilic and thermophilic bacteria, such as including *Pseudomonas sp., Xanthomonas sp., Bacillus sp.* and *Stearothermophilus sp.*, are often found in food wastes (Fujio and Mume 1991; Jeong et al. 1999) implementing their ability to utilize these wastes. Yi et al. (2006) reported that increase in temperature to 50°C during the fermentation process significantly improved the efficiency of thermophilic bacteria to grow and decompose food wastes. Shin et al. (2004) were of the view that thermophilic bacteria produce hydrogen by utilizing the food waste, which not only reduces the waste but also generates clean fuel. Kim and Lee (2011) were able to isolate thermophilic *Bacillus subtilis* from various organic materials for food waste treatment indicating the role of the species in waste degradation. In order to start bio-degradation with these bacteria, the first requirement is the starter culture that can be used in industrial setting. There are several methods for preparing microbial starters such as freeze-drying, fluidized bed drying and spray drying for industrial applications (Bayrock and Ingledew 1997; Luna-Solano et al. 2005; Ale et al. 2015). Recently, Lee et al. (2016) developed and optimized a procedure for producing *Saccharomyces* and non-*Saccharomyces* yeast starters at the industrial level using an air-blast drying method. These methods give us considerable amount of starter culture having superior quality which can be used to set up bio-degradation of food waste at a large scale.

9.3.2 Aerobic Wastewater Treatment

Traditionally, aerobic waste treatment relies on extra oxygen (provided by aeration) for degradation of biological waste. However, in case of certain chemicals, the degradation can be enhanced by increasing the temperature. The aerobic wastewater treatment done above 45°C comes under thermophilic aerobic wastewater treatment (Rozich and Colvin 1999). Thermophilic aerobic treatment systems are found to be useful for high-strength/low flow wastewater with toxicity concerns stemming from high salinity levels or the presence of hazardous compounds (Collivignarelli et al. 2019). There is a large number of the research and reports on thermophilic aerobic treatment that have largely focused on waste streams containing relatively low concentrations of biodegradable compounds (LaPara and Alleman 1999).

The biology of thermophilic aerobic wastewater treatment reactors differs from conventional activated sludge microflora in that nitrifying bacteria, floc-forming organisms, or protozoa and other life forms are not present. Nitrifying thermophiles have only been isolated once (Golovacheva 1976), but this result has never been confirmed. The reasons thermophilic bacteria fail to form discrete floc particles remain unknown; contributing factors may include: (1) a lack of floc-forming species (e.g., *Zooglea ramigera*), (2) failure to achieve proper physiological state conducive to floc formation, (3) physicochemical conditions inhibiting aggregation, and (4) improper conditions to selectively favor floc formers (especially since scavenging protozoa are not present) (LaPara and Alleman 1999). Many researchers have reported modest (Golovacheva 1976) to excellent (Gehm 1956; Stover and Samuel 1997) floc formation, although it is not clear which factors contributed to these results. Because the primary goal of thermophilic wastewater treatment processes is to reduce the level of organic compounds in the waste stream, the dominant microorganisms are almost certainly aerobic heterotrophs (LaPara and Alleman 1999).

Thermophilic aerobic processes can treat a broad range of wastewaters. This technology appears to be particularly well suited for high-strength waste streams because these wastewaters can fully benefit from the rapid substrate utilization and low sludge yields which are characteristic of thermophilic bacteria. High-strength waste streams also contain the energy content necessary for autothermal operation. Thermophilic reactors are also likely to be dominated by thermophilic bacilli which have relatively broad metabolic abilities, but have unique nutritional requirements that potentially complicate treatment. However, before designing the reactor, engineers should be careful about selecting the aeration system for accommodating enormous oxygen requirements (Barr et al. 1996). Despite the difficulties, the thermophilic wastewater treatment provides a better alternative to other wastewater treatment processes and is more environment friendly.

9.3.3 Agricultural Waste Treatment

With the development and expansion of human civilization and continuously growing human population, there has been an exponential increase in agriculture production, resulting in accumulation of a huge amounts of agricultural wastes (Tripathi et al. 2019). Every year, India alone generates a huge amount of agricultural waste (approx. 350–990 Mt/y) and is only next to China which contributes the maximum amount (Singh and Sidhu 2014). As a common practice, farmers burn up or leave agricultural waste in the field, which contributes significantly in environmental pollution (Kumar and Joshi 2013). Various techniques have been utilized for agri-residue disposal with thermophiles providing one of the ways to handle this environmental issue.

Microbes including bacteria having the ability to break down the cellulosic wastes play a pivotal role in bio-degradation of the agriculture waste and its conversion into usable fertilizer/compost (de Souza 2013). Composting of agriculture waste constitutes three stages during which different microorganisms (viz. actinomycetes, bacteria and fungi) break down cellulosic components and reduce the waste into humus. These processes involve not only mesophiles like *Streptomyces rectus* but also depend on the activity of thermophiles like *Actinobifida chromogena* and *Thermomonospora fusca* (Singh et al. 2019). The first stage involves the rise in temperature to a moderate range and depends on mesophiles which are followed by further increase in the temperature (around 70°C) through intervention of thermophilic microorganisms (Singh et al. 2019). Thermophilic phase is a critical step in this composting as it eliminates many of the pathogenic bacteria and fungi (due to high temperature). The process of bio-degradation is dependent on several parameters such as pH of the waste (< 7.0), moisture content present (~ 60%), amount of volatile ammonia emission (30–70%), temperature (30–60°C) and the composition of agriculture waste in terms of bio-molecules (such as polysaccharides, cellulose, hemicellulose, amino acids and fatty acids) (Zhang et al. 2003). Another advantage of direct composting of agricultural waste (crop residues) is that this transforms a large volume of lignocellulosic wastes into stable compost. This process yields compost with specific ingredients which have definite C:N ratio (Singh et al. 2019). Besides this advantage, the process which is mediated by microbes results in humification and transformation of waste. This process is not only cheaper but also sustainable (both in terms of resources needed and cost efficiency) (Singh et al. 2019).

9.4 LIMITATIONS AND FUTURE PERSPECTIVES

Thermophilic waste management is an interesting, resource saving and environment friendly approach. But like many processes, it does have its own limitations. A major limitation is the high temperature needed for the process. This is true in case of industrial water waste which is dispensed at higher temperature, but for others the slurry needs to be pre-heated. In those cases, the mesophilic treatment (Zhang et al. 2003) or a combination of mesophilic treatment followed by thermophilic treatment is more sustainable instead of thermophilic treatment. Another concern is the oxygen transfer and hydrophobicity of complex mixtures. Though, as a general rule the hydrophobicity decreases with temperature and oxygen transfer increases with temperature (Boogerd et al. 1990), but certain complex mixtures might not follow the trend. Therefore, if the composition of the waste changes, it becomes critical to analyze and study the waste mixture for these parameters before setting them up for processing. Another hurdle in implementing bio-degradation by microbes (including thermophiles) is that the legislation regarding solid waste disposal is not well defined and also not strictly applied in the majority of developing countries (Zhang et al. 2022). Also, though these techniques look good on paper, there are hardly reports or study done to see the cost-effectiveness of these techniques as well as the additional burden (time and economic) these techniques bring when the waste going to be treated is not at par with requirements (in terms of moisture, pH, etc.) (Zhang et al. 2022).

Despite these limitations, bio-degradation of wastes by thermophiles is a great alternative and has wide applications. If these methods are used by studying the type of waste needs to be dealt with, it can provide better solution to the problem of waste management. For instance, it has been reported that the methane production plant (depending on bio-degradation by thermophiles) is not cost-effective in low-population towns (Zhang et al. 2022). But this knowledge can be implemented to set up reactors in densely populated and urban societies, thereby solving the problem of waste management as well as generating fuel in form of methane. Similarly, many industries spent and discard liquid waste at high temperature. These conditions provide good chance to set up the thermophilic bio-treatment plant. However, more study needs to be carried out in order to optimize and harness the vast potential of bio-degradation by thermophiles and this can in the future provide a sustainable solution to constantly growing problem of waste management.

REFERENCES

Abedi, A.H.A.A., Arjunan, S., 2014. Molecular characterization of thermostable DNA polymerase of *Bacillus stearothermophilus* spp isolated from soil in Bangalore, India. *Euro. J. Exp. Biol.* 4, 67–72.

Ahring, B.K., Licht, D., Schmidt, A.S., Sommer, P., Thomsen, A.B., 1999. Production of ethanol from wet oxidised wheat straw by *Thermoanaerobacter mathranii*. *Biores. Technol.* 68, 3–9.

Akinosho, H., Yee, K., Close, D., Ragauskas, A., 2014. The emergence of *Clostridium thermocellum* as a high utility candidate for consolidated bioprocessing applications. *Front. Chem.* 2, 66.

Ale, C.E., Otero, M.C., Pasteris, S.E., 2015. Freeze-drying of wine yeasts and *Oenococcus oeni* and selection of the inoculation conditions after storage. *J. Bioproc. Biotech.* 5, 1000248.

Alisawi, H.A.O., 2020. Performance of wastewater treatment during variable temperature. *Appl. Water Sci.* 10, 89.

Amend, J.P., Shock, E.L., 2001. Energetics of overall metabolic reactions of thermophilic and hyperthermophilic archaea and bacteria. *FEMS Microbiol. Rev.* 25, 175–243.

An, J., Xie, Y., Zhang, Y., Tian, D., Wang, S., Yang, G., Feng, Y., 2015. Characterization of a thermostable, specific GH10 xylanase from *Caldicellulosiruptor bescii* with high catalytic activity. *J. Mol. Catal. B Enzym.* 117, 13–20.

Ando, S., Ishida, H., Kosugi, Y., Ishikawa, K., 2002. Hyperthermostable endoglucanase from *Pyrococcus horikoshii*. *Appl. Environ. Microbiol.* 68, 430–433.

Atomi, H., Sato, T., Kanai, T., 2011. Application of hyperthermophiles and their enzymes. *Curr. Opin. Microbiol.* 22, 618–626.

Aye, S.L., Fujiwara, K., Ueki, A., Doi, N., 2018. Engineering of DNA polymerase I from *Thermus thermophilus* using compartmentalized self-replication. *Biochem. Biophys. Res. Commun.* 499, 170–176.

Azlina, I.N., Norazila, Y., 2013. Thermostable alkaline serine protease from thermophilic *Bacillus* species. *Int. Res. J. Biol. Sci.* 2, 29–33.

Bacha, A.B., Al-Assaf, A., Moubayed, N.M.S., Abid, I., 2018. Evaluation of a novel thermo-alkaline *Staphylococcus aureus* lipase for application in detergent formulations. *Saudi. J. Biol. Sci.* 25, 409–417.

Balk, M., Weijma, J., Stams, A.J., 2002. *Thermotoga lettingae* sp. *nov.*, a novel thermophilic, methanol-degrading bacterium isolated from a thermophilic anaerobic reactor. *Int. J. Syst. Evol. Microbiol.* 52, 1361–1368.

Barr, T.A., Taylor, J.M., Duff, S.J.B., 1996. Effect of HRT, SRT and temperature on the performance of activated sludge reactors treating bleached kraft mill effluent. *Wat. Res.* 30, 799–810.

Bartlett, J.M., Stirling, D.A., 2003. Short history of the polymerase chain reaction. *Methods Mol. Biol.* 226, 3–6.

Bayrock, D., Ingledew, W.M., 1997. Mechanism of viability loss during fluidized bed drying of baker's yeast. *Food Res. Int.* 30, 417–425.

Bergquist, P.L., Gibbs, M.D., Morris, D.D., Teo, V.S.J., Saul, D.J., Morgan, H.W., 1999. Molecular diversity of thermophilic cellulolytic and hemicellulolytic bacteria. *FEMS Microbiol. Ecol.* 28, 99–110.

Bertoldo, C., Duffner, F., Jorgensen, P.L., Antranikian, G., 1999. Pullulanase type I from *Fervidobacterium pennavorans* Ven5: cloning, sequencing, and expression of the gene and biochemical characterization of the recombinant enzyme. *Appl. Environ. Microbiol.* 65, 2084–2091.

Bhardwaj, N., Kumar B., Agrawal, K., Verma, P., 2021. Current perspective on production and applications of microbial cellulases: a review. *Bioresour. Bioprocess.* 8, 95.

Bhosale, S.H., Rao, M.B., Deshpande, V.V., 1996. Molecular and industrial aspects of glucose isomerase. *Microbiol. Rev.* 60, 280–300.

Boers, S.A., Jansen, R., Hays, J.P., 2019. Understanding and overcoming the pitfalls and biases of next-generation sequencing (NGS) methods for use in the routine clinical microbiological diagnostic laboratory. *Eur. J. Clin. Microbiol. Infect. Dis.* 38, 1059–1070.

Bolshakova, E.V., Ponomariev, A.A., Novikov, A.A., Svetlichnyi, V.A., Velikodvorskaya, G.A., 1994. Cloning and expression of genes coding for carbohydrate degrading enzymes of *Anaerocellum thermophilum* in *Escherichia coli*. *Biochem. Biophys. Res. Commun.* 202, 1076–1080.

Boogerd, F.C., Bos, P., Kuenen, J.G., Heijnen, J.J., Van der Lans, R.G.J.M., 1990. Oxygen and carbondioxide mass transfer and the aerobic, autophilic cultivation of moderate and extreme thermophiles: A case study related to the microbial desulfurisation of coal. *Biotechnol. Bioeng.* 35, 1111–1119.

Boyce, A., Walsh, G., 2018. Expression and characterisation of a thermophilic endo-1,4-β-glucanase from *Sulfolobus shibatae* of potential industrial application. *Mol. Biol. Rep.* 45, 2201–2211.

Brito-Cunha, C.C.Q., Gama, A.R., Jesuino, R.S.A., Faria, F.P., Bataus, L.A.M., 2015. Production of cellulases from a novel thermophilic *Streptomyces thermocerradoensis* I3 using agricultural waste residue as substrate. *JAES.* 4, 90–99.

Brock, T.D., Freeze, H., 1969. *Thermus aquaticus* gen. n. and sp. n., a nonsporulating extreme thermophile. *J. Bacteriol.* 1969(98), 289–297.

Brock, T.D., 1986. Introduction, an overview of the thermophiles. In: Brock, T.D. (Ed.) *Thermophiles: General, Molecular and Applied Microbiology*. John Wiley & Sons, New York. pp. 1–16.

Brunecky, R., Chung, D., Sarai, N.S., Hengge, N., Russell, J.F., Young, J., et al., 2018. High activity CAZyme cassette for improving biomass degradation in thermophiles. *Biotechnol. Biofuels* 11, 22.

Cannio, R., Di Prizito, N., Rossi, M., Morana, A., 2004. A xylan degrading strain of *Sulfolobus solfataricus*: isolation and characterization of the xylanase activity. *Extremophiles.* 8, 117–124.

Carballeira, N., Nazabal, M., Brito, J., Garcia, O., 1990. Purification of a thermostable DNA polymerase from *Thermus thermophilus* HB8, useful in the polymerase chain reaction. *Biotechniques* 9, 276–281.

Cariello, N.F., Swenberg, J.A., Skopek, T.R., 1991. Fidelity of *Thermococcus litoralis* DNA polymerase (Ventt) in PCR determined by denaturing gradient gel electrophoresis. *Nucleic Acids Res.* 19, 4193–4198.

Chahed, H., Boumaiza, M., Ezzine, A., Marzouki, M.N., 2018. Heterologous expression and biochemical characterization of a novel thermostable *Sclerotinia sclerotiorum* GH45 endoglucanase in Pichia pastoris. *Int. J. Biol. Macromol.* 106, 629–635.

Choi, M., Chung, Y., Park, H., 1996. Effects of seeding on the microbial changes during thermophilic compositing of food waste. *J. Korean Org. Res. Recy. Assoc.* 4, 1–11.

Cline, J., Braman, J.C., Hogrefe, H.H., 1996. PCR fidelity of pfu DNA polymerase and other thermostable DNA polymerases. *Nucleic Acids Res.* 24, 3546–3551.

Cohen, G.N., Barbe, V., Flament, D., Galperin, M., Heilig, R., Lecompte, O., et al., 2003. An integrated analysis of the genome of the hyperthermophilic archaeon *Pyrococcus abyssi*. *Mol. Microbiol.* 47, 1495–1512.

Collivignarelli, M.C., Bertanza, G., Abbà, A., Torretta, V., Katsoyiannis, I.A., 2019. Wastewater treatment by means of thermophilic aerobic membrane reactors: respirometric tests and numerical models for the determination of stoichiometric/kinetic parameters. *Environ. Technol.* 40, 182–191.

Comfort, D.A., Chhabra, S.R., Conners, S.B., Chou, C.-J., Epting, K.L., Johnson, M.R., et al., 2004. Strategic biocatalysis with hyperthermophilic enzymes. *Green Chem.* 6, 459–465.

de Souza, W.R., 2013. Microbial degradation of lignocellulosic biomass. In: Chandel, A.K., da Silva, S.S. (Eds.) *Sustainable Degradation of Lignocellulosic Biomass- Techniques, Applications and Commercialization*. INTECH. DOI: 10.5772/54325.

Ding, S.Y., Xu, Q., Crowley, M., Zeng, Y., Nimlos, M., Lamed, R., Bayer, E.A., Himmel, M.E., 2008. A biophysical perspective on the cellulosome: new opportunities for biomass conversion. *Curr. Opin. Biotechnol.* 19, 218–227.

Drzyzga, O., Gerritse, J., Dijk, J.A., Elissen, H., Gottschal, J.C., 2001. Coexistence of a sulfate-reducing *Desulfovibrio* species and the dehalorespiring *Desulfitobactyerium frappieri* TCE1 in defined chemostat cultures grown with various combinations of sulfate and tetrachloroethene. *Environ. Microbiol.* 3, 92–99.

Du Plessis, C.A., Batty, J.D., Dew, D.W., 2007. Commercial applications of thermophile bioleaching. In: Rawlings, D.E., Johnson, D.B. (Eds.) *Biomining*. Springer-Verlag, Berlin Heidelberg. pp. 57–80.

Farazmandfar, T., Rafiei, A., Hashemi-Sotehoh, M.B., Valadan, R., Alavi, M., Moradian, F., 2013. A simplified protocol for producing Taq DNA polymerase in biology laboratory. *Res. Mol. Med.* 1, 23–26.

Fardeau, M.L., Magot, M., Patel, B.K., Thomas, P., Garcia, J.L., Ollivier, B., 2000. *Thermoanaerobacter subterraneus* sp. *nov*., a novel thermophile isolated from oilfield water. *Int. J. Syst. Evol. Microbiol.* 50, 2141–2149.

Fernando, S., Adhikari, S., Chandrapal, C., Murali, N., 2006. Biorefineries: Current status, challenges and future direction. *Energy Fuels* 20, 1727–1737.

Fiala, G., Stetter, K.O., 1986. *Pyrococcus furiosus* sp. *nov*. represents a novel genus of marine heterotrophic archaebacteria growing optimally at 100 degrees C. *Arch. Microbiol.* 145, 56–61.

Flores-Fernandez, C.N., Cardenas-Fernandez, M., Dobrijevic, D., Jurlewicz, K., Zavaleta, A.I., Ward, J.M., et al. 2019. Novel extremophilic proteases from *Pseudomonas aeruginosa* M211and their application in the hydrolysis of dried distiller's grain with solubles. *Biotechnol. Prog.* 35, e2728.

Foophow, T., Tanaka, S., Angkawidjaja, C., Koga, Y., Takano, K., Kanaya, S., 2010. Crystal structure of a subtilisin homologue, Tk-SP, from *Thermococcus kodakaraensis*: requirement of a C-terminal beta-jelly roll domain for hyperstability. *J. Mol. Biol.* 400, 865–877.

Fotouh, D.M.A., Bayoumi, R.A., Hassan, M.A., 2016. Production of thermoalkaliphilic lipase from *Geobacillus thermoleovorans* DA2 and application in leather industry. *Enz. Res.* 2016(2016), 9034364.

Frock, A.D., Kelly, R.M., 2012. Extreme thermophiles: moving beyond single-enzyme biocatalysis. *Curr. Opin. Chem. Eng.* 1, 363–372.

Fujio, Y., Mume, S., 1991. Isolation and identification of thermophilic bacteria from sewage sludge compost. *J. Ferment. Bioeng.* 72, 334–337.

Fukui, T., Atomi, H., Kanai, T., Matsumi, R., Fujiwara, S., Imanaka, T., 2005. Complete genome sequence of the hyperthermophilic archaeon *Thermococcus kodakaraensis* KOD1 and comparison with *Pyrococcus* genomes. *Genome Res.* 15, 352–363.

Gao, J., Huang, J.W., Li, Q., Liu, W., Ko, T.P., Zheng, Y., et al., 2017. Characterization and crystal structure of a thermostable glycoside hydrolase family 45 1,4-β-endoglucanase from *Thielavia terrestris. Enzyme Microb. Technol.* 99, 32–37.

Gao, X., Zeng, J., Yi, H., Zhang, F., Tang, B., Tanga, X.-F., 2016. Four inserts within the catalytic domain confer extra stability and activity to hyperthermostable pyrolysin from *Pyrococcus furiosus. Appl. Environ. Microbiol.* 83, e03228- 16.

Gaur, R., Tiwari, S., 2015. Isolation, production, purification and characterization of an organic-solvent-thermostable alkalophilic cellulase from *Bacillus vallismortis* RG-07. *BMC Biotechnol.* 15, 19.

Gehm, H.W., 1956. Activated sludge at high temperatures and pH values. In: McCabe, J., Eckenfelder, Jr. W.W., (Eds.) *Biological Treatment of Sewage and Industrial Wastes, Volume I: Aerobic Oxidation.* Reinhold Publishing Corporation, New York.

Ghosh, S., Lepcha, K., Basak, A., Mahanty, A.K., 2020. Thermophyles and thermophilic hydrolases. In: Salwan, R., Sharma, V. (Eds.) *Physiological and Biotechnological Aspects of Extremophiles*. Elsevier. pp. 219–236.

Gleeson, D., O'Connell, A., Jordan, K., 2013. Review of potential sources and control of thermoduric bacteria in bulk-tank milk. *Irish J. Agr. Food Res.* 52, 217–227.

Golovacheva, R.S., 1976. Thermophilic nitrifying bacteria from hot springs. *Mikrobiologiya* 45, 377–379.

Gomes, I., Gomes, J., Steiner, W., 2003. Highly thermostable amylase and pullulanase of the extreme thermophilic eubacterium *Rhodothermus marinus*: production and partial characterization. *Bioresour. Technol.* 90, 207–214.

Groenewald, W.H., Gouws, P.A., Witthuhn, R.C., 2009. Isolation, identification and typication of *Alicyclobacillus acidoterrestris* and *Alicyclobacillus acidocaldarius* strains from orchard soil and the fruit processing environment in South Africa. *Food Microbiol.* 26, 71–76.

Guagliardi, A., Martino, M., Iaccarino, I., De Rosa, M., Rossi, M., Bartolucci, S., 1996. Purification and characterization of the alcohol dehydrogenase from a novel strain of *Bacillus stearothermophilus* growing at 70°C. *Int. J. Biochem. Cell. Biol.* 28, 239–246.

Gumerov, V.M., Mardanov, A.V., Kolosov, P.M., Ravin, N.V., 2012. Isolation and functional characterization of lipase from the thermophilic alkali-tolerant bacterium *Thermosyntropha lipolytica. Appl. Biochem. Microbiol.* 48, 338–343.

Hamed, M.B., Karamanou, S., Ólafsdottir, S., Basílio, J., Simoens, K., Tsolis, K.C., et al., 2017. Large-scale production of a thermostable *Rhodothermus marinus* cellulase by heterologous secretion from *Streptomyces lividans. Microb. Cell. Fact.* 16, 232.

Harrell, R.A., Hart, R.P., 1994. Rapid preparation of *Thermus flavus* DNA polymerase. *PCR Methods Appl.* 3, 372–375.

Harris, J.M., Epting, K.L., Kelly, R.M., 2010. N-terminal fusion of a hyperthermophilic chitin-binding domain to xylose isomerase from *Thermotoga neapolitana* enhances kinetics and thermostability of both free and immobilized enzymes. *Biotechnol. Prog.* 26, 993–1000.

Huang, X.P., Monk, C., 2004. Purification and characterization of a cellulase (CMCase) from a newly isolated thermophilic aerobic bacterium *Caldibacillus cellulovorans* gen. *nov*., sp *nov. World J. Microbiol. Biotechnol.* 20, 85–92.

Huber, R., Langworthy, T.A., Konig, H., Thomm, M., Woese, C.R., Sletyr, U.B., Stetter, K.O., 1986. *Thermotoga maritima* sp. *nov.* represents a new genus of unique extremely thermophilic eubacteria growing up to 90°C. *Arch. Microbiol.* 144, 324–333.

Hung, K.S., Liu, S.M., Tzou, W.S., Lin, F.P., Pan, C.L., Fang, T.Y., Sun, K.H., Tang, S.J., 2011. Characterization of a novel GH10 thermostable, halophilic xylanase from the marine bacterium *Thermoanaerobacterium saccharolyticum* NTOU1. *Process Biochem.* 46, 1257–1263.

Hwang, K.Y., Song, H.K., Chang, C., Lee, J., Lee, S.Y., Kim, K.K., et al., 1997. Crystal structure of thermostable alpha-amylase from *Bacillus licheniformis* refined at 1.7 A resolution. *Mol. Cell* 7, 251–258.

Inagaki, K., Nakahira, K., Mukai, K., Tamura, T., Tanaka, H., 1998. Gene cloning and characterization of an acidic xylanase from *Acidobacterium Capsulatum. Biosci. Biotechnol. Biochem.* 62, 1061–1067.

Irwin, D., Jung, E.D., Wilson, D.B., 1994. Characterization and Sequence of a *Thermomonospora fusca* Xylanase. *Appl. Environ. Microbiol.* 60, 763–770.

Jeong, J., Jung, K., Park, W., 1999. Studies on the optimum condition for food waste composting by microorganism in food waste. *Korean J. Environ. Agr.* 18, 272–279.

Jung, J.H., Lee, S.B., 2006. Identification and characterization of *Thermoplasma acidopilum* glyceraldehydes dehydrogenase: a new class of NADP1-specific aldehyde dehydrogenase. *Biochem. J.* 397, 131–138.

Kaledin, A.S., Sliusarenko, A.G., Gorodetskiĭ, S.I., 1980. Isolation and properties of DNA polymerase from extreme thermophylic bacteria *Thermus aquaticus* YT-1. *Biokhimiia* 45, 644–651.

Kambourova, M., Mandeva, R., Fiume, I., Maurelli, L., Rossi, M., Morana, A., 2007. Hydrolysis of xylan at high temperature by co-action of the xylanase from *Anoxybacillus flavithermus* BC and the β-xylosidase/α-arabinosidase from *Sulfolobus solfataricus* Oα. *J. Appl. Microbiol.* 102, 1586–1593.

Kamm, B., Kamm, M., 2004. Principles of biorefineries. *Appl. Microbiol. Biotechnol.* 64, 137–145.

Karaguler, N.G., Sessions, R.B., Binay, B., Ordu, E.B., Clarke, A.R., 2007. Protein engineering applications of industrially exploitable enzymes: *Geobacillus steaarothermophilus* LDH and *Candida methylica* FDH. *Biochem. Soc. Trans.* 35, 1610–1615.

Keller, M., Loder, A.J., Basen, M., Izquierdo, J., Kelly, R.M., Adams, M.W., 2014. Production of lignofuels and electrofuels by extremely thermophilic microbes. *Biofuels* 5, 499–515.

Kengen, S.W.M., Breidenbach, C.G., Felske, A., Stams, A.J.M., Schraa, G., deVos, W.M., 1999. Reductive dechlorination of tetrachloroethene to cis21,2-dichloroethene by a thermophilic anaerobic enrichment culture. *Appl. Environ. Microbiol.* 65, 2312–2316.

Khasin, A., Alchanati, I., Shoham, Y., 1993. Purification and characterization of a thermostable xylanase from *Bacillus stearothermophilus* T-6. *Appl. Environ. Microbiol.* 59, 1725–1730.

Kim, C., Lee, S.A., 2011. Isolation of Bacillus subtilis CK-2 hydrolyzing various organic materials. *J. Life Sci.* 21, 1716–1720.

Kim, C.H., Nashiru, O., Ko, J.H. 1996. Purification and biochemical characterization of pullulanase type I from *Thermus caldophilus* GK-24. *FEMS Microbiol. Lett.* 138, 147–152.

Kiran, E.U., Trzcinski, A.P., Ng, W.J., Liu, Y., 2014. Bioconversion of food waste to energy: A review. *Fuel* 134, 389–399.

Kosaka, T., Toh, H., Toyoda, A., 2013. Complete genome sequence of a thermophilic hydrogenotrophic methanogen, *Methanothermobacter* sp strain CaT2. *Genome Announc.* 1, e00672- 13.

Kristjansson, J.K., Stetter, K.O., 1992. Thermophilic bacteria. In: Kristjansson, J.K. (Ed.) *Thermophilic Bacteria*. CRC Press, London. pp. 1–18.

Kumar, P., Joshi, L., 2013. Pollution caused by agricultural waste burning and possible alternate uses of crop stubble: a case study of Punjab. In: Nautiyal, S., Rao, K., Kaechele, H., Raju, K., Schaldach, R. (Eds.) *Knowledge Systems of Societies for Adaptation and Mitigation of Impacts of Climate Change. Environmental Science and Engineering*. Springer, Berlin, Heidelberg. pp. 367–385.

Kurr, M., Huber, R., Konig, H., Jannasch, H.W., Fricke, H., Trincone, A., Kristjansson, J.K., Stetter, K.O., 1991. *Methanopyrus kandleri*, gen. and sp. nov. represents a novel group of hyperthermophilic methanogens, growing at 110°C. *Arch. Microbiol.* 156, 239–247.

Kwon, B.G., Na, S., Lim, H., Lim, C., Chung, S., 2014. Slurry phase decomposition of food waste by using various microorganisms. *J. Korean Soc. Environ. Eng.* 36, 303–310.

Ladkau, N., Schmid, A., Bühler, B., 2014. The microbial cell- functional unit for energy dependent multistep biocatalysis. *Curr. Opin. Biotechnol.* 30, 178–189.

LaPara, T.M., Alleman, J.E., 1999. Thermophilic aerobic biological wastewater treatment. *Water Res.* 33, 895–908.

Lee, D.W., Koh, Y.S., Kim, K.J., Kim, B.C., Choi, H.J., Kim, D.S., et al., 1999. Isolation and characterization of a thermophilic lipase from *Bacillus thermoleovorans* ID-1. *FEMS Microbiol. Lett.* 179, 393–400.

Lee, S., Choi, W., Jo, H., Yeo, S., Park, H., 2016. Optimization of air-blast drying process for manufacturing *Saccharomyces cerevisiae* and non-Saccharomyces yeast as industrial wine starter. *AMB Express.* 6, 105–114.

Leggio, L.L., Kalogiannis, S., Bhat, M.K., Pickersgill, R.W., 1999. High resolution structure and sequence of *T. aurantiacus* xylanase I: implications for the evolution of thermostability in family 10 xylanases and enzymes with (beta) alpha-barrel architecture. *Proteins.* 36, 295–306.

Lehmann, M., Wyss, M., 2001. Engineering proteins for thermostability: The use of sequence alignments versus rational design and directed evolution. *Curr. Opin. Biotechnol.* 12, 371–375.

Lights, Z., 2019. 5 reasons why composting is the greenest thing you can do-one green planet. *One Green Planet.* 2019. Available online: https://www.onegreenplanet.org/lifestyle/5-reasons-why-composting-isthe-greenest-thing-you-can-do/ (accessed on 30 April 2020).

Lu, L., Wang, T.N., Xu, T.F., Wang, J.Y., Wang, C.L., Zhao, M., 2013. Cloning and expression of thermo-alkali-stable laccase of *Bacillus licheniformis* in *Pichia pastoris* and its characterization. *Bioresour. Technol.* 134, 81–86.

Luna-Solano, G., Salgado-Cervantes, M.A., Rodriguez-Jimenes, G.C., Garcia-Alvarado, M.A., 2005. Optimization of brewer's yeast spray drying process. *J. Food. Eng.* 68, 9–18.

Lundberg, K.S., Shoemaker, D.D., Adams, M.W., Short, J.M., Sorge, J.A., Mathur, E.J., 1991. High-fidelity amplification using a thermostable DNA polymerase isolated from *Pyrococcus furiosus. Gene.* 108, 1–6.

Maheshwari, R., Bharadwaj, G., Bhat, M.K., 2000. Thermophilic fungi: their physiology and enzymes. *Microbiol. Mol. Biol. Rev.* 64, 461–488.

Michelin, M., Silva, T.M., Bennasi, V.M., Peixoto-Nogueira, S.C., Moraes, L.A.B., Leao, J.M., et al., 2010. Purification and characterization of thermostable α-amylase produced by the fungus *Paecilomyces variotti. Carbohydr. Res.* 345, 2348–2353.

Nam, G.W., Lee, D.W., Lee, H.S., Lee, N.J., Kim, B.C., Choe, E.A., et al., 2002. Native-feather degradation by *Fervidobacterium islandicum* AW-1, a newly isolated keratinase-producing thermophilic anaerobe. *Arch. Microbiol.* 178, 538–547.

Ooteghem, S.A.V., Jones, A., van der Lelie, D., Dong, B., Mahajan, D., 2004. H_2 production and carbon utilization by *Thermotoga neapolitana* under anaerobic and microaerobic growth conditions. *Biotechnol. Lett.* 26, 1223–1232.

Paes, G., O'Donohue, M.J., 2006. Engineering increased thermostability in the thermostable GH-11 xylanase from *Thermobacillus xylanilyticus. J. Biotechnol.* 125, 338–350.

Panda, A.K., Bisht, S.S., De Mandal, S., Kumar, N.S., 2019. Microbial diversity of thermophiles through the lens of next generation sequencing. In: Das, S., Dash, H.R. (Eds.) *Microbial Diversity in the Genomic Era.* Elsevier. pp. 217–226.

Park, J.H., Kim, J.S., Kwon, S.T., Lee, D.S., 1993. Purification and characterization of *Thermus caldophilus* GK24 DNA polymerase. *Eur. J. Biochem.* 214, 135–140.

Podar, M., Reysenbach, A.L., 2006. New opportunities revealed by biotechnological explorations of extremophiles. *Curr. Opin. Biotechnol.* 17, 250–255.

Rainey, F.A., Donnison, A.M., Janssen, P.H., Saul, D., Rodrigo, A., Bergquist, P.L., Daniel, R.M., Stackebrandt, E., Morgan, H.W., 1994. Description of *Caldicellulosiruptor saccharolyticus* gen. *nov.*, sp. nov: An obligately anaerobic, extremely thermophilic, cellulolytic bacterium. *FEMS Microbiol. Lett.* 120, 263–266.

Rastogi, M., Nandal, M., Khosla, B., 2020. Microbes as vital additives for solid waste composting. *J. Exp. Biol. Ecol.* 6, e03343.

Reiss, R., Ihssen, J., Thony-Meyer, L., 2011. *Bacillus pumilus* laccase: a heat stable enzyme with a wide substrate spectrum. *BMC Biotechnol.* 11, 9.

Royter, M., Schmidt, M., Elend, C., Hobenreich, H., Schafer, T., Bornscheuer, U.T., Antranikian, G., 2009. Thermostable lipases from the extreme thermophilic anaerobic bacteria *Thermoanaerobacter thermohydrosulfuricus* SOL1 and *Caldanaerobacter subterraneus* subsp. *tengcongensis. Extremophiles* 13, 769–783.

Rozich, A.F., Colvin, R.J., 1999. Design and operational considerations for thermophilic aerobic reactors treating high strength wastes and sludges. *Proceedings of the 52nd Industrial Waste Conference*, Purdue University, Ann Arbor Press, USA.

Rüttimann, C., Cotorás, M., Zaldívar, J., Vicuña, R., 1985. DNA polymerases from the extremely thermophilic bacterium *Thermus thermophilus* HB-8. *Eur. J. Biochem.* 149, 41–46.

Sakaguchi, M., Osaku, K., Maejima, S., Ohno, N., Sugahara, Y., Oyama, F., Kawakita, M., 2014. Highly conserved salt bridge stabilizes a proteinase K subfamily enzyme, Aqualysin I, from *Thermus aquaticus* YT-1. *AMB Exp.* 4, 59.

Satyanarayana, T., Raghukumar, C., Shivaji, S., 2005. Extremophilic microbes: Diversity and perspectives. *Curr. Sci.* 89, 78–90.

Schlacher, A., Holzmann, K., Hayn, M., Steiner, W., Schwab, H., 1996. Cloning and characterization of the gene for the thermostable xylanase XynA from *Thermomyces lanuginosus*. *J. Biotechnol.* 49, 211–218.

Shao, X., Zhou, J., Olson, D.G., Lynd, L.R., 2016. A markerless gene deletion and integration system for *Thermoanaerobacter ethanolicus*. *Biotechnol. Biofuels* 9, 100.

Shin, H., Youn, J., Kim, S., 2004. Hydrogen production from food waste in anaerobic mesophilic and thermophilic acidogenesis. *Int. J. Hydrogen Energ.* 29, 1355–1363.

Sudharhsan, S., Senthilkumar, S., Ranjith, K., 2007. Physical and nutritional factors affecting the production of amylase from species of *Bacillus* isolated from spoiled food waste. *Afr. J. Biotechnol.* 6, 430–435.

Silva, T.M., Attili-Angeli, D., Carvalho, A.F., Silva, D.R., Boscolo, M., Gomes, E., 2005. Production of saccharogenic and dextrinogenic amylases by *Rhizomucor pusillus* A 13.36. *J. Microbiol.* 43, 561–568.

Singh, D.P., Prabha, R., Renu, S., Sahu, P.K., Singh, V., 2019. Agrowaste bioconversion and microbial fortification have prospects for soil health, crop productivity, and eco-enterprising. *Int. J. Recycl. Org. Waste Agricult.* 8, 457–472.

Singh, R.K., Tiwari, M.K., Singh, R., Lee, J.K., 2013. From protein engineering to immobilization: promising strategies for the upgrade of industrial enzymes. *Int. J. Mol. Sci.* 14, 1232–1277.

Singh, Y., Sidhu, H.S., 2014. Management of cereal crop residues for sustainable rice-wheat production system in the Indo-Gangetic plains of India. *Proc. Natl. Acad. Sci. India Sect. A (Phys. Sci.)* 80, 95–114.

Sriyapai, P., Kawai, F., Siripoke, S., Chansiri, K., Sriyapai, T., 2015. Cloning, expression and characterization of a thermostable esterase HydS14 from *Actinomadura* sp. strain S14 in Pichia pastoris. *Int. J. Mol. Sci.* 16, 13579–13594.

Stamford, T.L.M., Stanford, N.P., Coelho, L.C.B.B., Araujo, J.M., 2001. Production and characterization of a thermostable α-amylase from *Nocardiopsis* sp. endophyte of yam bean. *Bioresour. Technol.* 76, 137–141.

Steiner, K., Schwab, H., 2012. Recent advances in rational approaches for enzyme engineering. *Comput. Struct. Biotechnol. J.* 2, e201209010.

Stetter, K.O., 1996. Hyperthermophilic prokaryotes. *FEMS Microbiol. Rev.* 18, 149–158.

Stetter, K.O., 1999. Extremophiles and their adaptation to hot environments. *FEBS Lett.* 452, 22–25.

Stover, E.L., Samuel, G.J., 1997. High-rate thermophilic pretreatment of high strength industrial wastewaters. *Proceedings of the 52nd Industrial Waste Conference*, Purdue University, Ann Arbor Press, Ann Arbor, MI, U.S.A.

Swamy, N.V., Seenayya, G., 1996. Thermostable pullulanase and α-amylase activity from *Clostridium thermosulfurogenes* SV9-Optimization of culture conditions for enzyme production. *Process Biochem.* 31, 157–162.

Taguchi, H., 2017. The simple and unique allosteric machinery of *Thermus caldophilus* lactate dehydrogenase: Structure-function relationship in bacterial allosteric LDHs. *Adv. Exp. Med. Biol.* 925, 117–145.

Takahata, Y., Nishijima, M., Hoaki, T., Maruyama, T., 2001. *Thermotoga petrophila* sp. *nov.* and *Thermotoga naphthophila* sp. *nov.*, two hyperthermophilic bacteria from the Kubiki oil reservoir in Niigata, Japan. *Int. J. Syst. Evol. Microbiol.* 51, 1901–1909.

Taylor, M.P., Van Zyl, L., Tuffin, I.M., Leak, D.J., Cowan, D.A., 2011. Genetic tool development underpins recent advances in thermophilic whole-cell biocatalysts. *Microb. Biotechnol.* 4, 438–448.

Terpe, K., 2013. Overview of thermostable DNA polymerases for classical PCR applications: from molecular and biochemical fundamentals to commercial systems. *Appl. Microbiol. Biotechnol* 97, 10243–10254.

Toogood, H.S., Smith, C.A., Baker, E.N., Daniel, R.M., 2000. Purification and characterization of Ak.1 protease, a thermostable subtilisin with a disulphide bond in the substrate-binding cleft. *Biochem. J.* 350, 321–328.

Toplak, A., Wu, B., Fusetti, F., Quaedflieg, P.J.L.M., Janssen, D.B., 2013. Proteolysin, a novel highly thermostable and cosolvent-compatible protease from the thermophilic bacterium *Coprothermobacter proteolyticus*. *Appl. Environ. Microbiol.* 79, 5625–5632.

Tripathi, N., Hills, C.D., Singh, R.S., Atkinson, C.J., 2019. Biomass waste utilisation in low-carbon products, harnessing a major potential resource. *Clim. Atmos. Sci.* 2, 1–10.

Van Fossen, A.L., Lewis, D.L., Nichols, J.D., Kelly, R.M., 2008. Polysaccharide degradation and synthesis by extremely thermophilic anaerobes. *Ann. N. Y. Acad. Sci.* 1125, 322–337.

Verbeke, T.J., Zhang, X., Henrissat, B., Spicer, V., Rydzak, T., Krokhin, O.V., Fristensky, B., Levin, D.B., Sparling, R., 2013. Genomic evaluation of *Thermoanaerobacter* spp. for the construction of designer co-cultures to improve lignocellulosic biofuel production. *PLoS One* 8, 1–18.

Verhaart, M.R.A., Bielen, A.A.M., van der Oost, J., Stams, A.J.M., Kengen, S.W.M., 2010. Hydrogen production by hyperthermophilic and extremely thermophilic bacteria and archaea: mechanisms for reductant disposal. *Environ. Technol.* 31, 993–1003.

Verma, A., Shirkot, P., 2014. Purification and characterization of thermostable laccase from thermophilic *Geobacillus thermocatenulatus* MS5 and its applications in removal of textile dyes. *J. Biosci.* 2, 479–485.

Vieille, C., Zeikus, G.J., 2001. Hyperthermophilic enzymes: Sources, uses, and molecular mechanisms for thermostability. *Microbiol. Mol. Biol. R.* 65, 1–43.

Voorhorst, W.G., Warner, A., de Vos, W.M., Siezen, R.J., 1997. Homology modelling of two subtilisin-like proteases from the hyperthermophilic archaea *Pyrococcus furiosus* and *Thermococcus stetteri*. *Protein Eng.* 10, 905–914.

Walia, A., Mehta, P., Guleria, S., Shirkot, C.K., 2015. Modification in the properties of paper by using cellulase-free xylanase produced from alkalophilic *Cellulosimicrobium cellulans* CKMX1 in bio-leaching of wheat straw pulp. *Can. J. Microbiol.* 61, 671–681.

Wang, Q., Wang, Q., Tong, W., Bai, X., Chen, Z., Zhao, J., Zhang, J., Liu, S., 2012. Regulation of enzyme activity of alcohol dehydrogenase through its interactions with pyruvate-ferredoxin oxidoreductase in *Thermoanaerobacter tengcongensis*. *Biochem. Biophys. Res. Commun.* 417, 1018–1023.

Wu, W.L., Chen, M.Y., Tu, I.F., Lin, Y.C., EswarKumar, N., Chen, M.Y., Ho, M.C., Wu, S.H., 2017. The discovery of novel heat-stable keratinases from *Meiothermus taiwanensis* WR-220 and other extremophiles. *Sci. Rep.* 7, 4658.

Yadav, P., Maharajan, J., Korpole, S., Prasad, G.S., Sahni, G., Bhattarai, T., Sreerama, L., 2018. Production, purification, and characterization of thermostable alkaline xylanase from *Anoxybacillus kamchatkensis* NASTPD13. *Front. Bioeng. Biotechnol.* 6, 65.

Yang, S.J., Kataeva, I., Wiegel, J., Yin, Y., Dam, P., Xu, Y., Westpheling, J., Adams, M., 2010. Classification of 'Anaerocellum thermophilum' strain DSM 6725 as Caldicellulosiruptor bescii sp. nov. *Int. J. Syst. Evol. Microbiol.* 60, 2011–2015.

Yang, W., He, Y., Xu, L., Zhang, H., Yan, Y., 2016. A new extracellular thermo-solvent-stable lipase from *Burkholderia ubonensis* SL-4: identification, characterization and application for biodiesel production. *J. Mol. Catal. B Enzym.* 126, 76–89.

Yi, H., Jeong, J., Park, Y., Deul, K., Ghim, S., 2006. Effect of thermophilic bacteria on degradation of food wastes. *Korean J. Microbiol. Biotechnol.* 34, 363–367.

Zamost, B.L., Nielsen, H.K., Starnes, R.L., 1991. Thermostable enzymes for industrial applications. *J. Ind. Microbiol.* 8, 71–82.

Zhang, M., Tashiro, Y., Ishida, N., Sakai, K., 2022. Application of autothermal thermophilic aerobic digestion as a sustainable recycling process of organic liquid waste: Recent advances and prospects. *Sci. Total Environ.* 828, 54187.

Zhang, T., Liu, H., Fang, H.H.P., 2003. Biohydrogen production from starch in wastewater under thermophilic conditions. *J. Environ. Manag.* 69, 149.

Zhou, Q., Ji, P., Zhang, J., Li, X., Han, C., 2017. Characterization of a novel thermostable GH45 endoglucanase from *Chaetomium thermophilum* and its biodegradation of pectin. *J. Biosci. Bioeng.* 124, 271–276.

10 Extremophiles

Biofactories for Bioremediation

Rachna Chaturvedi
Amity Institute of Biotechnology, Amity University Uttar Pradesh, Lucknow Campus, Lucknow, India

CONTENTS

10.1 INTRODUCTION

Extremophiles are ubiquitous living entities that thrive in extreme environments like high-level of salty conditions, presence of contaminants, high temperature, low temperature, high UV or on substrates having acidic or alkaline pH. These types of organisms have modified their way of life and all their cellular machinery is suitable for survival under extremely stressful environments. In this captivating sphere of different microbes, extremophiles are the greatest cryptic group of life on the planet Earth (Rothschild and Mancinelli 2001) and may be on different other planets as well (Navarro-Gonzalez et al. 2009).

Both prokaryotes and eukaryotes can stay in extreme environmental conditions. Extremophilic microorganisms are classified into different categories on the basis on their capability to adapt to extreme environmental conditions, for example, alkaliphiles, acidophiles, thermophile, endoliths, hyperthermophiles, piezophiles, oligotrophs, hypolith, metallotolerant, psychrophiles, toxin tolerant, radioresistant and xerophilies (Cowan et al. 2015) (Figure 10.1). Acidophiles survive at pH below 3.0, while alkaliphiles live in alkaline conditions (pH 9–11). Endoliths live inside the rocks, corals and animal shells. Hyperthermophiles develop at temperatures between 80 and 122°C. Hypolith lives underneath the rocks in cold deserts. Metallotolerant organisms are able to tolerate high levels of noxious metals, while oligotrophs grow in nutritionally deficient environments.

DOI: 10.1201/9781003324706-14

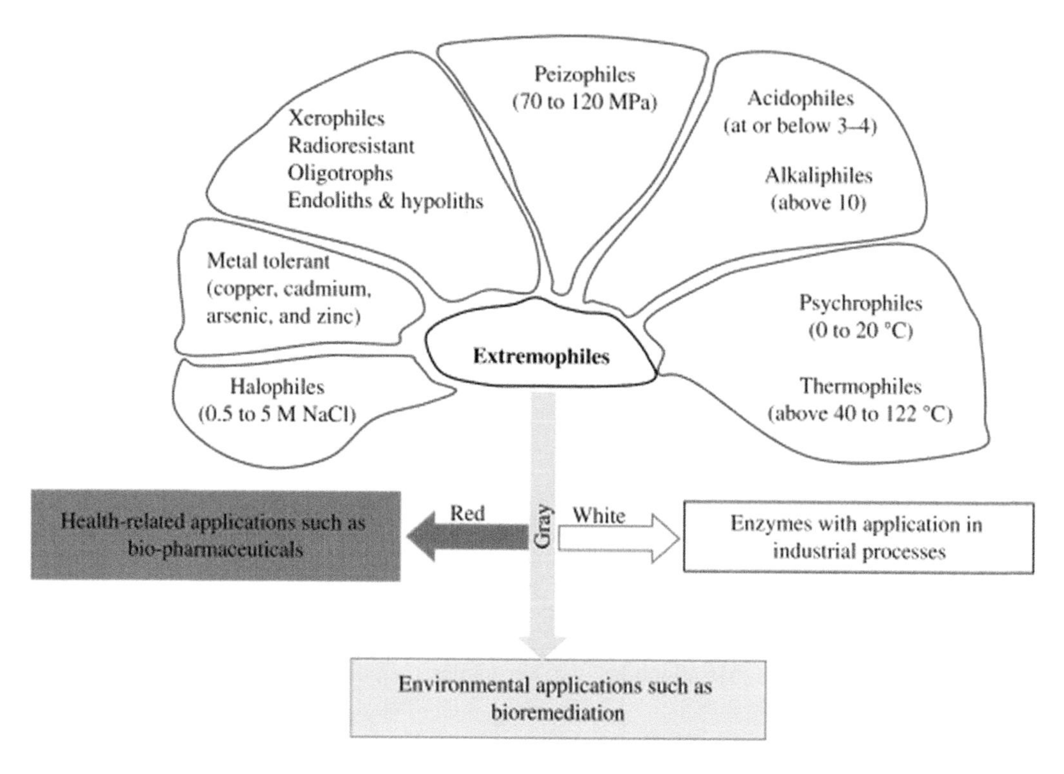

FIGURE 10.1 Extremophilic microorganisms and their applications in diverse fields (Reprinted with permission from Salwan and Sharma 2022).

Piezophiles flourish in high hydrostatic pressure. Psychrophiles grow at temperatures of around −10 to −20°C. The halophilic *Haloarchaea* and the bacteria *Deinococcus radiodurans* can tolerate high levels of ionizing radiation of ~3000 Gray and 1000 Gray, respectively, without loss of viability (Bruckbauer and Cox 2021).

Most of the extremophiles have the capability to synthesize exopolysaccharides that are high molecular weight carbohydrate biopolymers with diverse structures and functions (Figure 10.2). These exopolysaccharides are in great demand as natural polymers in the industrial sector,

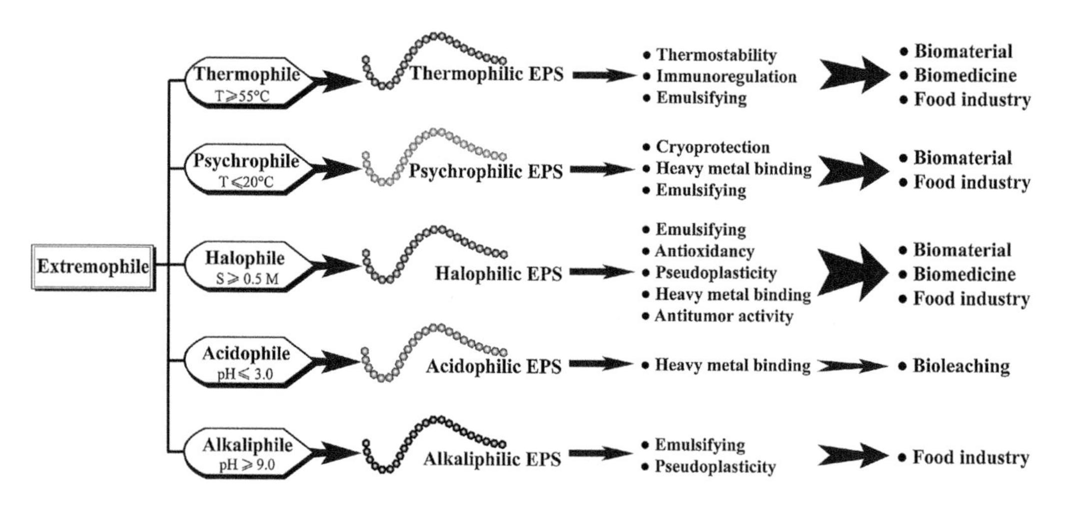

FIGURE 10.2 The exopolysaccharides from diverse extremophiles and potential applications (Reprinted with permission from Wang et al. 2019).

enzymes, such as primary and secondary metabolites which have applications in food, pharmaceutical, leather, textile, paper and pulp and agricultural sector.

Owing to indiscriminate industrial activity and other human activities, the air and community health sectors are susceptible to the enormous volume of lethal impurities that have collected in the atmosphere, on land and in water bodies. Hence, maintaining the ecological environment from noxious toxins has become a big task for human beings over the earlier decades. In recent times, a variety of methods have been intensively established for the safety of the environment for stopping the spread of toxic contaminants. Several approaches have been identified for the elimination of several noxious materials (Gebreeyessus et al. 2019; Ouyang et al. 2019; Xiang et al. 2019; Muddemann et al. 2019). However, these processes may have disadvantages such as excessive cost, development of lethal byproducts and less efficient removal effectiveness. Hence, there is an urgent requirement for alternative approaches to get rid of the contaminants from contaminated environments. Microbial bioremediation has caught the attention of environmentalists as a hopeful technology that can beat the weaknesses of the presently used physicochemical methods (Mishra 2017; Diep et al. 2018; Igiri 2018). Bioemulsification, bioreduction, bioprecipitation, biosorption and bioaccumulation are techniques wherein microbes use proteins to fascinate metal ions inside their intracellular spaces. Extremophilic microorganisms have recently been extensively researched for the management of diverse lethal contaminants (Azubuike et al. 2016; Waigi et al. 2017; Kiadehi et al. 2018) since these are not only useful in detoxifying lethal contaminants with the help of microbial cellular metabolism, but can also resist extremely harsh environments.

Extremophiles habitually live-in environments that have high concentrations of metals, in acidic environments where solubility of metal is quite higher than neutrophilic environments or are found in hot springs that contain metals from geological sources. Extremophiles acquire numerous adaptations that are familiar to all types of microbes to withstand pollutant toxicity, but a lot of them remain mysterious and to be detected. Apart from utilization of tremendous energy in countering stressful environments, the extremophiles have unique metabolic networks and regulatory circuits in order to survive and flourish in extreme environments (Salwan and Sharma 2022).

10.2 TYPES OF EXTREMOPHILES

Extremophiles are classified according to the environment in which they are found. Thus, we have a number of types that have been discussed below.

10.2.1 Acidophiles and Alkaliphiles

Acidophiles are the microbes that can live under exceptionally low pH (less than pH 3) conditions and protect themselves by managing proton leakage (Siliakus et al. 2017) Alkaliphiles are the microorganisms which exist in the alkaline environments such as alkaline soda lakes, alkaline hydrothermal vents, carbonate-rich soils and even hind gut of insects (Jones et al. 1998; Thongaram et al. 2003, 2005). Alkaliphiles are further subcategorized into the following two types (Guffanti et al. 1986; Preiss et al. 2015):

a. Obligate alkaliphies: These grow only at pH values of ~pH 9 and above.
b. Facultative alkaliphiles: These can grow optimally both in stringent alkaline conditions as well as near neutral pH.

One of the maximum extraordinary properties of such types of organisms is the use of proton pumps to give a neutral pH internally. For pH reworking, alkaliphiles and acidophiles make use of numerous approaches. Acidophiles have molecular mechanisms for familiarizing extreme condition as potassium antiporter releases protons concerning the extracellular medium, ATP synthase and chaperones. Likewise, alkaliphiles have electrochemical gradient of Na^+ and H^+ by antiporters for gathering of protons, Na^+-solute receiving system, cytochromes a-, b- and c-type, higher concentration of membrane lipids, and acidic polymers like galacturonic acid, glucuronic acid, glutamic acid, aspartic acid, phosphoric acid and teichuronopeptides in the cell wall (Borkar 2015).

10.2.2 Thermophiles and Psychrophiles

Thermophiles are heat-loving microbes which have optimal growth temperatures between 50 and 55°C (Singh et al. 2011). They can be usually categorized into mild thermophiles (optimum growth temperature: 50–60°C), intense thermophiles (optimum growth temperature: 60–80°C) and hyperthermophiles (optimum growth temperature: 80–110°C). Thermophiles have been isolated from the different environmental zones viz. hot springs as well as deep-sea zones of the earth. *Pyrolobus fumarii* and *Geogemma barossii* are two well-known hyperthermophilic archaea which have the ability to tolerate temperatures up to 121°C. Similarly, the hyperthermophilic archaeon *Methanopyrus kandleri* isolated from deep-sea hydrothermal field withstands temperature of 122°C which is the highest recorded for any living form.

The microorganisms which survive at temperatures of around −10 to −20°C are known as psychrophiles or psychrotolerant. These are the cold atmosphere-loving microbes and live in the low-temperature zones of the planet generally in the Arctic and Antarctic oceans. Their diverse habitats comprise polar regions, glaciers, ocean deeps, upper atmosphere, refrigerated appliances, shallow subterranean regions and inside plants and animals inhabiting cold areas. One example of psychrophile is *Psychrobacter cryopegellain* which was discovered in the Siberian permafrost and has the capability to survive in temperatures as low as −20°C.

10.2.3 Halophiles and Piezophiles

Halophiles can thrive in a high-salt atmosphere which hinders normal organism's existence due to osmolar imbalance and metabolic challenges (Gunde-Cimerman, et al. 2018). Halophiles are classified as extreme halophiles that show best growth in media containing 1.5–5.2 M, moderate halophiles grow best in media with 0.2–0.5 M and light halophiles grow best in media containing 0.2–0.5 M NaCl (Kumar and Khare 2012).

Microbes that choose always high-pressure requirements are called as piezophiles or barophiles inhabiting in oceans or deep-sea. Barophilic bacteria are those that display optimal growth at pressures >40 MPa, whereas barotolerant bacteria display optimal growth at pressure <40 MPa but can also grow at atmospheric pressure (Horikoshi 1998). Typical examples of barophiles include the thermophilic methanogens, *Methanocaldococcus jannaschii* and *Methanothermococcus thermolithotrophicus* which have been discovered in high-pressure niches of deep-sea beds. It is believed that enzymes obtained from piezophiles are steady at high-pressure condition and do not require pressure-associated modifications. The molecular basis of piezophiles is now being examined significantly. The emphasis has essentially been on the detection of pressure-controlled operons indicating the connection between pressure and microbial growth and purpose of specific proteins.

10.2.4 Radiophiles, Metallophiles and Xerophiles

Radiophiles are the microbes that are exceptionally resistant to high level of ionizing and non-ionizing radiations. Such type of radiation-resistant microbes showing elevated capability in the treatment of radioactive wastes of environment. Radiophiles are getting a lot of consideration in recent times, due to their ability to live in environments of oxidative stress, starvation and elevated amounts of DNA destruction. Earlier studies on how they can fit and live in elevated dose of radiation and oxidative stress requirements have shown that these microbes acquire robust DNA-repair structures and antioxidation methods to tolerate serious irradiation trauma (Liu et al. 2017; Srinivasan et al. 2017; Park et al. 2018; Tanner et al. 2020).

Metallophiles are the microorganisms that can survive in the habitat of high metal concentrations. Heavy metal pollution has become a serious problem to community healthiness, fishery and wildlife (Bhargava et al. 2012; Srivastava and Bhargava 2022). Hence it is need of the hour to recognize rising significance of metallophiles in bioremediation of the noxious heavy metals from soils, sediments, environments and wastewaters. To this end, a considerable variety of microorganism-metal connections are being manipulated, varying from decrease in anaerobic respiration to decrease in decontamination, as well as bioaccumulation, bioleaching, biosorption and biomineralization (Gadd 2010). Heavy metals usually exert a suppressing act on microbes by stopping vital operations, replacing important metal ions or altering the effective shape of biotic particles (Rawlings 2002). Though, specific microbes are capable to nullify such effects and showing endurance by either collecting the metals in the kind of specific protein-metal association or by heavy metal efflux methods (Naz et al. 2005).

The microbes that can survive in exceptionally dry environments or at extremely low water activity (a_w) levels are called xerophiles. Only a few specific genera amongst bacteria, yeasts, fungi, lichens and algae can live in such type of environments (Rothschild and Mancinelli 2001). Xerophiles are accountable for destroying of dry foods and stored grains, spices, nuts and oilseeds. A notable example is the food-spoilage fungi *Xeromycesbisporus*, which is capable of growth at a_w 0.61–0.62

10.3 BIOREMEDIATION

Bioremediation involves the use of a biological system, either living or dead, to decontaminate polluted sites. Several bioremediation methods are available that include bioaugmentation, bioattenuation, biostimulation, biopiling and bioventing. Bioaugmentation is the technique that incorporates the add-on of contaminant-demeaning microbes (natural/exotic/engineered) to complement the biodegradative ability of native populations of microorganisms on the polluted site. Since, normal types are not adequate for breaking down certain compounds, sometimes genetically modified or extremophilic microbes are utilized for bioremediation of soil and groundwater (Thapa et al. 2012). Bioattenuation or natural attenuation is the extermination of contaminants from the environments where the risk of exposure to contaminants is within acceptable limits. Bioattenuation involves procedures such as anaerobic and aerobic biodegradation, ion exchange, complexation, abiotic transformation and physical singularities such as dilution, volatilization, advection, dispersion and sorption (Mulligana and Yong 2004). Biostimulation involves stimulation of native occurring bacterial and fungal populations at the contaminated site via the infusion of specialized nutrients, supplements, electron acceptors, oxygen and other significant mixes. These supplements, which are the fundamental developing units of life, improve the (co) metabolic activities of the microflora and allows microorganism to increase in number and aid in the removal of contaminants (Madhavi and Mohini 2012; Muthukumaran 2022). Biopiling is an ex situ treatment technology that aims to reduce aerobically remediable petroleum pollutants using biological processes. In this process, air is supplied to the biopile structure all through a method of pumps and fluting to improve microbial action via microbial breathing (Emami et al. 2012). Bioventing is an aerobic bioremediation technique that involves venting of oxygen via a low-level air flow rate all through soil to simulate the growth of indigenous microorganisms, thus enhancing the rate of organic constituents (Agarry and Latinwo 2015).

Bioremediation is categorized into two types viz. *in situ* where the decontamination is carried out directly on the polluted sites and *ex situ* that refers to cleaning carried out on excavated materials (Table 10.1).

TABLE 10.1
Conventional Bioremediation Approaches for Taking Into Consideration Different Factors with Pros and Cons

Type	Technique	Factors	Pros	Cons
Ex situ	Landfilling	Biodegradative capabilities of native microbes	Low cost	Requirement of space
	Bio piles	Biodegradability and dissemination of contaminants	Cost-efficient	Necessary to control abiotic loss
	Composting	Chemical solubility	Optimized environ- mental parameters	Expanded treat- ment time
	Aqueous bioreactor	Ecological parameters	Rapid degradation kinetic	Mass transfer trouble
	Slurry bioreactors	Existence of metals and other inorganics	Enhance mass transfer	Bioavailability limitation
		Bioaugmentation	Effective use of inoculants and surfactants	Soil requires excavation
		Toxicity concentration of contaminants	-	High-cost capital
		Toxicity of amendments	-	High operating cost
In situ	*In situ* bioremediation	Biodegradative skills of indigenous microorganisms	Highly cost-efficient	Environmental limitations
	Bioventing	Biodegradability and dissemination of pollutants	Comparatively passive	Expanded treatment time
	Bioaugmentation	Chemical solubility	Pure attenuation processes	Monitoring problems
	Biosparging	Environmental parameters	Soil and water treatment	
		Occurrence of metals and other inorganics	Noninvasive	

10.3.1 *In situ* Bioremediation

In situ methods are generally considered more efficient due to their cost and use of natural microbes for degradation of contaminants. These methods are broadly classified into intrinsic and engineered types. The former utilizes the inbuilt capacity of naturally occurring microorganism for reducing contaminants while the latter envisages boosting the growth of existing microbes as well as introduction of specific microbial species in the polluted site to accelerate degradation process. The engineered approach is generally costly but yields comprehensive transformation of pollutants to harmless constituents such as CO_2 and water.

10.3.2 *Ex-situ* Bioremediation

In the *ex situ* form of bioremediation, contaminated soil is excavated, placed on a coated aboveground area and aerated to improve the degradation of pollutants by encouraging growth of indigenous microbial populations. The *ex situ* method is appropriate for a wide range of pollutants but shows less efficacy against heavy metal pollutants and some chlorinated hydrocarbons. This method is additionally classified into two types viz. slurry-phase and solid-phase bioremediation. Slurry-phase bioremediation envisages excavation of polluted soil, mixing it with water to form a slurry and further treatment in a bioreactor (Woodhull and Jerger 1994; Grosse et al. 2000). Slurry-phase method is quite attractive as compared to the solid-state method due to requirement of a small area, efficient control over operating conditions and higher treatment efficiency which leads to enhanced degradation. Solid-phase type of bioremediation is a technology in which the contaminated soil is unearthed and laid into piles. Bacterial growth is accelerated all the way through a system of pipes that are spread out across the piles. Required aeration is offered for microbial respiration all through the pipes by dragging air. This method needs huge amount of space, and additional time for cleaning to achieve the desired goal. Some solid-phase treatment processes include biopiling, composting and land farming.

10.4 EXTREMOPHILES IN BIOREMEDIATION

The capacity of extremophilic microbes to survive in a extensive range of environmental conditions establishes them as exceptional entities for cleanup of polluted soil and water (Figure 10.3). Bioremediation of organic pollutants and the immobilization, mobilization and/or alteration of

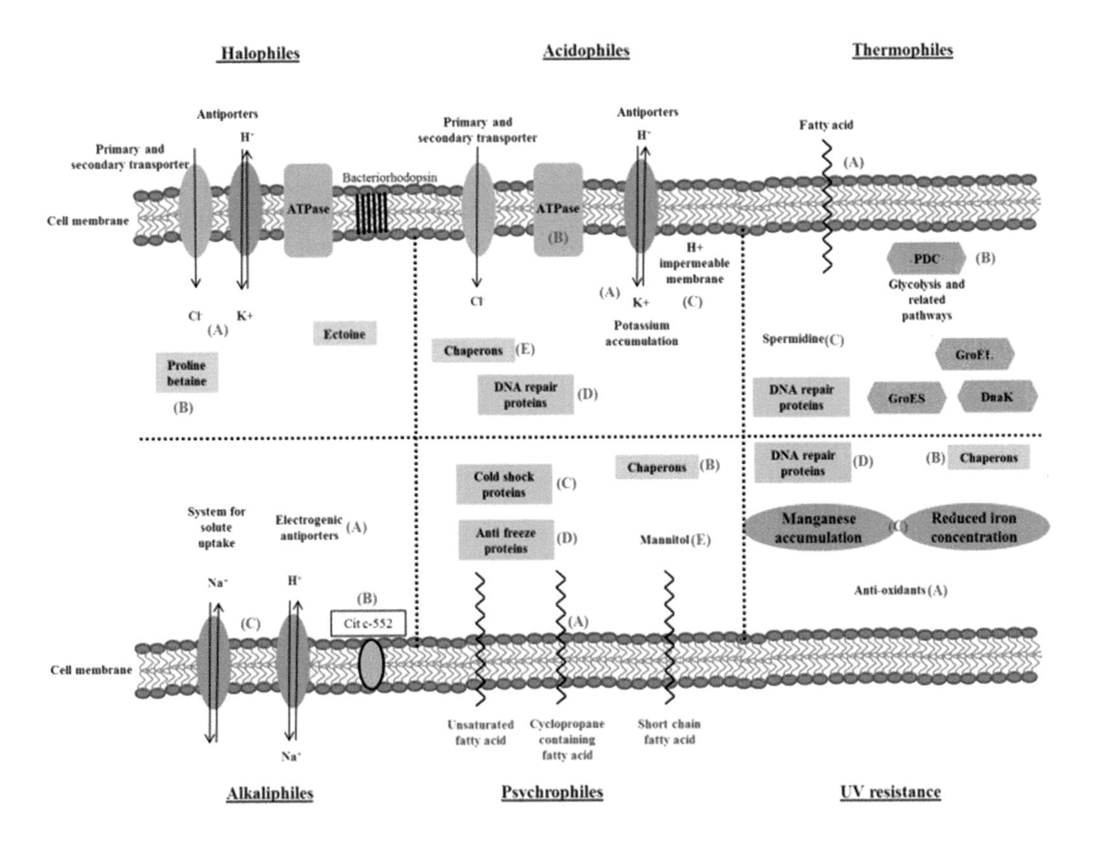

FIGURE 10.3 Molecular mechanisms of extremophiles that are useful for bioremediation of polluted soils and waters: [I] *Halophiles*: (A) High salt-in mechanism (i) potassium uptake into cells by concerted action of bacteriorhodopsin and ATP synthase, (ii) chloride transporters and (B) low-salt strategy (i) synthesis or uptake of osmoprotectants (proline-betaine, ectoine) to maintain osmotic pressure; [II] *Acidophiles*: (A) potassium antiporter, (B) ATP synthase, (C) proton impermeable membrane, (D) DNA-repair proteins, and (E) chaperones; [III] *Thermophiles*: (A) long chain dicarboxylic fatty acids, (B) upregulated glycolysis proteins and (C) polyamines (spermidine); [IV] *Alkaliphiles*: (A) electrogenic antiporters for proton accumulation, (B) Cytochrome *c*-552 enhanced terminal oxidation function by electron and H^+ accumulation and (C) Na^+-solute uptake system; [V] *Psychrophiles*: (A) cyclopropane containing fatty acids and short chain fatty acids, (B) chaperones, (C) cold shock proteins, (D) antifreeze proteins and (E) manniol or other solutes in cytoplasm as cryo-protectants to prevent protein aggregation; and [VI] *UV resistance*: (A) antioxidants, (B) chaperones, (C) manganese accumulation and reduced iron levels and (D) DNA-repair proteins (Reprinted with permission from Koul et al. 2021).

metal(loid)s are the chief remediation actions that can be enabled by the act of numerous microbes particularly those extremophiles existing in aggressive atmospheres with elevated concentrations of pollutants (Donati et al. 2019). Removal and sanitization of these pollutants and wastes can be attained with the assistance of extremozymes which have excellent properties, for example, thermostability and ability to denaturing some of the agents like detergents, organic solvents and extreme pH. Haloarchaea have been efficiently confirmed for biotechnological purposes (Arora et al. 2014). In recent times, extremophiles have been developed as microfactories/biofactories which are capable of providing metabolic or genetic alterations leading to more efficient cleanup of environmental pollution (Arora et al. 2014; Marques 2018).

The extremophilic microbes have firmly established themselves as suitable candidates for bioremediation purposes. There have been increased efforts in the optimization of bioremediation methodologies in high-salt environments, high temperature and extreme pH ranges since a large amount of waste products and contaminants generated from numerous industrial actions, mining events for oil spills warrant urgent attention (Sivaperumal et al. 2017).

10.4.1 Petroleum Bioremediation and Extremophiles

Hydrocarbon pollutants comprise of petroleum and its derivatives, polycyclic aromatic hydrocarbons (PAHs) and halogenated hydrocarbon compounds that are released into the environment due to human activities (Giovanlla et al. 2020). These are found in a wide range of habitats and negatively affect the health of organisms along with creating an imbalance in the ecosystem (Arulazhagan et al. 2017). Degradation is more difficult in compounds like the PAHs that have complex chemical structures (Fathepure 2014). Bioremediation of severely polluted sites involves the use of extremophiles that are most suited to these environments. The extremophilic microorganisms are capable of performing significant remediation of the surroundings and have been developed for microbial degradation of crude oil and other types of refined petroleum pollutants (Khemili-Talbi et al. 2015). Extremophilic microbes from Archaea domain as well from extreme conditions have been found as probable agents for the bioremediation of hydrocarbons (Giovanlla et al. 2020). Hydrocarbons can be mineralized or converted via the biodegradation process that happens in a variety of extreme habitats (Park and Park 2018). An extremophilic microbe *Stenotrophomonas maltophilia* strain AJH1 isolated from a mineral mining location in Saudi Arabia was found to be highly efficient in degrading both high and low molecular weight contaminants at extremely low pH (Arulazhagan et al. 2017). The archaea named as *Natrialba* sp. C21 identified from oil-polluted saline water in AinSalah (Algeria) was competent to survive in high-salt concentrations of up to 25% sodium chloride-containing aromatic hydrocarbons (Khemili-Talbi et al. 2015). This strain had a great capacity for degrading naphthalene (3% v/v) and pyrene (3% v/v) after 7 days at 40°C, pH 7.0 and high salinity situations. Hassan and Aly (2018) evaluated three haloalkaliphilic *Pseudomonas* strains (HA10, HA12 and HA14) for biodegradation of BTEX (benzene, toluene, ethylbenzene and xylene) and obtained encouraging results at pH 9 in the presence of NaCl (7% w/v). *Haloarchaea* 1M1011 isolated from Changlu Tanggu saltern (China) was capable of degrading 57% of hexadecane in the presence of NaCl in 24 days at 37°C (Zhao et al. 2017).

10.4.2 Pesticide Bioremediation and Extremophiles

Pesticides are mixtures that are widely used to control pests most of which affect the agricultural fields. The best-known pesticides belonging the organophosphorus group have firmly established themselves for agricultural use since the 1950s, but are extremely noxious compounds. Although pesticides play a significant role in the protection of crops, their disproportionate and persistence use has resulted in severe soil contamination and depreciated soil quality. Various methodologies have been followed for decontamination and remediation of these chemicals, but most are unsuited with respect to ecologically approachable remediation and severe situations (Jacquet et al. 2016).

Biodegradation of chlorpyrifos was successfully accomplished by Verma et al. (2020) using *Sphingobacterium* sp. C1B (Family: Sphingobacteriaceae), a Gram-negative, non-lactose-fermentative psychrotolerant bacterium strain isolated from apple orchard in Himachal Pradesh of India. The pesticide was degraded ≥42 ppm and ≥36 ppm within 14 days at 20°C and 15°C, respectively, which pointed toward the efficiency and possible use of the bacterium in cold environment. Verma et al. (2021) carried out biodegradation of malathion, an organophosphorus pesticide using the psychrotolerant bacteria *Ochrobactrum* sp. M1D isolated from a soil sample of peach orchards. The microbe utilized malathion as the sole source of carbon and energy and had the capability of degrading 100% of 100 mg l^{-1} malathion at 20°C, pH 7.0 within a period of 12 days.

10.4.3 Heavy Metal Bioremediation and Extremophiles

The accumulation of heavy metals and metalloids in different systems has triggered severe health worries globally since the heavy metals cannot be reduced into non-toxic types and persist in the ecosystems for a long period (Ayangbenro and Babalola 2017). Extremophiles have been depicted as excellent sources for bioremediation of heavy metals (Figure 10.4). Extremophilic microbes such as *Metallosphaera sedula* and *Leptospirillum ferriphilum* (Mi et al. 2011), along with *Sulfolobus solfataricus* (Schelert et al. 2013) contain gene merA that confers resistance to mercury. *Acidithiobacillus ferrooxidans* SUG 2–2 has been reported to volatilize mercury from contaminated acidic soils. Extremophiles have been focused upon for heavy metal and radionuclide contamination (Figueroa et al. 2018). Several halophilic archaea have established their potential in bioremediation of heavy metals since they frequently absorb heavy metals (Zhuang et al. 2010). *Halobacterium sp.* NRC-1 has exhibited high-level endurance to arsenic. Existence of genes has been reported for antimonite and arsenite specific structure on plasmids (Wang et al. 2012). Haloarchaeal approaches of change to high metal concentration of heavy metals such as copper, zinc, manganese iron, nickel cobalt, using *Halobacterium sp.* NRC-1 as a universal microbe has been extensively researched. Intracellular synthesis of silver nanoparticles has been accomplished using the haloarchaeal isolate *Halococcus salifodinae* BK3 as soon as the microbial cells were placed in the medium comprising silver nitrate (Srivastava et al. 2013). Likewise, selenium nanoparticles are also created in the presence of sodium selenite. Tolerance to 0.5, 1, 2 and 4 mM cadmium concentrations has been observed in haloarchaeal strains from salterns in India (Chaudhary et al. 2014). Biosorption of heavy metals by the microbe at the periphery or by the exopolysaccharides facilitates microorganisms to survive in the presence of toxic heavy metals (Srivastava and Kowshik 2013). *Halobacterium sp.* GUSF was described to be capable to take up Mn at high-level intensity as well as on high levels (Naik and Furtado 2014). *Halobacterium noricense* has been reported to adsorb cadmium (Showalter et al. 2016), while *Haloferax* St. BBK2 accumulates Cd intracellularly (Das et al. 2014). Therefore, members of Haloarchaea can serve as excellent candidates for remediation of hypersaline heavy metals contaminated areas.

10.4.4 Radionuclides Bioremediation and Extremophiles

The widespread use of radioactive substances at different research labs, industrial spots and biomedical organizations has generated a huge collection of radioactive litter. Radionuclide exposure or radiation triggers severe wellbeing impacts with different apparent symptoms (Kamiya et al. 2015). Since the physiochemical approaches are quite costly, bioremediation has been taken into consideration as the environmentally friendly accountable substitute for removal of these ecologically harmful substances. Extremophilic microbes have been utilized to remediate environmental pollutants as radionuclides. *Rhodanobacter sp.* and *Desulfuromusa ferrireducens* have the ability to interact with these pollutants which start solubility of transmuted radionuclides by accumulation or elimination of electrons, and permits their removal from the atmosphere (Green et al. 2012).

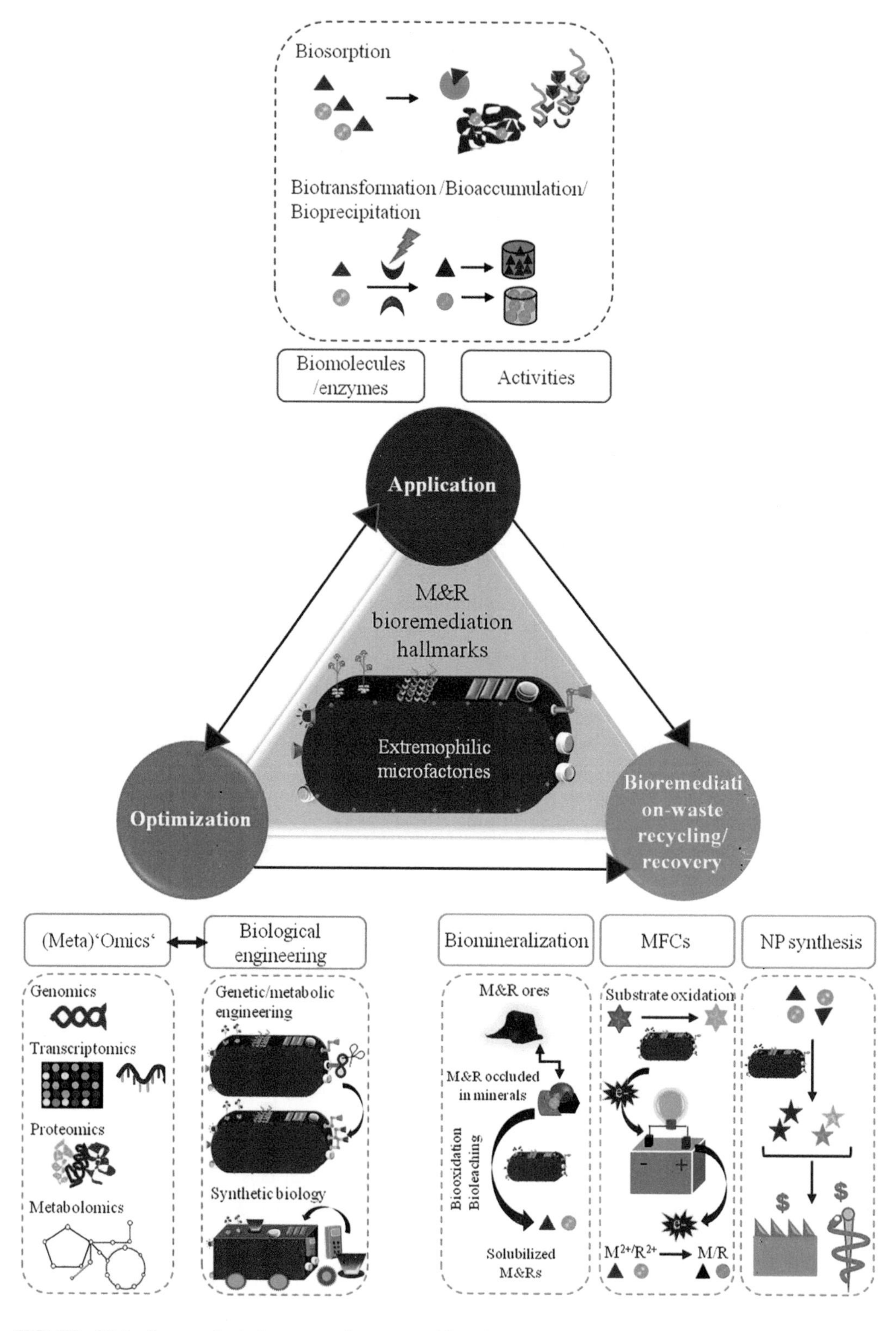

FIGURE 10.4 Image depicting use of extremophiles in bioremediation of metals and radionuclides (Reprinted from Marques 2018).

Biosorption requires the appropriation of positively charged metal ions to the negatively charged cell membranes and polysaccharides produced on the surface of bacteria because of capsule and slime formation (Prakash et al. 2013). Biosorption on its own might not be adequate to eliminate radionuclides without an increase in the ground biomass. Biostimulation, utilizing populations of microbes, is also known to enhance the bioremediation of radionuclides. Research on molecular techniques with the extremophilic microbial alteration of radionuclides and manipulating them in the process of bioremediation might be helpful for tracking the accountable microbial metabolic products for cell-free bioremediation and additional contribution in effective elimination of radionuclides from the polluted atmospheres.

10.4.5 Wastewater Bioremediation and Extremophiles

The principal purpose of wastewater treatment is to decrease the pollutants to the concentration at which the effluents are not harmfully affecting the biosphere and not causing a threat for health. The extremophilic microbes which are capable of degrading ammonia are currently one of the best candidates for wastewater treatment along with other microbes. Since the industrial discharges are rich in salts along with organic mixtures and heavy metals, polyextremophilic microbes highly resistant to metals, complex dyes as well as high-level salt concentration can be utilized for wastewater treatment. Such type of polyextremophilic microorganisms can be detected and isolated from industrial discharge or waste locations. In the current scenario, haloarchaea have been effectively evaluated for bioremediation and different biotechnological treatments (Arora et al. 2012; Nájera-Fernández et al. 2012). Most of the species from Haloferacease and Halobacteriaceae families are able to flourish in anaerobic environments in different conditions of salt concentrations (Torregrosa-Crespo et al. 2016). Therefore, these microbes may be utilized for bioremediation process in saline and hypersaline wastewater treatments since they have high tolerance to salt, metals and organic pollutants (Bonete et al. 2015). The production of pesticides, herbicides and explosive normally produce discharges including complex mixtures of salts and nitrate or nitrite leading to development of tolerance to extremely high-level nitrate and nitrite concentrations in some species of *Haloferax*. Consequently, it might be beneficial for bioremediation purposes in sewage plants with high salts, nitrate and nitrite concentrations. Halophilic extremophile *Haloferax mediterranei* is capable to bring out denitrification process, therefore offering outstanding types to discover significant bioremediation methods to eliminate nitrogen mixtures from sea waters and salty waters (Torregrosa-Crespo et al. 2019).

10.5 CONCLUSION

Bioremediation provides a technique for eliminating pollutants form natural environment and is one of the extremely desirable options to deal with contaminated environments. Extremophiles have enormous uses in bioremediation and biodegradation of numerous toxic compounds, pesticides and many other types of wastes. However, usage of extremophiles in the process of bioremediation is still not clear due to a lack in information about the genome-level structures, their metabolic pathways and kinetics. Hence, evolving knowledge about discovery of microbial microenvironments and a thorough knowledge of the metabolic pathways under various climatic and extreme environments is required to be further explored. Consequently, additional investigations from molecular biology and biochemical perspectives are needed to appropriately understand the metabolism regulation in extremophiles. Novel niches and extreme microecosystems in terms of pH, salt concentration and temperature should be discovered to recognize and separate extremophilic microbes for effective use in remediation of pollutants such as heavy metals, hydrocarbons and chlorinated compounds. In the coming years, it is anticipated that metagenomic studies along with genome sequencing will pave the way for the detection of new-found extremozymes from extremophilic microbes for bioremediation. Utilizing high-throughput sequencing techniques with metaproteomic and metabolomics studies will permit the detection of genes and metabolites accountable for the creation of biomolecules to be utilized in bioremediation. These multiomics tools are additionally stuffing gaps in the understanding of gene representation, metabolism and ecology of extremophilic microbes which might allow the developments in understanding associated to their use in the area of bioremediation.

REFERENCES

Agarry, S., Latinwo, G.K., 2015. Biodegradation of diesel oil in soil and its enhancement by application of bioventing and amendment with brewery waste effluents as bio stimulation-bioaugmentation agents. *J. Ecol. Eng.* 13, 82–91.

Arora, S., Trivedi, R., Rao, G.G., 2012. Bioremediation of coastal and inland salt affected soils using halophilic soil microbes. *Salinity News* 18, 3.

Arora, S., Vanza, M., Mehta, R., Bhuva, C., Patel, P., 2014. Halophilic microbes for bio-remediation of salt affected soils. *Afr. J. Microbiol. Res.* 8, 3070–3078.

Arulazhagan, P., Al-Shekri, K., Huda, Q., Godon, J.J., Basahi, J.M., Jeyakumar, D., 2017. Biodegradation of polycyclic aromatic hydrocarbons by an acidophilic *Stenotrophomonas maltophilia* strain AJH1 isolated from a mineral mining site in Saudi Arabia. *Extremophiles* 21, 163–174.

Ayangbenro, A.S., Babalola, O.O., 2017. A new strategy for heavy metal polluted environments: A review of microbial biosorbents. *Int. J. Environ. Res. Public Health* 14, 1–14.

Azubuike, C.C., Chikere, C.B., Okpokwasili, G.C., 2016. Bioremediation techniques-classification based on site of application: Principles, advantages, limitations and prospects. *World J. Microbiol. Biotechnol.* 32, 180.

Bhargava, A., Carmona, F.F., Bhargava, M., Srivastava, S., 2012. Approaches for enhanced phytoextraction of heavy metals. *J. Environ. Manag.* 105, 103–120.

Bonete, M., Bautista, V., Esclapez, J., Garcia-Bonete, M., Pire, C., Camacho, M., 2015. New uses of haloarchaeal species in bioremediation processes. In: Shiomi, M., (Ed.) *Advances in Bioremediation of Wastewater and Polluted Soil.* InTech, Shanghai. pp. 23–49.

Borkar, S., 2015. Alkaliphilic bacteria: Diversity, physiology and industrial applications. In: Borkar, S. (Ed.) *Bioprospects of Coastal Eubacteria.* Springer, Cham. pp. 59–83.

Bruckbauer, S.T., Cox, M.M., 2021. Experimental evolution of extremophile resistance to ionizing radiation. *Trends Genet.* 37, 830–845.

Chaudhary, A., Pasha, M.I., Salgaonkar, B.B., Braganca, J.M., 2014. Cadmium tolerance by Haloarchaeal strains isolated from solar salterns of Goa, India. *Intl. J. Biosci. Biochem. Bioinform.* 4, 1–6.

Cowan, D.A., Ramond, J.B., Makhalanyane, T.P., De Maayer, P., 2015. Metagenomics of extreme environments. *Curr. Opin. Microbiol.* 25, 97.

Das, D., Salgaonkar, B.B., Mani, K., Braganca, J.M., 2014. Cadmium resistance in extremely halophilic archaeon *Haloferax* strain BBK2. *Chemosphere* 112, 385–392.

Diep, P., Mahadevan, R., Yakunin, A.F., 2018. Heavy metal removal by bioaccumulation using genetically engineered microorganisms. *Front. Bioeng. Biotechnol.* 6, 157.

Donati, E.R., Sani, R.K., Goh, K.M., Chan, K.G., 2019. Recent advances in bioremediation/ biodegradation by extreme microorganisms. *Front. Microbiol.* 10, 1851.

Emami, S., Pourbabaee, A.A., Alikhani, H.A., 2012. Bioremediation principles and techniques on petroleum hydrocarbon contaminated soil. *Tech. J. Eng. Appl. Sci.* 2, 320–323.

Fathepure, B.Z., 2014 Recent studies in microbial degradation of petroleum hydrocarbons in hypersaline environments. *Front. Microbiol.* 5, 1–16.

Figueroa, M., Fernandez, V., Arenas-Salinas, M., Ahumada, D., Muñoz-Villagran, C., Cornejo, F., 2018. Synthesis and antibacterial activity of metal(loid) nanostructures by environmental multi-metal(loid) resistant bacteria and metal(loid)-reducing flavoproteins. *Front. Microbiol.* 9, 959.

Gadd, G.M., 2010. Metals, minerals, and microbes: Geomicrobiology and bioremediation. *Microbiology* 156, 609–643.

Gebreeyessus, G.D., 2019. Status of hybrid membrane-ion-exchange systems for desalination: A comprehensive review. *Appl. Water. Sci.* 9, 135.

Giovanlla, P., Gabriela, A.L.V., Igor, V.R.O., Elisa, P.P., Bruno de Jesus, F., Lara, D.S., 2020. Metal and organic pollutants bioremediation by extremophile microorganisms. *J. Hazard. Mater.* 382, 121024.

Green, S.J., Prakash, O., Jasrotia, P., Overholt, W.A., Cardenas, E., Hubbard, D., 2012. Denitrifying bacteria from the genus *Rhodanobacter* dominate bacterial communities in the highly contaminated subsurface of a nuclear legacy waste site. *Appl. Environ. Microbiol.* 78, 1039–1047.

Grosse, D.W., Sahle-Demessie, E., Bates, E.R. 2000. Bioremediation treatability studies of contaminated soils at wood preserving sites. *Remed. J.* 10, 67–84.

Guffanti, A.A., Finkelthal, O., Hicks, D.B., Falk, L., Sidhu, A., Garro, A., Krulwich T.A., 1986. Isolation and characterization of new facultatively alkalophilic strains of *Bacillus* species. *J. Bacteriol.* 167, 766–773.

Gunde-Cimerman, N., Plemenitaš, A., Oren, A., 2018. Strategies of adaptation of microorganisms of the three domains of life to high salt concentrations. *FEMS Microbiol. Rev.* 42, 353–375.

Hassan, H.A., Aly, A.A., 2018. Isolation and characterization of three novel catechol 2,3-dioxygenase from three novel haloalkaliphilic BTEX-degrading Pseudomonas strains. *Int. J. Biol. Macromol.*, 106, 1107–1114.

Horikoshi, K., 1998. Barophiles: deep-sea microorganisms adapted to an extreme environment. *Curr. Opin. Microbiol.* 1, 291–295.

Igiri, B.E., Okoduwa, S.I.R., Idoko, G.O., Akabuogu, E.P., Adeyi, A.O., Ejiogu, I.K., 2018. Toxicity and bioremediation of heavy metals contaminated ecosystem from Tannery wastewater: A review. *J. Toxicol.* e2568038.

Jacquet, P., Daude, D., Bzderenga, J., Masson, P., Elias, M., Chabriere, E., 2016. Current and emerging strategies for organophosphate decontamination: Special focus on hyper stable enzymes. *Environ. Sci. Pollut. Res.* 23, 8200–8218.

Jones, B.E., Grant, W.D., Duckworth, A.W., Owenson, G.G., 1998. Microbial diversity of soda lakes. *Extremophiles* 2, 191–200.

Kamiya, K., Ozasa, K., Akiba, S., Niwa, O., Kodama, K., Takamura, N., Zaharieva, E.K., Kimura, Y., Wakeford, R., 2015. Long-term effects of radiation exposure on health. *Lancet* 386, 469–478.

Khemili-Talbi, S., Kebbouche-Gana, S., Akmoussi-Toumi, S., Angar, Y., Gana, M.L., 2015. Isolation of an extremely halophilic archaeon *Natrialba* sp. C21 able to degrade aromatic compounds and to produce stable biosurfactant at high salinity. *Extremophiles* 19, 1109–1120.

Kiadehi, M.S.H., Amoozegar, M.A., Asad, S., Siroosi M., 2018. Exploring the potential of halophilic archaea for the decolorization of azo dyes. *Water Sci. Technol.* 77, 1602–1611.

Koul, B., Chaudhary, R., Taak P., 2021. Extremophilic microbes and their application in bioremediation of environmental contaminants. In: Kumar, A., Singh, V.K., Singh, P., Mishra, V.K. (Eds.) *Microbe Mediated Remediation of Environmental Contaminants*. Woodhead Publishing, Elsevier. pp. 115–128.

Kumar, S., Khare. S.K., 2012. Purification and characterization of malt oligosaccharide-forming á-amylase from moderately halophilic *Marinobacter* sp. EMB8. *Biores. Technol.* 116, 247–251.

Liu, Z., Kim, M.C., Wang, L., Zhu, G., Zhang, Y., Huang, Y., Wei, Z., Danzeng, W., Peng, F., 2017. *Deinococcus taklimakanensis* sp. *nov*. isolated from desert soil. *Int. J. Syst. Evol.* 67, 4311–4316.

Madhavi, G.N., Mohini, D.D., 2012. Review paper on parameters affecting bioremediation. *Int. J. Life Sci. Pharma Res.* 2, 77–80.

Marques, C.R., 2018 Extremophilic micro factories: Applications in metal and radionuclide bioremediation. *Front. Microbiol.* 9, 1191.

Mi, S., Song, J., Lin, J., Che, Y., Zheng, H., Lin, J., 2011. Complete genome of *Leptospirillum ferriphilum* ML-04 provides insight into its physiology and environmental adaptation. *J. Microbiol.* 49, 890–901.

Mishra, G.K., 2017. Microbes in heavy metal bioremediation: A review on current trends and patents. *Recent Pat. Biotechnol.* 11, 188–196.

Muddemann, T., Haupt, D., Sievers, M., Kunz, U., 2019. Electrochemical reactors for wastewater treatment. *ChemBioEng. Rev.* 6, 142–156.

Mulligana, C.N., Yong, R.N., 2004. Natural attenuation of contaminated soils. *Environ. Int.* 30, 587–601.

Muthukumaran, M., 2022. Advances in bioremediation of nonaqueous phase liquid pollution in soil and water. In: Kumar, S., Hashmi, M.Z., (Eds.) *Biological Approaches to Controlling Pollutants*. Elsevier. pp. 191–231.

Naik, S., Furtado, I., 2014. Equilibrium and kinetics of adsorption of Mn^{+2} by Haloarchaeon *Halobacterium* sp. GUSF (MTCC3265). *Geomicrobiol. J.* 31, 708–715.

Nájera-Fernández, C., Zafrilla, B., Bonete, M.J., Martínez-Espinosa, R.M., 2012. Role of the denitrifying Haloarchaea in the treatment of nitrite-brines. *Intern. Microbiol.* 15, 111–119.

Navarro-Gonzalez, R.I., Niguez, E., de la Rosa, J., McKay, C.P., 2009. Characterization of organics, microorganisms, desert soil, and Mars-like soils by thermal volatilization coupled to mass spectrometry and their implications for the search for organics on Mars by Phoenix and future space missions. *Astrobiology* 9, 703–715.

Naz, N., Young, H.K., Ahmed N., Gadd G.M., 2005. Cadmium accumulation and DNA homology with metal resistance genes in sulfate-reducing bacteria. *Appl. Environ. Microbiol.* 71, 4610–4618.

Ouyang, W., Chen, T., Shi, Y., Tong, L., Chen, Y., Wang, W., Yang, J., Xue, J., 2019. Physico-chemical process. *Water Environ. Res.* 91, 1350–1377.

Park, C., Park, W., 2018. Survival and energy producing strategies of alkane degraders under extreme conditions and their biotechnological potential. *Front. Microbiol.* 9, 1–15.

Park, M.R., Song, J.H., Nam, G.G., Joung, Y.C., Zhao, L., Kim, M.K., Cho, J.C., 2018. *Deinococcus lacus* sp. *nov.*, a gamma radiation-resistant bacterium isolated from an artificial freshwater pond. *Int. J. Syst. Evol.*, 68, 1372–1377.

Prakash, D., Gabani, P., Chandel, A.K., Ronen, Z., Singh, O.V., 2013. Bioremediation: A genuine technology to remediate radionuclides from the environment. *Microbiol. Biotechnol.* 6, 349–360.

Preiss, L., Hicks, D.B., Suzuki, S., Meier, T., Krulwich, T.A., 2015. Alkaliphilic bacteria with impact on industrial applications, concepts of early life forms, and bioenergetics of ATP synthesis. *Front. Bioeng. Biotechnol.* 3, 75.

Rawlings, D.E., 2002. Heavy metal mining using microbes. *Ann. Rev. Microbiol.* 56, 65–91.

Rothschild, L.J., Mancinelli, R.L., 2001. Life in extreme environments. *Nature* 409, 1092–10101.

Salwan, R., Sharma V., 2022. Genomics of prokaryotic extremophiles to unfold the mystery of survival in extreme environments. *Microbiol. Res.* 264, 127–156.

Schelert, J., Rudrappa, D., Johnson, T., Blum, P., 2013 Role of MerH in mercury resistance in the archaeon *Sulfobolus solfataricus*. *Microbiology* 159, 1198–1208.

Showalter, A.R., Szymanowski, J.E.S., Fein, J.B., Bunker, B.A., 2016. An x-ray absorption spectroscopy study of Cd binding onto a halophilic archaeon. *J. Phys. Conf. Ser.* 712, 012079.

Siliakus, M.F., Van der Oost, J., Kengen, S.W.M., 2017. Adaptations of archaeal and bacterial membranes to variations in temperature, pH and pressure. *Extremophiles* 21, 651–670.

Singh, S.K., Singh, S.K., Tripathi, V.R., Khare, S.K., Garg, S.K., 2011. A novel sychrotrophic, solvent tolerant *Pseudomonas putida* SKG-1 and solvent stability of its psychro-thermoalkalistable protease. *Process Biochem.* 46, 1430–1435.

Sivaperumal, P., Kamala, K., Rajaram, R., 2017. Bioremediation of industrial waste through enzymes producing marine microorganisms. *Adv. Food Nutr. Res.* 80, 165–179.

Srinivasan, S., Lim, S.Y., Lim, J.-H., Jung, H.Y., Kim, M.K., 2017. *Deinococcus rubrus* sp. *nov.*, a bacterium isolated from Antarctic coastal sea water. *J. Microbiol. Biotechnol.* 27, 535–541.

Srivastava, P., Kowshik, M. 2013. Mechanisms of metal resistance and homeostasis in haloarchaea. *Archaea* 1, 1–16.

Srivastava, S., Bhargava, A., 2022. Heavy metal remediation- the microbial approach. In: Molina, G., Usmani, Z., Sharma, M., Yasri, A., Gupta, V.K., (Eds.) *Microbes in Agri-Forestry Biotechnology*. CRC Press, Boca Raton, USA. pp. 197–219.

Srivastava, P., Bragança, J., Ramanan, S.R., Kowshik, M., 2013. Synthesis of silver nanoparticles using haloarchaeal isolate *Halococcus salifodinae* BK3. *Extremophiles* 17, 821–831.

Tanner, K., Molina-Menor, E., Latorre-Pérez, A., Vidal-Verdú, À., Vilanova, C., Peretó, J., Porcar, M., 2020. Extremophilic microbial communities on photovoltaic panel surfaces: A two-year study. *Microb. Biotechnol.*, 13, 1819–1830.

Thapa, B., Kumar, A.K.C., Ghimire, A., 2012. A review on bioremediation of petroleum hydrocarbon contaminants in soil. *Kathmandu Univ. J. Sci. Eng. Technol.* 8, 164–170.

Thongaram, T., Kosono, S., Ohkuma, M., Hongoh, Y., Kitada, M., Yoshinaka, T., Trakulnaleamsai, S., Noparatnaraporn, N., Kudo, T., 2003. Gut of higher termites as a niche for alkaliphiles as shown by culture-based and culture-independent studies. *Microb. Environ.* 18, 152–159.

Thongaram, T., Hongoh, Y., Kosono, S., Ohkuma, M., Trakulnaleamsai, S., Noparatnaraporn, N., Kudo, T. 2005. Comparison of bacterial communities in the alkaline gut segment among various species of higher termites. *Extremophiles* 9, 229–238.

Torregrosa-Crespo, J., Martinez-Espinosa, R., Esclapez, J., Bautista, V., Pire, C., Camacho, M., 2016. Anaerobic metabolism in *Haloferax* genus: Denitrification as case of study. In: Poole, R.K., (Ed.) *Advances in Microbial Physiology*, Oxford Academic Press, Oxford. pp. 41–85.

Torregrosa-Crespo, J., Pire, C., Martínez-Espinosa, R.M., Bergaust, L., 2019. Denitrifying haloarchaea within the genus *Haloferax* display divergent respiratory phenotypes, with implications for their release of nitrogenous gases. *Environ. Microbiol.* 21, 427–436.

Verma, S., Singh, D., Chatterjee, S., 2020. Biodegradation of organophosphorus pesticide chlorpyrifos by *Sphingobacterium* sp. C1B, a psychrotolerant bacterium isolated from apple orchard in Himachal Pradesh of India. *Extremophiles* 24, 897–908.

Verma, S., Singh, D., Chatterjee, S., 2021. Malathion biodegradation by a psychrotolerant bacteria *Ochrobactrum* sp. M1D and metabolic pathway analysis. *Lett. Appl. Microbiol.* 73, 326–335.

Waigi, M.G., Sun, K., Gao, Y.Z., 2017, Sphingomonads in microbe-assisted phytoremediation: Tackling soil pollution. *Trends Biotechnol.* 35, 883–899.

Wang, J., Feng, X., Anderson, C.W., Xing, Y., Shang, L., 2012. Remediation of mercury contaminated sites-a review. *J. Hazard. Mater.* 221–222, 1–18.

Wang, J., Salem, D.R., Sani, R.K., 2019. Extremophilic exopolysaccharides: A review and new perspectives on engineering strategies and applications. *Carbohydr. Polym.* 205, 8–26.

Woodhull, P.M., Jerger, D.E., 1994. Bioremediation using a commercial slurry phase biological treatment system: Site-specific applications and costs. *Remediation* 4, 353–363.

Xiang, Q., Nomura, Y., Fukahori, S., Mizuno, T., Tanaka, H., Fujisawa, T., 2019. Innovative treatment of organic contaminants in reverse osmosis concentrate from water reuse: A mini review. *Curr. Pollut. Rep.* 5, 294–307.

Zhao, D., Kumar, S., Zhou, J., Wang, R., Li, M., Xiang, H., 2017. Isolation and complete genome sequence of *Halorientalis hydrocaebonoclasticus* sp. *nov.*, a hydrocarbon-degrading haloarchaon. *Extremophiles* 21, 1081–1090.

Zhuang, X., Han, Z., Bai, G., Zhuang, G., Shim, H., 2010. Progress in decontamination by halophilic microorganisms in saline wastewater and soil. *Environ. Poll.* 158, 1119–1126.

Section V

Bioinformatics

11 Role of Bacterial Infection in Cancer Genomics

Suchitra Singh, Piyush Kumar Yadav, and Ajay Kumar Singh
Department of Bioinformatics, School of Life Sciences, Central University of South Bihar, Bihar, India

CONTENTS

11.1 INTRODUCTION

Neoplasia or "new growth" refers to a condition wherein normal cells undergo irreversible genetic variations and renders them unresponsive to ordinary controls on growth (Newkirk et al. 2017). Tumor, nodule and mass are other commonly used terms for neoplasms. The term tumor may refer to either benign or malignant growths, but the term "cancer" always denotes a malignant growth. Cancer is the change of a healthy cell into a malignant cell. When the body's regular control mechanism malfunctions, cancer develops. Instead of dying, old cells expand out of control, generating new and aberrant cells. These additional cells may create a tumor, which is a mass of tissue made up of abnormal cells. Tumor development is a multistage process having potentially pre-neoplastic changes that include hyperplasia, hypertrophy, metaplasia and dysplasia (Figure 11.1).

Cancers are malignant tumors that escape from the primary site, spread in the body through the circulatory or lymphatic systems, escape deadly combat with immune cells and infiltrate surrounding normal tissue through a process known as metastasis (Figure 11.2). Metastatic cancers are tough to cure and account for considerable number of cancer-related deaths (Stoletov et al. 2020). The cells continue to divide, causing damage to adjacent normal cells and diminishing the function of the organ. Cancer can appear anywhere in the body. There are various types of cancer such as breast cancer, lung cancer, liver cancer, mouth cancer and many more. Changes in the cells' DNA due to mutations are a common cause of cancer. Mutations can be induced by a variety

DOI: 10.1201/9781003324706-16

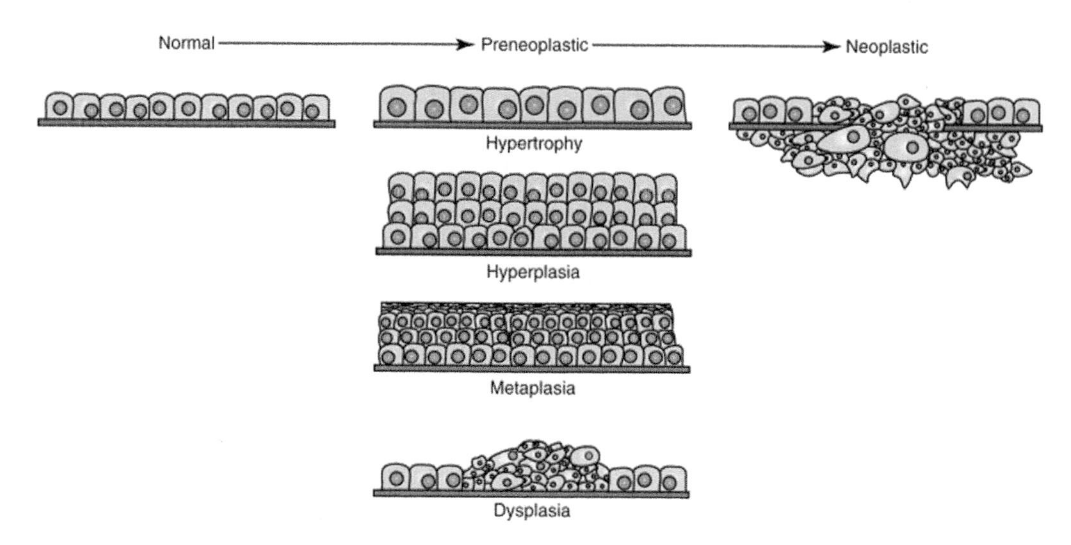

FIGURE 11.1 Tumor development (Reprinted with permission from Newkirk et al. 2017).

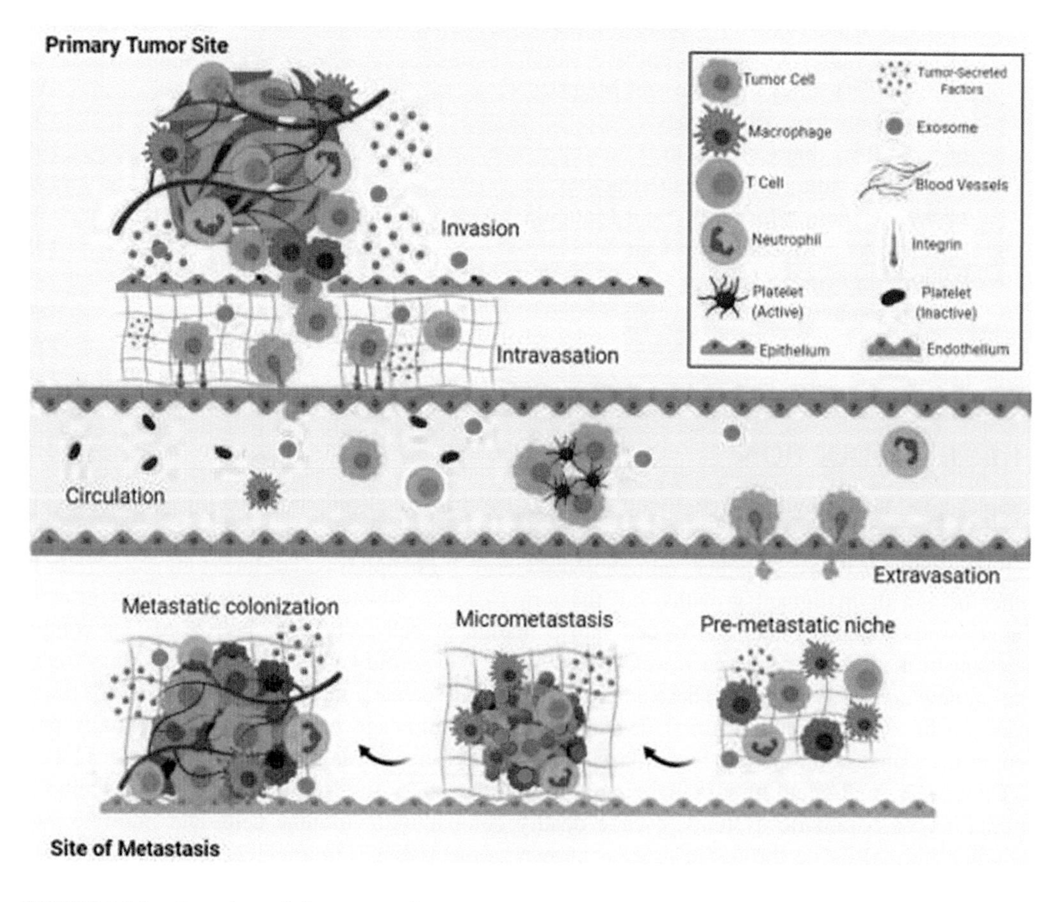

FIGURE 11.2 Overview of the metastatic cascade depicting five key steps of metastasis (Reprinted from Fares et al. 2020).

of chemicals, radiation sources, viruses and bacteria, all of which have been linked to cancer growth. Cancer is a multi-step process that occurs over time. The cells get additional capacities as they become more aberrant. The growth and size of cancerous tissue are detected by the staging system. A staging method detects the progression and size of malignant tissue. The staging approach is crucial since it identifies the type of treatment that is necessary. Cancer staging is of two kinds of methods, i.e., TNM (tumor, node, metastasis) and number system. The TNM system employs both numbers and letters to describe the stage. T stands for tumor size and how far it has progressed into neighboring tissue; it can be 1, 2, 3 or 4, with 1 indicating a small tumor and 4 indicating a large tumor. The number N indicates whether cancer has spread to the lymph nodes; it can range from 0 (no cancer cells in lymph nodes) to 3 (all lymph nodes contain cancer cells).

There are various types of cancer, which are distinguished by their names, organs or tissue where cancer forms. Cancer can be caused by a variety of factors, including nicotine, alcohol and viral infections. Early studies have reported that systemic inflammation caused by bacteria leads to the development of cancer. Table 11.1 provides information on the role of bacteria involved in. It has evidence of bacteria can cause cancer by inducing chronic inflammation, up-regulating or secreting anti-apoptotic enzymes, producing carcinogenic substances and by evading the immune system.

In contrast, conventional cancer treatments such as chemotherapy, surgery and radiotherapy have improved cancer survival rates in different countries around the globe (Spratt et al. 2013) Chemotherapy is a form of cancer treatment that is used all over the world, but because most of these medications are non-specific and injure all bodily cells, they are unable to penetrate the tumor site (damaged and normal cells). As a result, it's crucial to adopt alternative cancer treatment methods (Nair et al. 2014) because radiation penetration is determined by the oxygen level in the cancer environment, surgery eliminates the solid tumor and ongoing chemotherapy causes resistance and ultimately an oxygen shortage. Most of these aspects have prompted researchers to investigate and propose a different strategy cancer for treatment (Pucci et al. 2019).

11.2 BACTERIAL INVOLVEMENT IN CANCER GENETICS

Cancer is a genetic disease caused by genetic alterations in genes. Changes in the genes evade normal function and growth of the cells which results in the development of cancer. Changes in the gene occur due to mutation. Infection caused by several bacteria influences cancer genetics. A significant number of hallmarks of cancer have been involved in various cancer as shown in Table 11.1.

11.3 SOMATIC MUTATIONS INDUCED BY BACTERIA IN CANCER

There are two types of mutation in cancer cells: somatic and germline mutation. Somatic cells are those cells that are not a part of germ cells. Somatic cells contribute to an organism's growth and development. Germ cells give rise to reproductive cells known as gametes which help in the genetic inheritance. So, unlike somatic cells, germ cell mutations can be passed down from generation to generation. Cancer due to somatic mutation is also known as sporadic cancer, whereas cancer that occurs due to germ cell mutation is known as hereditary cancer (Eldridge 2021). Bacteria have been found to incorporate into the somatic cells and lead to somatic mutations in cancer (Akimova et al. 2022). By producing carcinogens, bacteria mutate the genetic material of the somatic cell. Butyrate-producing bacteria promote tumor genesis in colon cancer patients (Okumura et al. 2021). Acetaldehyde a well-known carcinogen produced by many bacterial species promotes somatic cell mutation in gastric cancer (Salaspuro 2003).

11.4 BACTERIA-RELATED MUTATIONS IN CANCER-CRITICAL GENES

Not all of the genes found in the organism are mutated in cancer. There are two main classes of cancer-critical genes (i) proto-oncogene and (ii) tumor suppressor gene. Proto-oncogenes are

TABLE 11.1
Important Cancer-causing Bacteria and Their Clinical Manifestations

Intestinal Bacteria	Bacterial Mechanism	Hallmark Affected	Mouse Models	References
Enterotoxigenic *Bacteroides fragilis* (ETBF)	*B. fragilis* toxin (BFT)	Sustaining proliferative signaling	WT mice	Rhee et al. (2009)
		Genome instability and mutations	$Apc^{Min/+}$	Goodwin et al. (2011)
	Unknown mechanism	Tumor-promoting inflammation	$Apc^{Min/+}$	Wu et al. (2009)
Fusobacterium nucleatum	FadA adhesin	Sustaining proliferative signaling	Xenograft model	Rubinstein et al. (2013)
	Fap2 adhesin	Avoiding immune destruction	$Apc^{Min/+}$	Kostic et al. (2013); Gur et al. (2015)
pks+Escherichia coli	Colibactin	Genome instability and mutations	In vitro cellular assays	Nougayrède et al. (2006)
			AOM/$Il10^{-/-}$	Arthur et al. (2012)
		Sustaining proliferative signaling	AOM/DSS xenograft model	Cougnoux et al. (2014)
Enterococcus faecalis	Unknown mechanism	Genome instability and mutations	Allograft model	Wang et al. (2015)
Alistipes spp.	Unknown mechanism	Tumor-promoting inflammation	$Il10^{-/-}$ $Lcn2^{-/-}$	Moschen et al. (2016)
Bifidobacterium spp.	Unknown mechanism	Inhibits avoiding immune destruction	Subcutaneous B16.SIY melanoma	Sivan et al. (2015)
Bacteroides thetaiotamicron and *B. fragilis*	Unknown mechanism	Inhibits avoiding immune destruction	MCA205 sarcoma, Ret melanoma and MC38 CRC xenograft	Vétizou et al. (2015)

Source: Reprinted from Fulbright et al. (2017).

normal cellular genes that help in the growth, division and differentiation of cells (Newkirk et al. 2017). These genes also play an important role in the stimulation of cellular proliferation during tissue regeneration. Mutation in proto-oncogene results in a gain of function and they are known as oncogenes which leads to uncontrollable cell growth. The tumor suppressor gene slows down the cell division, repairs DNA damage and helps in apoptosis of the cell. Mutation in these genes causes loss in function and can lead to cancer. Mutation events can be dominant or recessive. A gain in function in a single copy of proto-oncogenes is a dominant event whereas inactivation of both copies of tumor suppressor gene is a recessive event. A variety of bacteria has been associated with the mutation of a huge number of tumor suppressors and proto-oncogenes. Microbial virulence factors (fimbriae, lipopolysaccharides), enzymes (protease and collagenase) and their by-products (ammonia, butyric acid, fatty acid and hydrogen sulfide) directly induce mutation in tumor suppressor and proto-oncogenes, which eventually promotes the activation of oncogenic signaling pathways.

11.5 BACTERIA INFLUENCING HALLMARKS OF CANCER

The hallmarks of cancer describe the complexities of a cancer cell in an organized way. It provides insights into the biological aspects of the cancer cell acquired by changes in the genetic constituents of the cell. There are seven hallmarks of cancer described. Many investigations have reported that bacteria influence the hallmarks of cancer.

11.5.1 Sustaining Proliferative Signaling

Cancer cells can grow without external signals. They grow and divide by producing their signaling known as autocrine signaling. These signals help in constitutive cell division, growth and survival of tumors. As seen in Table 11.1, the crucial hallmarks for bacterial infection-causing cancer are shown in Figure 11.3. In colorectal cancer patients, *Fusobacterium nucleatum* induces the level of interleukin-8 (IL-8) and C-X-C motif chemokine ligand 1 (CXCL1)

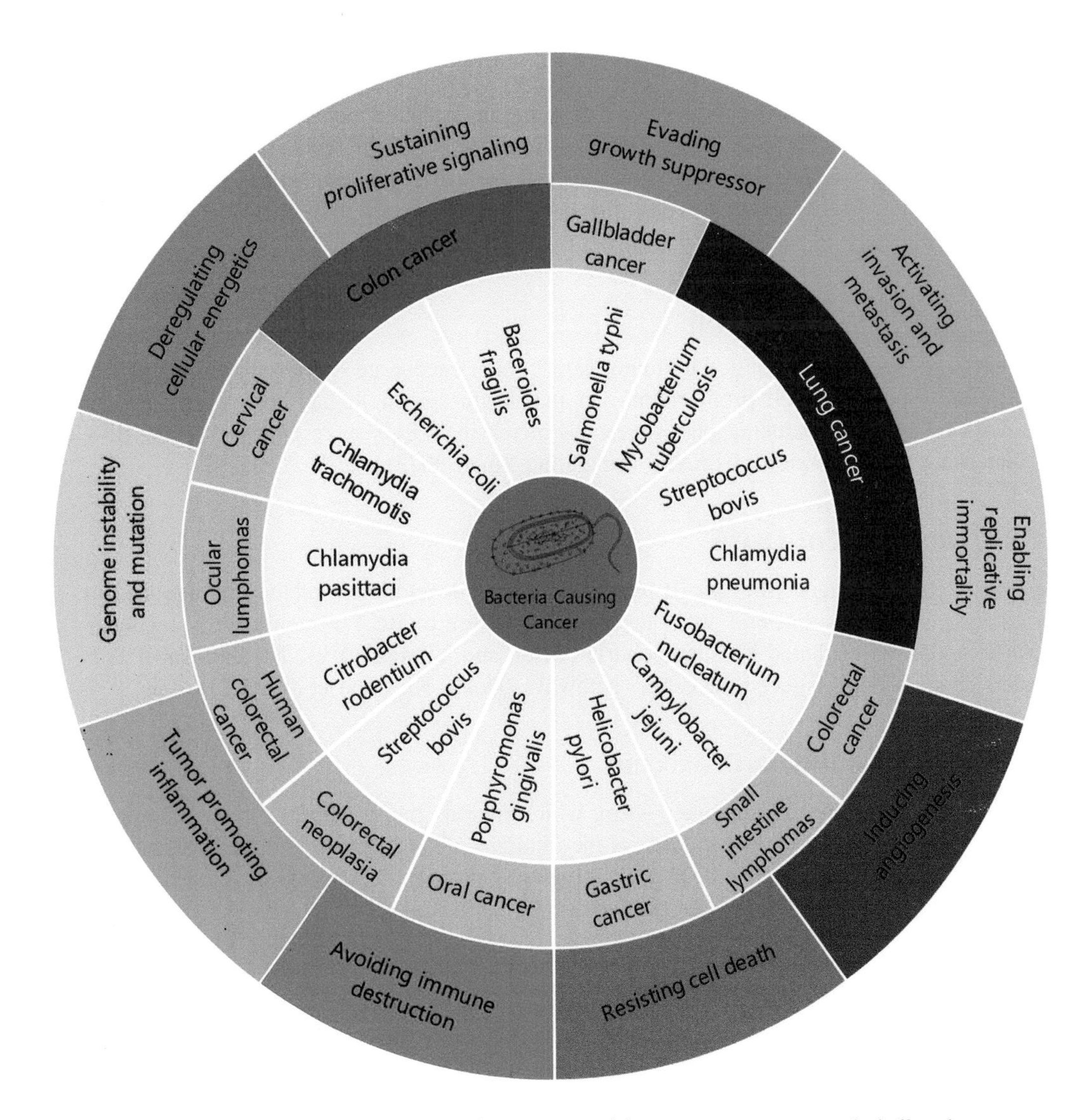

FIGURE 11.3 Bacteria causing their respective cancer and important cancer genomic hallmarks.

which helps rapid cell proliferation by autocrine signaling (Grivennikov and Karin 2008). *Helicobacter pylori* up-regulate the IL-8 level which induces uncontrollable cell division (Lee et al. 2013).

11.5.2 Evading Growth Suppressor

Tumor suppressor is a cancer-critical gene mutation that inactivates its growth suppression function in a variety of cancer. In a normal cell, the tumor suppressor gene controls the cell proliferation, but in the cancer cell, this control is lost due to the inactivation of the tumor suppressor gene and hence the cells proliferate continuously by evading the antigrowth signaling. *H. pylori* inactivates PTEN a tumor suppressor protein by increasing the level of PLK1 which inhibits the PTEN activity by phosphorylating it (Fu and Xie 2019). *Porphyromonas gingivalis* diminished the activity of P53 which helps in the regulation of cell growth (Whitmore and Lamont 2014).

11.5.3 Activating Invasion and Metastasis

A tumor that spreads to nearby tissues and distant organs is called metastasis. Metastasis occurred in an advanced stage of cancer in which patients have the risk of reoccurrence of tumors after surgical treatment of primary tumors (Kawada and Taketo 2011). *F. nucleatum* increased the level of mir-21, which further promotes tumorigenesis and metastasis of colorectal cancer (Fu-Mei Shang 2018).

11.5.4 Inducing Angiogenesis

Nutrients and oxygen are required by tumor cells, which encourages tumor growth. Angiogenesis is therefore crucial for cell proliferation and metastatic spread. It has been suggested that *P. gingivalis* is capable of inducing angiogenesis by increasing VEGF secretion via IL-6 and IL-8 and by increasing the expression of *EENB2* (Singh and Singh 2022).

11.5.5 Resisting Cell Death

Apoptosis plays a crucial role in cell retention and the development of the immune system. *P. gingivalis* a pathogenic bacteria promotes the upregulation of Bcl-2 protein and dysregulates the intrinsic apoptosis mediated by mitochondria (Olsen and Yilmaz 2019). The antigen of *S. bovis* inhibits apoptosis and induced the cancer pathways (Dokht Khosravi et al. 2022).

11.5.6 Avoiding Immune Destruction

Cancer is controlled by the immune system, which inhibits the growth of tumors. Carcinogenic antigen produced by *F. nucleatum* destroys the immune cells and protects themselves and cancer cell from being killed (Li et al. 2021). *P. gingivalis* promotes apoptosis of T-cells caused by activating program death ligands in OSCC cells (Groeger et al. 2011).

11.5.7 Genomic Instability and Mutation

Cancer cell promotes their survival by installing normal cell control. This tendency is carried out by damaging the DNA, epigenetics modification and telomerase damage. Colibactin a toxin substance secreted by *E. coli* causes DNA damage and assists tumorigenesis (Li et al. 2021).

11.6 BACTERIA AS A BIOMARKER IN CANCER

Since cancer occurred due to the imbalance of the normal function of the cells in the body, this is due to genetic changes. Several indicators are used in the detection of cancer, these are biological molecules known as a biomarker. A biomarker has been shown as one of the powerful prognostic tools in cancer. Biomarkers can be DNA, RNA, protein and metabolites. More than seven hundred bacterial species are found in the saliva of the oral cavity, which is present in a balanced amount in healthy organisms. Pathogenic bacteria disrupt this balance and can promote cancer. Pathogenic bacteria have been present in abundance amount in the saliva of a diseased person (Zhang et al. 2016). Pathogenic bacterial DNA has been found in the saliva of Cancer patients (Pushalkar et al. 2011). Thus, bacterial DNA in saliva may act as a promising non-invasive prognostic biomarker. Using bacterial biomarkers, cancer can be detected early.

11.7 BACTERIA IN CANCER THERAPY

Several biomarkers are used for cancer treatment as discussed in Table 11.2. Cancer is a complicated disease that necessitates a multifaceted approach for effective treatment (Parsonnet 1995). For the first time, doctors used a living bacterium viz. *Clostridia* and *Streptococci* to treat cancer. This is now being done with genetically modified bacteria. Bacteria can be used to treat cancer in a variety of ways (Hajitou et al. 2006; Mengesha et al. 2007). Native bacterial toxicity, in conjunction with other medicines, bacteria that can influence anticancer agent development, expression of tumor-specific antigens, gene transfer, RNA interference and prodrug cleavage are among these techniques (Forbes 2010). In numerous cancer experimental models, the use of whole live, attenuated and/or genetically engineered bacteria alone or in combination with conventional treatments has been tested. *Salmonella, Lactobacillus, Pseudomonas, Listeria, Clostridium, Escherichia, Proteus, Bifidobacterium, Caulobacter* and *Streptococcus* are by far the most prevalent bacteria utilized in this domain (Heimann and Rosenberg 2003; Nemunaitis et al. 2003;

TABLE 11.2
Potential Uses for Cancer Biomarkers

Use	Example	References
Estimate risk of developing cancer	BRCA1 germline mutation (breast and ovarian cancer)	Easton et al. 1995, Hall et al. 1990
Screening	Prostate-specific antigen (prostate cancer)	Lin et al. 2008
Differential diagnosis	Immunohistochemistry to determine tissue of origin	
Determine prognosis of disease	21 gene recurrence score (breast cancer)	Paik et al. 2004
Predict response to therapy	KRAS mutation and anti-EGFR antibody (colorectal cancer)	Allegra et al. 2009
	HER2 expression and anti-Her2 therapy (breast and gastric cancer)	Bang et al. 2010, Piccart-Gebhart et al. 2005, Romond et al. 2005
	Estrogen receptor expression (breast cancer)	EBCTCG 2011
Monitor for disease recurrence	CEA (colorectal cancer)	Locker et al. 2006
	AFP, LDH, βHCG (germ cell tumor)	Gilligan et al. 2010
Monitor for response or progression in metastatic disease	CA15-3 and CEA (breast cancer)	Harris et al. 2007

Source: Reprinted from Henry and Hayes (2012); published under a Creative Commons Attribution (CC BY) License.

Nallar et al. 2017). Many clinical studies have shown significant responses, indicating that more research on humans is required. Cancer therapy using genetically modified bacteria genetic engineering therapy is an innovative method of cancer treatment. The main advantages of genome editing are selective targeting and tumor cell elimination (Nallar et al. 2017). Bacterial strains that have been genetically modified may be able to reduce host pathogenicity while increasing anticancer efficacy (Nuyts et al. 2001). It's been discovered that genetically engineered bacteria can multiply more rapidly in tumors in comparison to normal tissues (Dang et al. 2001).

Despite many drawbacks like insufficient drug concentrations in tumors, occurrence of systemic toxicity (gastrointestinal, heart, hematological, alopecia and skin toxicity) in many types of cancer, and almost inevitable induction of drug resistance, chemotherapy is still the mainstay of treatment for incurable cancer (Barber 1965). Radiation treatment is also one of the efficacious ways of treating a diverse range of malignancies. Even though, normal tissue damage cannot be prevented, which is a significant constraint to the success of this cancer treatment technique (Burdelya et al. 2012). The prevalence of ischemic hypoxic (i.e., inadequately vascularized) zones that seem immune to radiation is one of the prime reasons for radiotherapy's lack of efficacy in some solid tumors. Even though few studies have been conducted employing microorganisms to increase radiotherapy, this subject may potentially become a viable method in clinical radiation oncology. Salmonella bacteria that have been genetically altered have been shown to exhibit the desirable characteristics of an anticancer vector. Salmonella which addresses tumors from a remote injection site can decrease tumor growth (Pawelek et al. 1997). Multiple novel techniques for reducing radiotherapy have been presented. (Abdollahi 2015). Bacteria can also be employed to minimize natural cellular injury during or after radiotherapy. Certain bacterial strains (especially *Bifidobacteria* and *Lactobacilli*) had indeed been recommended to minimize the negative effects of radiotherapy. Many clinical and preclinical investigations have demonstrated the utility of probiotics to protect normal tissue during or after radiation (Delia et al. 2002).

11.8 CONCLUSION AND FUTURE ASPECTS

There are numerous well-known factors of cancer however these elements are still unable to explain all cancer cases that occurred around the world. So, our study emphasized that factors of cancer are not well discovered. Recent investigations have revealed that bacterial infection has a role in cancer, so we emphasized this fact and try to relate it to cancer genetics and cancer hallmarks. Levels of cancer-associated bacteria have been seen in a higher abundance in cancer patients, they can be utilized as a marker and can serve as a therapeutic goal for cancer treatment. Cancer-associated bacteria express virulence factors or proteins that have the transforming ability, which can affect several important cellular types of machinery like cell division, growth and apoptosis, etc. Therefore, identification and characterization of molecular mechanisms in cancer associated with bacteria will help in the development of therapeutic targets which further have a role in preventive and diagnostic implications in cancer.

ACKNOWLEDGMENTS

The author(s) would like to acknowledge the Department of Bioinformatics, Central University of South Bihar, Gaya, Bihar.Author Contributions: All authors have contributed equally to the chapter. The authors have declared no conflicts of interest.

REFERENCES

Abdollahi, H., 2015. Beneficial effects of cellular autofluorescence following ionization radiation: hypothetical approaches for radiation protection and enhancing radiotherapy effectiveness. *Med. Hypothesis* 84, 194–198.

Akimova, E., Gassner, F. J., Greil, R., Zaborsky, N., Geisberger, R., 2022. Detecting bacterial-human lateral gene transfer in chronic lymphocytic leukemia. *Int. J. Mol. Sci.* 23, 1094.

Allegra, C.J., Jessup, J.M., Somerfield, M.R., Hamilton, S.R., Hammond, E.H., Hayes, D.F., McAllister, P.K., Morton, R.F., Schilsky, R.L., 2009. American Society of Clinical Oncology provisional clinical opinion: Testing for KRAS gene mutations in patients with metastatic colorectal carcinoma to predict response to anti-epidermal growth factor receptor monoclonal antibody therapy. *J. Clin. Oncol.* 27, 2091–2096.

Arthur, J.C., Perez-Chanona, E., Mühlbauer, M., Tomkovich, S., Uronis, J.M., Fan, T.J., et al, 2012. Intestinal inflammation targets cancer-inducing activity of the microbiota. *Science* 338, 120–123.

Bang, Y.J., Van Cutsem, E., Feyereislova, A., Chung, H.C., Shen, L., Sawaki, A., et al., 2010. Trastuzumab in combination with chemotherapy versus chemotherapy alone for treatment of HER2-positive advanced gastric or gastro-oesophageal junction cancer (ToGA): A phase 3, open-label, randomised controlled trial. *Lancet* 376, 687–697.

Barber, M., 1965. Drug combinations in antibacterial chemotherapy. *Proc. Royal Soc. Med.* 58, 990–995.

Burdelya, L.G., Gleiberman, A.S., Toshkov, I., Aygun-Sunar, S., Bapardekar, M., Manderscheid-Kern, P., 2012. Toll-like receptor 5 agonist protects mice from dermatitis and oral mucositis caused by local radiation: Implications for head-and-neck cancer radiotherapy. *Int. J. Rad. Oncol.* 83, 228–234.

Cougnoux, A., Dalmasso, G., Martinez, R., Buc, E., Delmas, J., Gibold, L., et al., 2014. Bacterial genotoxin colibactin promotes colon tumour growth by inducing a senescence-associated secretory phenotype. *Gut* 63, 1932–1942.

Dang, L.H., Bettegowda, C., Huso, D.L., Kinzler, K.W., Vogelstein, B., 2001. Combination bacteriolytic therapy for the treatment of experimental tumors. *Proc. Natl. Acad. Sci. USA* 98, 15155–15160.

Delia, P., Sansotta, G., Donato, V., Messina, G., Frosina, P., Pergolizzi, S., De Renzis, C., Famularo, G., 2002. Prevention of radiation-induced diarrhea with the use of VSL#3, a new high-potency probiotic preparation. *Am. J. Gastroenterol.* 97, 2150–2152.

Dokht Khosravi, A., Seyed-Mohammadi, S., Teimoori, A., Asarehzadegan Dezfuli, A., 2022. The role of microbiota in colorectal cancer. *Folia Microbiol.* 67, 683–691.

Easton, D.F., Ford, D., Bishop, D.T., 1995. Breast and ovarian cancer incidence in BRCA1-mutation carriers. Breast Cancer Linkage Consortium. *Am. J. Human Genet.* 56, 265–271.

Early Breast Cancer Trialists' Collaborative Group (EBCTCG), Davies, C., Godwin, J., Gray, R., Clarke, M., Cutter, D., Darby, S., McGale, P., Pan, H. C., Taylor, C., Wang, Y. C., Dowsett, M., Ingle, J., Peto, R., 2011. Relevance of breast cancer hormone receptors and other factors to the efficacy of adjuvant tamoxifen: Patient-level meta-analysis of randomised trials. *Lancet* 378, 771–784.

Eldridge, L., 2021. Hereditary and Acquired Gene Mutations: Differences in Cancer. https://www.verywellhealth.com/hereditary-vs-acquired-gene-mutations-in-cancer-4691872

Fares, J., Fares, M.Y., Khachfe, H.H., Salhab, H.A., Fares, Y., 2020. Molecular principles of metastasis: a hallmark of cancer revisited. *Signal Transduct. Target. Ther.* 5, 28.

Forbes, N. S., 2010. Engineering the perfect (bacterial) cancer therapy. *Nature Rev. Cancer* 10, 785–794.

Fu-Mei Shang, H.-L.L., 2018. *Fusobacterium nucleatum* and colorectal cancer: A review. *World J. Diab.* 9, 157–164.

Fu, L., Xie, C., 2019. A lucid review of *Helicobacter pylori*-induced DNA damage in gastric cancer. *Helicobacter* 24, e12631.

Fulbright, L.E., Ellermann, M., Arthur, J.C., 2017. The microbiome and the hallmarks of cancer. *PLoS Pathog.* 13, e1006480.

Gilligan, T.D., Seidenfeld, J., Basch, E.M., Einhorn, L.H., Fancher, T., Smith, D.C., and American Society of Clinical Oncology, 2010. American Society of Clinical Oncology Clinical Practice Guideline on uses of serum tumor markers in adult males with germ cell tumors. *J. Clin. Oncol.* 28, 3388–3404.

Goodwin, A.C., Destefano Shields, C.E., Wu, S., Huso, D.L., Wu, X., Murray-Stewart, T.R., et al., 2011. Polyamine catabolism contributes to enterotoxigenic *Bacteroides fragilis*-induced colon tumorigenesis. *Proc. Natl. Acad. Sci. USA*, 15354–15359.

Grivennikov, S., Karin, M., 2008. Autocrine IL-6 Signaling: A key event in tumorigenesis? *Cancer Cell* 13, 7–9.

Groeger, S., Domann, E., Gonzales, J. R., Chakraborty, T., and Meyle, J., 2011. B7-H1 and B7-DC receptors of oral squamous carcinoma cells are upregulated by *Porphyromonas gingivalis*. *Immunobiology* 216, 1302–1310.

Gur, C., Ibrahim, Y., Isaacson, B., Yamin, R., Abed, J., Gamliel, M., et al., 2015. Binding of the Fap2 protein of *Fusobacterium nucleatum* to human inhibitory receptor TIGIT protects tumors from immune cell attack. *Immunity* 42, 344–355.

Hajitou, A., Trepel, M., Lilley, C. E., Soghomonyan, S., Alauddin, M. M., Marini, F. C. 3rd, et al., 2006. A hybrid vector for ligand-directed tumor targeting and molecular imaging. *Cell* 125, 385–398.

Hall, J.M., Lee, M.K., Newman, B., Morrow, J.E., Anderson, L.A., Huey, B., King, M.C., 1990. Linkage of early-onset familial breast cancer to chromosome 17q21. *Science* 250, 1684–1689.

Harris, L., Fritsche, H., Mennel, R., Norton, L., Ravdin, P., Taube, S., Somerfield, M.R., Hayes, D. F., Bast, R.C.J., 2007. American Society of Clinical Oncology 2007 update of recommendations for the use of tumor markers in breast cancer. *J. Clin. Oncol.* 25, 5287–5312.

Heimann, D.M., Rosenberg, S.A., 2003. Continuous intravenous administration of live genetically modified *Salmonella typhimurium* in patients with metastatic melanoma. *J. Immunother.* 26, 179–180.

Henry, N. L., Hayes, D.F., 2012. Cancer biomarkers. *Molecular Oncol.* 6, 140–146.

Kawada, K., Taketo, M.M., 2011. Significance and mechanism of lymph node metastasis in cancer progression. *Cancer Res.* 71, 1214–1218.

Kostic, A.D., Chun, E., Robertson, L., Glickman, J.N., Gallini, C.A., Michaud, M., et al., 2013. *Fusobacterium nucleatum* potentiates intestinal tumorigenesis and modulates the tumor-immune microenvironment. *Cell Host Microb.* 14, 207–215.

Lee, K.E., Khoi, P.N., Xia, Y., Park, J.S., Joo, Y.E., Kim, K.K., Choi, S.Y., Jung, Y. Do., 2013. *Helicobacter pylori* and interleukin-8 in gastric cancer. *World J. Gastroenterol.* 19, 8192–8202.

Li, S., Liu, J., Zheng, X., Ren, L., Yang, Y., Li, W., Fu, W., Wang, J., Du, G., 2021. Tumorigenic bacteria in colorectal cancer: mechanisms and treatments. *Cancer Biol. Med.* 19, 147–162.

Lin, K., Lipsitz, R., Janakiraman, S., 2008. Benefits and harms of prostate-specific antigen screening for prostate cancer: An evidence update for the u.s. preventive services task force. *Agency for Healthcare Research and Quality (US).*

Locker, G.Y., Hamilton, S., Harris, J., Jessup, J.M., Kemeny, N., Macdonald, J.S., Somerfield, M.R., Hayes, D.F., Bast, R.C.J., 2006. ASCO 2006 update of recommendations for the use of tumor markers in gastrointestinal cancer. *J. Clin. Oncol.* 24, 5313–5327.

Mengesha, A., Dubois, L., Chiu, R.K., Paesmans, K., Wouters, B.G., Lambin, P., Theys, J., 2007. Potential and limitations of bacterial-mediated cancer therapy. *Front. Biosci.* 12, 3880–3891.

Moschen, A.R., Gerner, R.R., Wang, J., Klepsch, V., Adolph, T.E., Reider, S.J., et al., 2016. Lipocalin 2 protects from inflammation and tumorigenesis associated with gut microbiota alterations. *Cell Host Microbe.* 19, 455–469.

Nair, N., Kasai, T., Seno, M., 2014. Bacteria: Prospective savior in battle against cancer. *Antic. Res.* 34, 6289–6296.

Nallar, S.C., Xu, D.-Q., Kalvakolanu, D.V., 2017. Bacteria and genetically modified bacteria as cancer therapeutics: Current advances and challenges. *Cytokine* 89, 160–172.

Nemunaitis, J., Cunningham, C., Senzer, N., Kuhn, J., Cramm, J., Litz, C., et al., 2003. Pilot trial of genetically modified, attenuated *Salmonella* expressing the *E. coli* cytosine deaminase gene in refractory cancer patients. *Cancer Gene Ther.* 10, 737–744.

Newkirk, K.M., Brannick, E.M., Kusewitt, D.F., 2017. Neoplasia and tumor biology. In: Zachary, J.F., (Ed.) *Pathologic Basis of Veterinary Disease*. Elsevier. pp. 286–321.

Nougayrède, J.P., Homburg, S., Taieb, F., Boury, M., Brzuszkiewicz, E., Gottschalk, G., et al., 2006. *Escherichia coli* induces DNA double-strand breaks in eukaryotic cells. *Science* 313, 848–851.

Nuyts, S., Van Mellaert, L., Theys, J., Landuyt, W., Lambin, P., Anné, J. 2001. The use of radiation-induced bacterial promoters in anaerobic conditions: A means to control gene expression in *Clostridium*-mediated therapy for cancer. *Rad. Res.*, 155, 716–723.

Okumura, S., Konishi, Y., Narukawa, M., Sugiura, Y., Yoshimoto, S., Arai, Y., et al., 2021. Gut bacteria identified in colorectal cancer patients promote tumourigenesis via butyrate secretion. *Nature Commun.* 12, 5674.

Olsen, I., Yilmaz, Ö., 2019. Possible role of *Porphyromonas gingivalis* in orodigestive cancers. *J. Oral Microbiol.* 11, 1563410.

Paik, S., Shak, S., Tang, G., Kim, C., Baker, J., Cronin, M., et al., 2004. A multigene assay to predict recurrence of tamoxifen-treated, node-negative breast cancer. *The New Eng. J. Med.* 351, 2817–2826.

Parsonnet, J., 1995. Bacterial infection as a cause of cancer. *Environ. Health Persp.* 103 Suppl, 263–268.

Pawelek, J.M., Low, K.B., Bermudes, D., 1997. Tumor-targeted *Salmonella* as a novel anticancer vector. *Cancer Res.* 57, 4537–4544.

Piccart-Gebhart, M.J., Procter, M., Leyland-Jones, B., Goldhirsch, A., Untch, M., Smith, I., …Herceptin Adjuvant (HERA) Trial Study Team, 2005. Trastuzumab after adjuvant chemotherapy in HER2-positive breast cancer. *N. Engl. J. Med.*, 353, 1659–1672.

Pucci, C., Martinelli, C., Ciofani, G., 2019. Innovative approaches for cancer treatment: current perspectives and new challenges. *Ecancermedicalscience* 13, 961.

Pushalkar, S., Mane, S. P., Ji, X., Li, Y., Evans, C., Crasta, O. R., Morse, D., Meagher, R., Singh, A., Saxena, D., 2011. Microbial diversity in saliva of oral squamous cell carcinoma. *FEMS Immunol. Med. Microbiol.* 61, 269–277.

Rhee, K.J., Wu, S., Wu, X., Huso, D.L., Karim, B., Franco, A.A., et al., 2009. Induction of persistent colitis by a human commensal, enterotoxigenic *Bacteroides fragilis*, in wild-type C57BL/6 mice. *Infect. Immun.* 77, 1708–1718.

Romond, E. H., Perez, E. A., Bryant, J., Suman, V. J., Geyer, C. E., Jr, Davidson, N. E., et al., 2005. Trastuzumab plus adjuvant chemotherapy for operable HER2-positive breast cancer. *N. Engl. J. Med.*, 353(16), 1673–1684.

Rubinstein, M.R., Wang, X., Liu, W., Hao, Y., Cai, G., Han, Y.W., 2013. *Fusobacterium nucleatum* promotes colorectal carcinogenesis by modulating E-cadherin/β-catenin signaling via its FadA adhesin. *Cell Host Microbe* 14, 195–206.

Salaspuro, M.P., 2003. Acetaldehyde, microbes, and cancer of the digestive tract. *Crit. Rev. Clin. Lab. Sci.* 40, 183–208.

Singh, S., Singh, A.K., 2022. *Porphyromonas gingivalis* in oral squamous cell carcinoma: A review. *Microb. Infect.* 24, 104925.

Sivan, A., Corrales, L., Hubert, N., Williams, J.B., Aquino-Michaels, K., Earley, Z.M., et al., 2015. Commensal *Bifidobacterium* promotes antitumor immunity and facilitates anti-PD-L1 efficacy. *Science* 350, 1084–1089.

Spratt, D.E., Pei, X., Yamada, J., Kollmeier, M.A., Cox, B., Zelefsky, M.J., 2013. Long-term survival and toxicity in patients treated with high-dose intensity modulated radiation therapy for localized prostate cancer. *Intern. J. Radiat. Oncol.* 85, 686–692.

Stoletov, K., Beatty, P.H., Lewis, J.D., 2020. Novel therapeutic targets for cancer metastasis. *Expert Rev. Anticancer Ther.* 20, 97–109.

Vétizou, M., Pitt, J.M., Daillère, R., Lepage, P., Waldschmitt, N., Flament, C., et al., 2015. Anticancer immunotherapy by CTLA-4 blockade relies on the gut microbiota. *Science* 350, 1079–1084.

Wang, X., Yang, Y., Huycke, M.M., 2015. Commensal bacteria drive endogenous transformation and tumour stem cell marker expression through a bystander effect. *Gut* 64(3), 459–468.

Whitmore, S.E., Lamont, R.J., 2014. Oral bacteria and cancer. *PLoS Pathog.* 10, 1–3.

Wu, S., Rhee, K.J., Albesiano, E., Rabizadeh, S., Wu, X., Yen, H.R., et al., 2009. A human colonic commensal promotes colon tumorigenesis via activation of T helper type 17 T cell responses. *Nat. Med.* 15, 1016–1022.

Zhang, C.Z., Cheng, X.Q., Li, J.Y., Zhang, P., Yi, P., Xu, X., Zhou, X.D., 2016. Saliva in the diagnosis of diseases. *Intern. J. Oral Sci.* 8, 133–137.

12 Structural Recognition and Cleavage Mechanism of SARS-CoV-2 Spike Protein

N. Dhingra
Department of Agriculture, Medi-Caps University, Indore, Madhya Pradesh, India

U. Bhardwaj
School of Sciences, Noida International University, Yamuna Expressway, Gautam Budh Nagar, Uttar Pradesh, India

R. Bhardwaj
Socorro, Bardez, North Goa, India

S. Sharma
Department of Biology, College of Arts and Sciences, Georgia State University, Atlanta, GA, USA

CONTENTS

12.1 INTRODUCTION

In December 2019, the World Health Organization (WHO) was informed about an outbreak of an infection in the respiratory tract of patients with pneumonia in Wuhan, Hubei Province, China which was then characterized as a newly identified β-coronavirus (nCoV). WHO classified the severe acute respiratory syndrome coronavirus 2 (SARS-CoV-2) epidemic as a public health emergency of international concern on January 30, 2020. As per WHO data, there were more than 224,122,263 confirmed cases with COVID-19, including 4,622,559 deaths. The United States, India and Brazil are the three countries in the world with the highest cumulative number of cases.

DOI: 10.1201/9781003324706-17

These features indicate that SARS-CoV-2 evades the human immune surveillance more effectively than the other viruses. Contribution to the widespread disease due to SARS-CoV-2 is believed to be a combination of immune evasion and high infectivity (Kirtipal et al. 2020; Hu et al. 2021).

Human coronaviruses like HCoV-229E and HCoV-OC43 have long been known to circulate in the community, and they, together with the more recently discovered HCoV-NL63 and HCoV-HKU1, induce seasonal and typically mild respiratory tract infections that mimic the symptoms of the "common cold". SARS-CoV, Middle East respiratory syndrome coronavirus (MERS-CoV) and SARS-CoV-2, which have all evolved in the human population in the last 20 years, are all extremely pathogenic. MERS-CoV, SARS-CoV and SARS-CoV-2 infect various respiratory parts causing life-threatening lung injuries for which till date no specific treatment has been approved (Sanyal 2020).

Despite the fact that several studies and clinical trials on COVID-19 are underway around the world, there is no proof from randomized clinical trials that any prospective medication improves patient outcomes. As the pandemic grows, finding a particular COVID-19 treatment is important, and vaccines targeting several SARS-CoV-2 proteins are now being developed.

12.2 VIRAL DISEASES IN HUMANS

Throughout history, periodic emergence and spread of infectious diseases such as Plaque, Cholera, Flu and MERS-CoV have been well documented and are responsible for millions of human lives lost during the past century. In the last two decades, several coronaviruses have caused major issues in humans and animals. The severe acute respiratory SARS-CoV, MERS-CoV and porcine epidemic diarrhea virus (PEDV) are the most well-known examples (Piret et al. 2021). The origin and re-emergence of several of these viruses may have been aided by urbanization and the greater frequent mingling of diverse species in densely populated areas. Coronaviruses, on the other hand, are known to have rapid mutation and recombination rates, which may allow them to adapt to new hosts and traverse species barriers (Stone et al. 2021).

The SARS pandemic in 2003 made scientists and the rest of the world aware of coronaviruses' capacity to transmit from animal to human. Zoonotic origin of these viral agents and their transmission to humans through consumption, contacts, breeding and hunting of animals have become a serious concern (Lindahl and Grace 2015). Zoonotic transmission of pathogen from animals to human (cross-species transmission), and its ability to afflict human involves five key steps; 1. Pathogen specifically infect the host animal in natural condition; 2. pathogen transmitted to human but are unable to withstand human to human transmission; 3. pathogen use host information to few cycles of secondary transmission between humans; 4. pathogens adapt to afflict human without the requirement of animal host; 5. finally, humans are the exclusive host for the pathogen (El-Sayed and Kamel 2020, White and Razgour 2020). Rapid change in climatic conditions and availability of host plays an important role in transmission and emergence of pathogen, with potential for zoonotic transmission at human–animal interface.

Only two coronaviruses, human coronavirus (HCoV)-229E and HCoV-OC43, were known to infect humans prior to the SARS pandemic. Two other human coronaviruses, HCoV-NL63 and HCoV-HKU1, were discovered shortly after the SARS outbreak, despite the fact that these viruses had likely been circulating in people for a long period before their detection (Kahn and McIntosh 2005). The WHO reported the discovery of a novel coronavirus, MERS-CoV, in the Middle East on September 23, 2012. *Tylonycteris* bat coronavirus HKU4 and *Pipistrellus* bat coronavirus HKU5 were discovered to be the most closely related lineage C betacoronaviruses at the time (Ye et al. 2020; Liu et al. 2021).

Phylogenetic analysis of beta-coronavirus indicates that beta-coronavirus is identical to SARS virus found among bats. While 96% of the WGS resemble to the virus found in horseshoe bats (*Rhinolophus affinis*) from Yunnan province, China. Similarly, 91% of beta-coronavirus genome resemble coronavirus found in Malayan pangolin (*Manis javanica*). However, bioinformatics analysis indicated the absence of a unique peptide insertion in pangolins coronavirus. Genetic variability and adaptability of beta-coronavirus could have resulted from natural selection and may have jumped into humans through zoonotic transfer, with human being the most suitable host. MERS appears to have a greater fatality rate (>35%) than SARS (9.6%), which could be explained in part by the high frequency of comorbid conditions among infected patients. Apart from SARS-CoV and MERS-CoV, which are relatively new, the other four human-pathogenic coronaviruses, HCoV-229E, HCoV-OC43, HCoV-NL63 and HCoV-HKU1, have been circulating in humans for decades or centuries. Although they are commonly associated with mild respiratory ailments, accumulating evidence suggests that they can cause serious infections, particularly in patients who are elderly or have comorbidities (Boni et al. 2020; Ye et al. 2020; Liu et al. 2021).

12.3 SEVERE ACUTE RESPIRATORY SYNDROME CORONAVIRUS 2 (SARS-CoV-2)

Coronaviruses (CoVs) are members of the Coronavirus genus, Coronaviridae family and Nidovirales order. Coronaviruses are a positive sense, single-stranded RNA virus belonging to the beta-coronavirus family and is known to infect a wide range of hosts, including humans, horses, pigs, cats, etc., characterized by pneumonia and the common cold (Piret et al. 2021). In humans, SARS-CoV-2 infection results in pneumonia, acute respiratory distress, lung injury and neurological disorder such as loss of taste, smell and memory, ultimately leading to death. Severity and pathogenicity of CoV-2 infection depend upon the susceptibility and permissiveness of host cells. While transmission primarily occurs through large droplets or aerosols, and through respiratory route into human host. Proximity and repeated exposure to an infected person increase the pathogenicity of infection. During the early emergence of SARS-CoV-2 in 2019, R0 = 2.2–3.58, R0 is total number of secondary infections occurred from a case; any value more than 1 implies proclivity for spread. Pathogenesis of SARS-CoV-2 is determined by the structural complexity of envelop-based spike protein (S protein) (Mohanty et al. 2020). Structural recognition and cleavage mechanism of SARS-COV-2 spike protein are discussed in the later part of the chapter.

Clinical samples from SARS-COV-2-infected samples indicate severe acute respiratory distress syndrome (ARDS) to mild flu-like symptoms. The larger population remains undiagnosed, a symptomatic and subclinical level. WHO identifies that 40% of infected persons display mild to moderate symptoms, 15% sever and 5% critical (https://www.who.int/docs/default-source/coronaviruse/risk-comms-updates/update-18-epi-win%2D%2Dcovid-19.pdf?sfvrsn=cfb0471f_2). Patient admitted to hospitals in China indicate that approximately 88% of the patient display very high fever, while 68% display fatigue, sore throat and shortness of breath (https://www.who.int/docs/default-source/coronaviruse/who-china-joint-mission-on-covid-19-final-report.pdf). Advance cases generally had advance cardiovascular, respiratory and renal failure leading up to multiorgan failure (Yan et al. 2020a, 2020b). Gross anatomy of SARS-COV-2-infected human lungs was heavy, with interstitial edema, congestion, pulmonary embolid, patchy lung parenchyma and thrombosis in prostatic vein. Noteworthy changes in pulmonary architecture have been linked to clinical to severe SARS-COV-2 infections; however, long-term studies of SARS-COV-2 infection indicate widespread multiorgan pathogenicity of SARS-COV-2, including neuropathogenesis (Harrison et al. 2020). Therefore, it is essential to understand the structural and virulence pathogenic nature of spike protein and its mechanism of action.

12.4 STRUCTURE OF SARS-CoV-2

12.4.1 Structure of Virus

SARS-CoV-2 is a long RNA-containing envelope virus surrounded by a protective capsid which is repeated protein referred as coat or capsid (Figure 12.1). Furthermore, it is surrounded by an outer

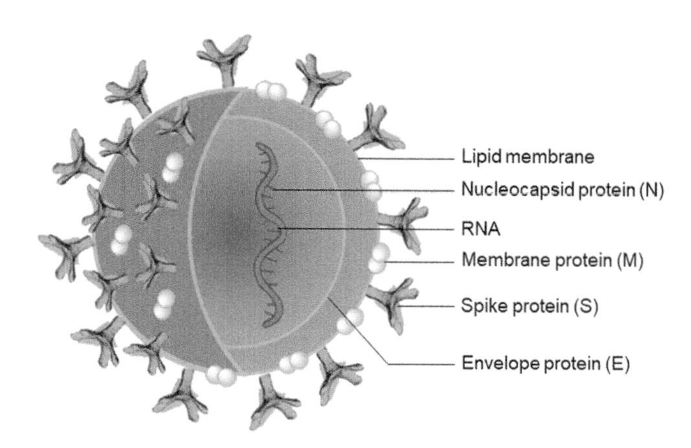

FIGURE 12.1 Structure of SARS-CoV-2.

membrane made up of lipids with protein inserted. SARS-CoV-2 is made up of four primary structural proteins: spike (S), small envelope (E), membrane (M) glycoprotein and nucleocapsid (N), as well as a number of auxiliary proteins. Fusion is based on the cleavage of the SARS-CoV-2 S1 and S2 subunits. Spike protein facilitates binding of envelope viruses to host cell by angiotensin-converting enzyme 2 (ACE2). The nucleocapsid known as N protein which is structurally bound to the nucleic acid of virus is involved in processes like viral replication cycle and the cellular response of host cells to viral infections (Figure 12.1). Membrane protein plays an important role in determining the shape of virus envelope and can bind to all other structural proteins. Binding of N protein with M protein helps its stabilization and promotes completion of viral assembly by stabilizing N protein-RNA complex. Envelope protein, the smallest protein, plays role in the production and maturation of this virus (Figure 12.1) (Astuti 2020; Cueno and Imai 2021; Philip et al. 2021).

12.4.2 Structure of Spike Protein

The total length of SARS-CoV-2 S protein is 1273 amino acids long and consists of a signal peptide (amino acids 1–13) located at the N-terminus, the S1 subunit (14–685 residues) and the S2 subunit (686–1273 residues). Spike protein has an extracellular N-terminus, a transmembrane (TM) domain anchored in the viral membrane, and a brief intracellular C-terminal portion with a size of 180–200 kDa. Once the virus interacts with host cell, there occurs extensive structural conformation changes in S protein which allows the fusion of virus to host cell membrane. S1 subunit is responsible for receptor binding whereas, S2 subunit helps in the membrane fusion. There is N-terminal domain (14–305 residues) and receptor-binding domain (RBD, 319–541 residues) in the S1 subunit. The fusion peptide (FP) (788–806), heptapeptide repeat sequence 1 (HR1) (912–984 residues), HR2 (1163–1213 residues), TM domain (1213–1237 residues) and cytoplasmic domain (1237–1273 residues) make up the S2 subunit. (Figure 12.2) (Brian and Baric 2005; Li 2016; Bangaru et al. 2020; Letarov et al. 2021, Pruimboom 2021).

The RBD which is situated in S1 subunit binds to ACE2 in the aminopeptidase region for viral fusion and entry FP, HR1, HR2, TM domain and cytoplasmic domain fusion (CT) of S2 are responsible. Membrane fusion occurs by disrupting and connecting lipid bilayer of the host cell membrane takes place by FP. The S2 subunits HR1 and HR2, which constitute the six-helical bundle (6-HB), are critical for viral fusion and entrance. While binding of RBD to ACE2, there is conformation change in S2 by inserting FP into the target cell membrane which exposes the prehairpin coiled-coil of HR1 and triggering interaction between HR2 and HR, thus brings the viral envelope and cell membrane into proximity for viral fusion and entry (Bosch et al. 2003; Taylor

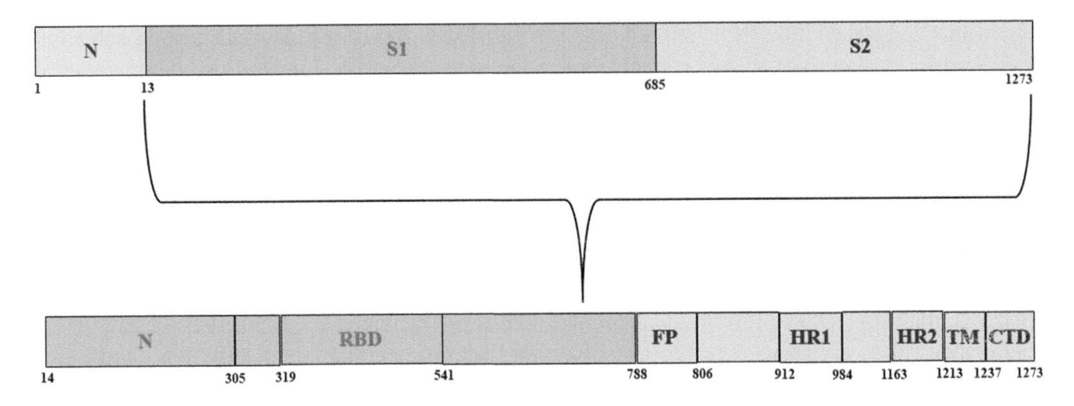

FIGURE 12.2 Structural domains of spike protein of SARS-CoV-2. N, N-terminal domain; RBD, receptor-binding domain; FB, fusion peptide, HR1, heptapeptide repeat sequence 1; HR2 heptapeptide repeat sequence 2; TM, transmembrane; CTD, cytoplasmic domain.

and Tom 2012; Miao et al. 2017; Daniel et al. 2020; Lim et al. 2020; Yu et al. 2020; Kalra and Kandimalla 2021; Koley et al. 2021; Belouzard et al. 2012).

12.5 CLEAVAGE DYNAMICS

SARS CoV-2 enters the host cell through the attachment of the S glycoprotein to the receptor. Virus enters the host cell through ACE2 receptors which are found in various organs such as kidneys, lungs, heart and gastrointestinal tract, thus facilitating viral entry into target cells. Attachment occurs at RBD which are present at 331–524 residues of S protein of SARS-CoV-2 and ACE2 receptors. Fusion is based on the cleavage of the SARS-CoV-2 S1 and S2 subunits. Host proteases cleave S protein into two parts, the S1 subunit and S2 subunit, and the subunits exist in a noncovalent form until viral fusion occurs. It has been found that the specific furin cleavage site is located in the cleavage site of SARS-CoV-2. The furin cleavage site is positioned at the intersection of the S1 and S2 subunits and contains four residues (P681, R682, R683 and A684). Transmembrane protease, serine 2 (TMPRSS2), in addition to host cell proteases is essential for S protein priming and has shown to be activated in the entry of SARS-CoV-2. After fusion occurs, the type II transmembrane serine protease (TMPRSS2) that is present on the surface of the host cell will activate the receptor-attached spike-like, S proteins and will clear the ACE2. When the S proteins are activated, conformational changes occur, allowing the virus to enter the cells. TMPRSS2 and ACE2 both proteins are the main determinants of the entry of this virus (Belouzard et al. 2009; Bestle et al. 2020; Jaimes et al. 2020; Shang et al. 2020; Xia et al. 2020; Yan et al. 2020a, 2020b; Thomas et al. 2021).

12.6 DESIGN OF MOLECULAR BLOCKERS

Viral attachment and entry are of particular interest among possible therapeutic targets in the life cycle of viruses because they represent the first steps in the replication cycle and take place at a relatively accessible extracellular site. For both SARS-CoV and SARS-CoV-2, this involves binding to human ACE2 followed by proteolytic activation by human proteases. Hence, blockade of the RBD–hACE2 protein–protein interaction (PPI) can disrupt infection, and most vaccines and neutralizing antibodies (nAbs) aim to abrogate this interaction. Peptide's blockers are considered to be highly target specific but in addition to this, they are hindered by difficulties related to their solubility, unsuitability for administration and immunogenicity. By being foreign proteins, they themselves can act as antigens and elicit strong immune responses in certain patients and this is only further exacerbated by their long elimination half-lives. Hence, an alternative approach could be offered by designing peptide or small molecular for SARS-CoV-2 infectivity (Campos et al. 2020; Duan et al. 2020; Wang et al. 2020; Xuhua 2021).

12.6.1 Peptide Blockers of Cleavage Site

Peptides are considered to be more effective and selective than small-molecule medications, giving them a viable alternative to small-molecule SARS-CoV-2 therapy. Thus, synthesis of peptide can be quickly implemented and altered in place of small molecules. In contrast, strategies for synthesizing small molecules are time-consuming and costly (Schutz et al. 2020; Pedro et al. 2021). Due to their relatively large size and proteolytic degradation, peptides frequently have low (oral) bioavailability and short half-lives and thus to translate them into clinical applications peptides need to be stable, potent, bioavailable and safe. However, peptides can be modified to improve activity and stability by shortening their sequences, changing amino acids or adding moieties that increase their affinities to the respective partner as often determined by computational studies. Antiviral peptides majorly target the structures responsible for viral replication. For instance, they may target or interfere the components essential for virus infection such as enzymatic activity, modulate conformational changes or viral proteins (Huang et al. 2020a, 2020b; Schutz et al. 2020; Zhao et al. 2021).

Some of them, including the endogenous peptide inhibitor of CXCR4 EPI-X4 or the virus inhibitory peptide VIRIP, are released from abundant precursor proteins by proteolytic cleavage. Based on the structure and function of viral proteins alternatively, bioactive peptides can be rationally designed. For the design of antiviral peptides, the structural moiety is designed on the analysis of charges, steric effects, polarity and biomolecular simulations. A straightforward approach to peptide design is to adopt sequences from the interaction sites of proteins. These peptides have the potential to block key PPIs. Peptides are frequently engineered to operate extracellularly, that is, to target early stages of viral replication like viral envelope glycoprotein activation, receptor attachment or fusion. This offers the advantage of removing the need for the therapeutic peptide to permeate the cell membrane, as well as reducing the risk of detrimental interactions between the viral pathogen and host cells (Schutz et al. 2020; Surid et al. 2020; Gilg et al. 2021; Sokkar et al. 2021).

There are various targets where peptide can act *viz.* prevention of ACE2 binding of the SARS-CoV-2 spike protein, targeting ACE2, targeting proteolytic S protein activation (furin, TMPRSS2 and cathepsin L) and fusion mechanism (Table 12.1).

12.6.2 Small-Molecule Inhibitors

Various pharmacological companies are actively involved in developing small-molecule inhibitors to prevent the entry of the virus into human hosts although, physical isolation is the ideal way of limiting the spread (Beeraka et al. 2020; Bojadzic et al. 2021). Small molecules could target in various ways against SARS-CoV-2 *viz.* as viral entry inhibitors, kinase inhibitors, membrane protease inhibitors and interferon therapy (Table 12.2). There are various small-molecule inhibitors which have been computationally and experimentally proven to be potent target against SARS-CoV-2 (Yanxiao and Petr 2020).

Antibodies, like all protein therapeutics, face challenges such as solubility, suitability for oral or inhaled administration, and immunogenicity, in addition to being overly target specific. Because they are foreign proteins, they can behave as antigens and provoke powerful immune responses in some patients, which is made worse by their extended elimination half-lives. Even among treatments approved by the US Food and Drug Administration (FDA), biologics have more postmarket safety problems than small-molecule medications. As a result, peptides or tiny compounds may provide an alternative. There have been some reports of peptide disruptors for this PPI, but none

TABLE 12.1
Various Peptide and Peptide Inhibitors with Their Target and Sequence

Target Protein	Name	Sequence
Spike Protein	SBP1	IEEQAKTFLDKFNHEAEDLFYQS
	AHB1	EVAKEAKDASRRGD DERAKEQMERAMRLF DQVFELAQELQE KQTDGNRQKATHLDKA VKEAADELYQR VRELEEQVMHVLDQVSEL AHELLHKLT GEELERAAYFNWWATEMML ELIKSDDEREIREIEEEAR RILEHLEELARK
	LCB1	DKEWILQKIYEIMRLLDELGHAEASMRVSDLIYEF MKKGDERLLEEAERLLEEVER
	CRM197	
ACE2	SARS-CoV-2	Unknown
α5β1 integrin	ATN-161	Ac-PHSCN-NH2
TMPRSS2	Aprotinin	RPDFC LEPPY TGPCK ARIIR YFYNA KAGLC QTFVY GGCRA KRNNF KSAED CMRTC GGA
	MI-432	(S)-3-(3-(4-(2-Aminoethyl) piperidin-1-yl)-2-((2′,4′-dichloro-[1,1′-biphenyl])-3-sulfonamido)-3-oxopropyl) benzimidamide
	MI-1900	(S)-4-(3-(3-Carbamimidoylphenyl)-2-((2′,4′-dimethoxy-[1,1′-biphenyl])-3-sulfonamido) propanoyl)-N-cyclohexylpiperazine-1-carboxamide
Cathepsin L	Teicoplanin	antibiotic glycopeptide
	P9	NGAICWGPCPTAFRQIGNCGHFKVRCCKIR
	P9R	NGAICWGPCPTAFRQIGNCGRFRVRCCRIR
	8P9R	8 x NGAICWGPCPTAFRQIGNCGRFRVRCCRIR
Furin	dec-RVKR-cmk	dec-RVKR-cmk
	MI-1851	(S)-N-((S)-1-((4-Carbamimidoylbenzyl) amino)-4-(guanidinooxy)-1-oxobutan-2-yl)-2-((S)-2-(2-(4-(guanidinomethyl)phenyl) acetamido)-4-(guanidinooxy) butanamido)-3,3-dimethylbutanamide
Spike S2 (membrane fusion)	EK1	SLDQINVTFLDLEYEMKKLEEAIKKLEESYIDLKEL
	2019-nCoV-HR2P	DISGINASVVNIQKEIDRLNEVAKNLNESLIDLQEL
	[SARSHRC-PEG4] 2-chol	[DISGINASWNIQKEIDRLNEVAKNLNESLIDLQEL-PEG4]2-chol
	EK1-C4	Ac-SLDQINVTFLDLEYEMKKLEEAIKKLEESYIDLKELGSGSG-amino-PEG3-acetyl-Cys (cholesteryloxycarbonylmethyl)-NH acetate salt
	IPB01-IPB-09	IBP02: ISGINASVVNIQKEIDRLNEVAKNLNESLIDLQELK(Chol)

TABLE 12.2
Small-Molecule Inhibitors with Their Structures

Small Molecule	Structure	Small Molecule	Structure
SiRNA		PD98059	
GRL0617	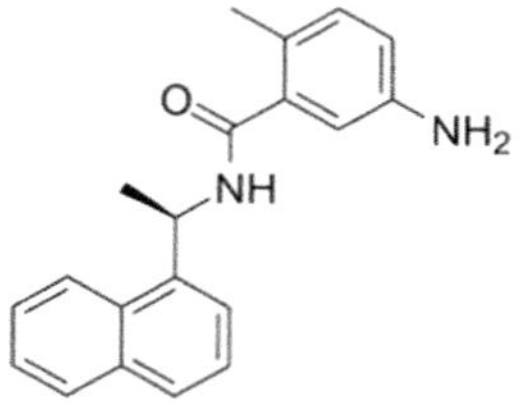	SB308520	-
Benzodioxolane derivatives	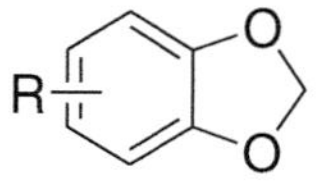	SP600125	-
5-chloropyridinyl indolecarboxylate	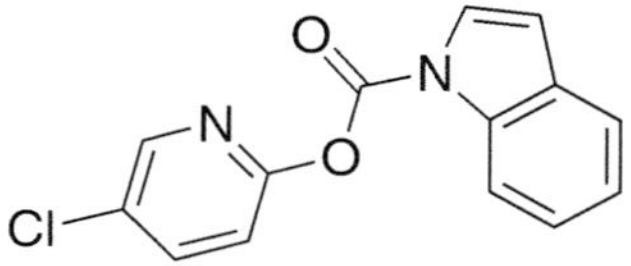	Chloroquine	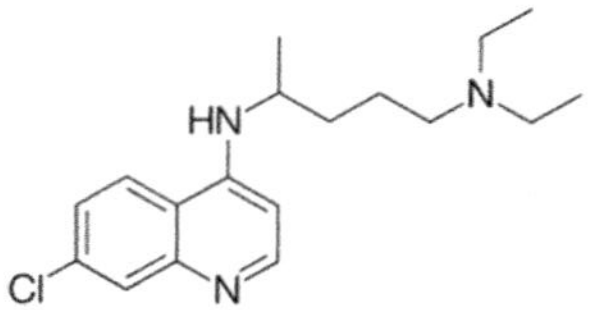
2978/10 humanized antibodies	-	Hydroxychlorquine	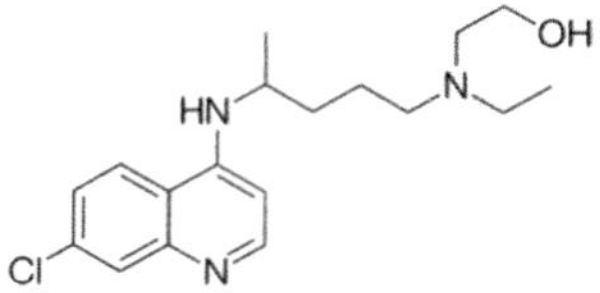

Amiodarone	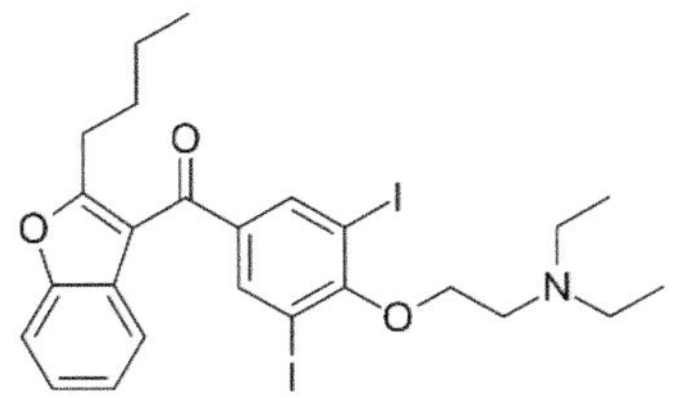	EGCG	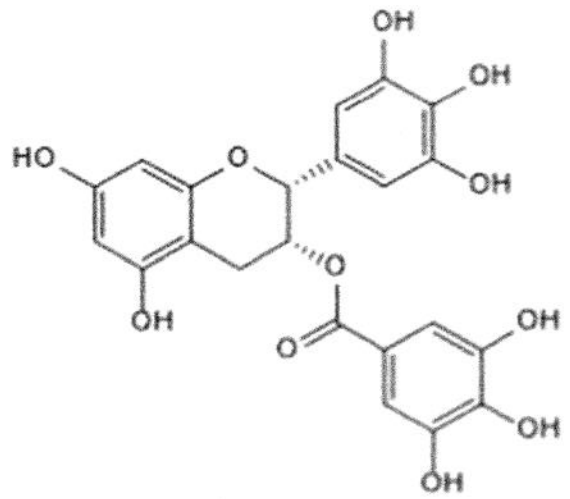
Arbidol	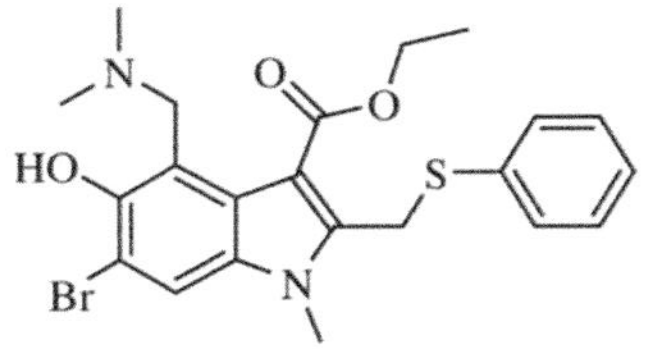	SFN	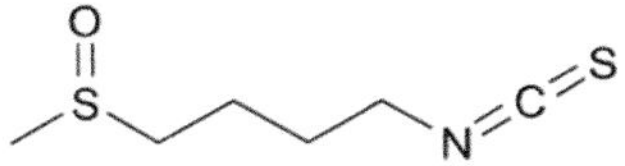
TSL-1	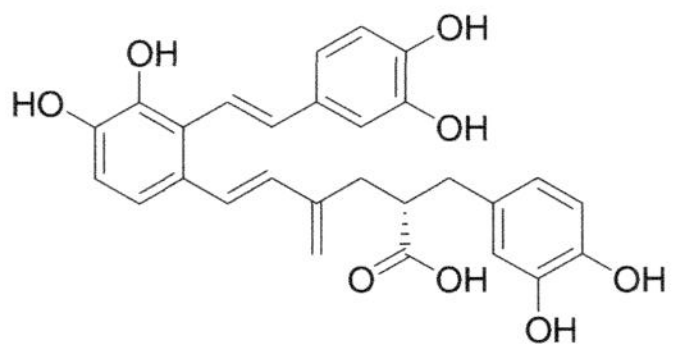	a-luminol (monosodium a-luminol	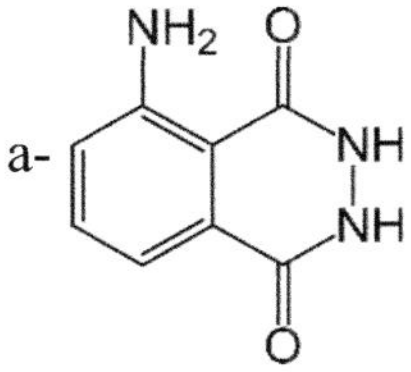
TACE inhibitor (TAPI-2)	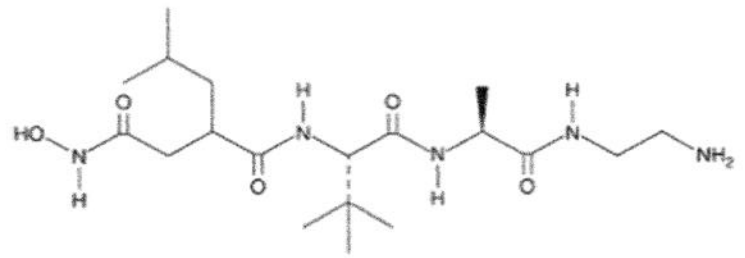	Tanshinone IIA	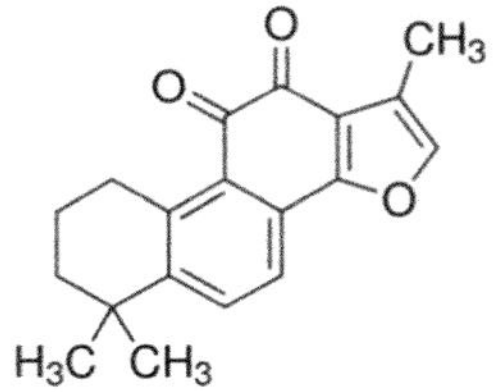

(Continued)

TABLE 12.2 (Continued)
Small-Molecule Inhibitors with Their Structures

Small Molecule	Structure	Small Molecule	Structure
IFN—a B/D	-	Lucidone	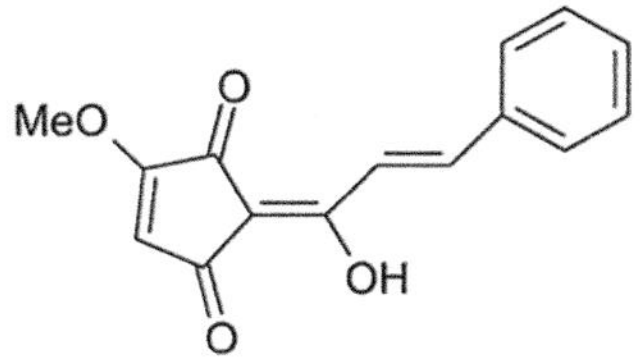
IFN-b and-g	-	Celastrol (quinone methide triterpene)	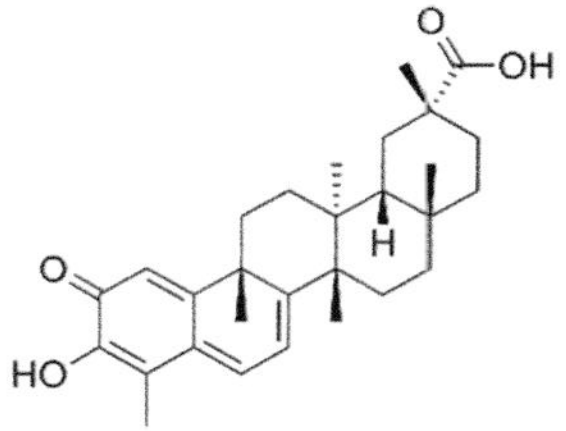
Camostat	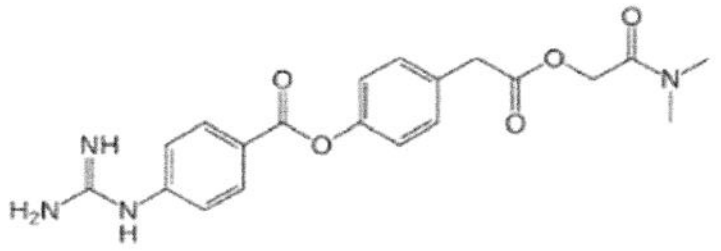	Bakuchiol (phenolic isoprenoid)	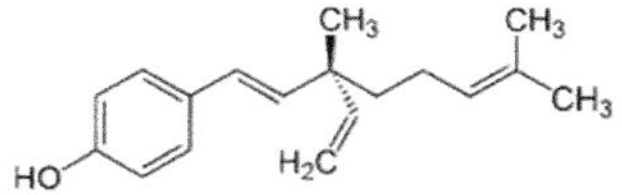
Nafamostat	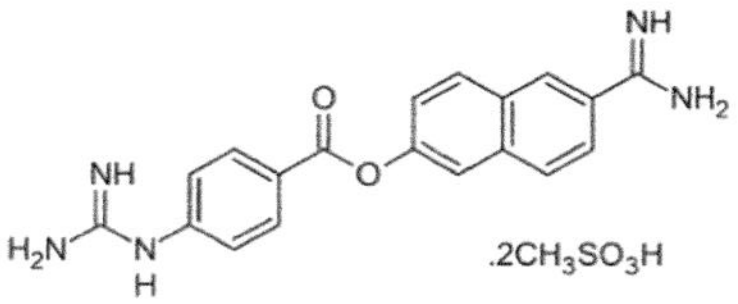	Rupestonic acid (sesquiterpene)	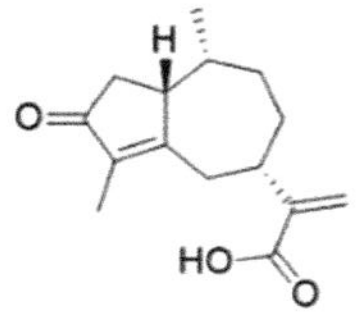

Lopinavir	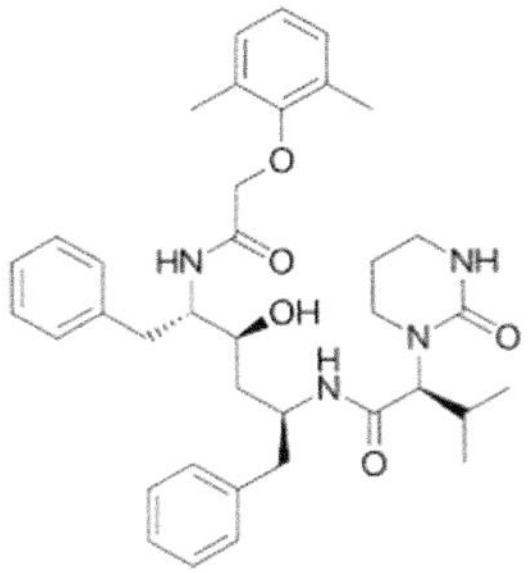	Curcumin	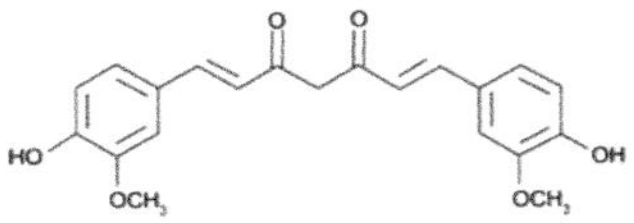
Nelfinavir	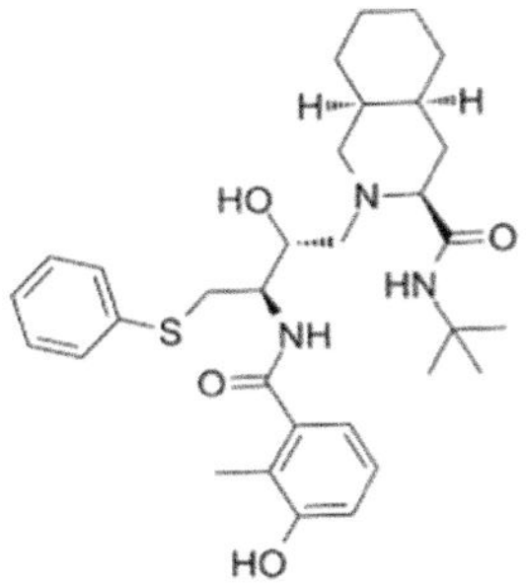	Baicalin (a flavonoid)	(a 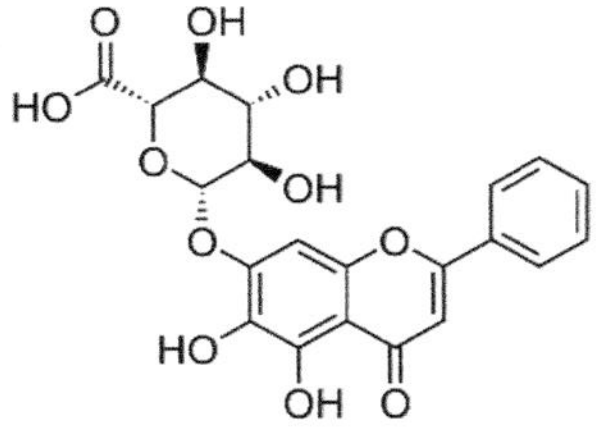
[[(Z)-1-thiophen-2-ylethylideneamino]thiourea]	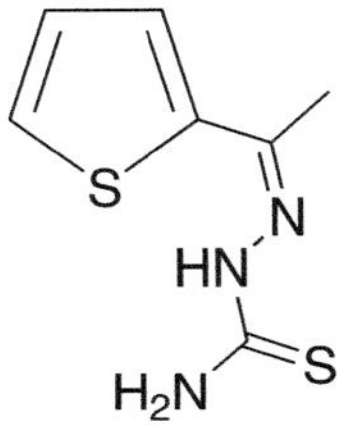	Scutellarin (a flavone glycoside)	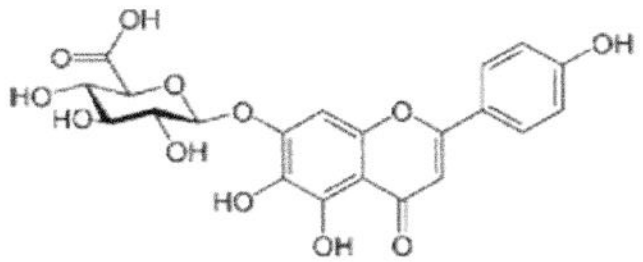

(Continued)

TABLE 12.2 (Continued)
Small-Molecule Inhibitors with Their Structures

Small Molecule	Structure	Small Molecule	Structure
N-[[4-(4-methylpiperazin-1-yl)phenyl]methyl]-1,2-oxazole-5-carboxamide	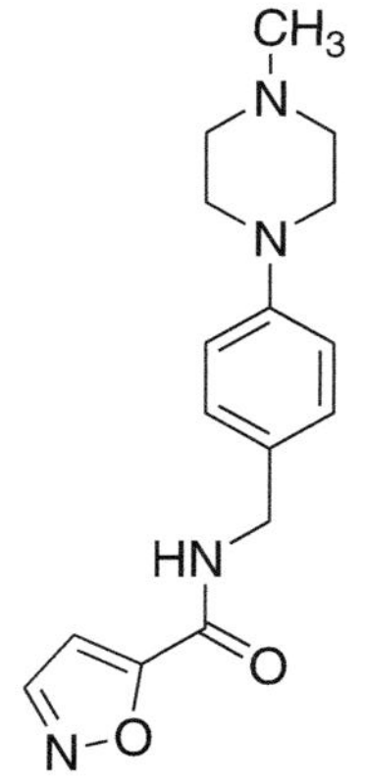	Nicotianamine	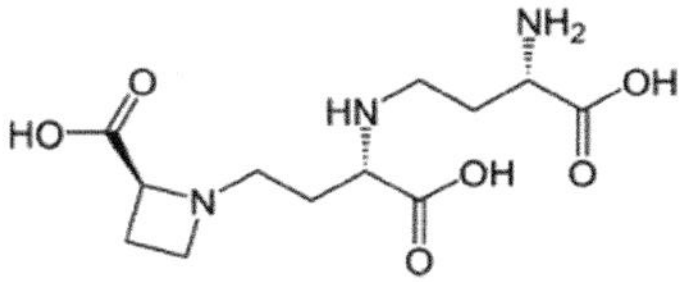
[N-(9,10-dioxo-9,10-dihydroanthracen-2-yl)benzamide]	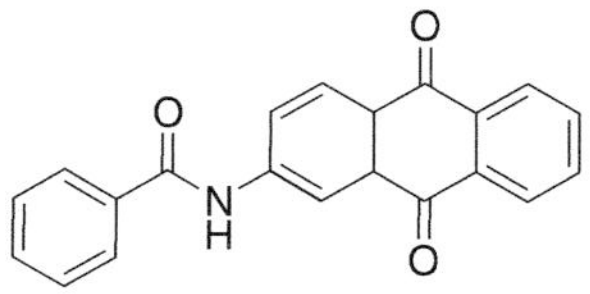	Glycyrrhizin	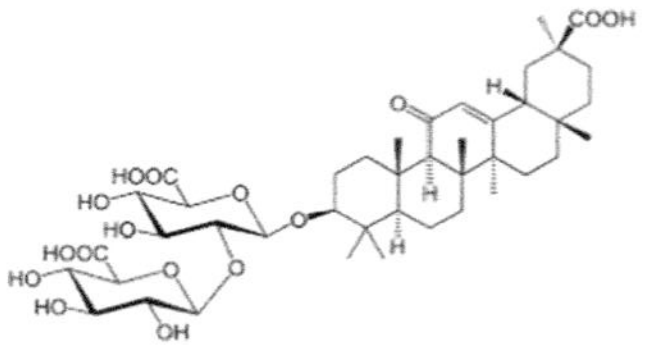
Trametinib	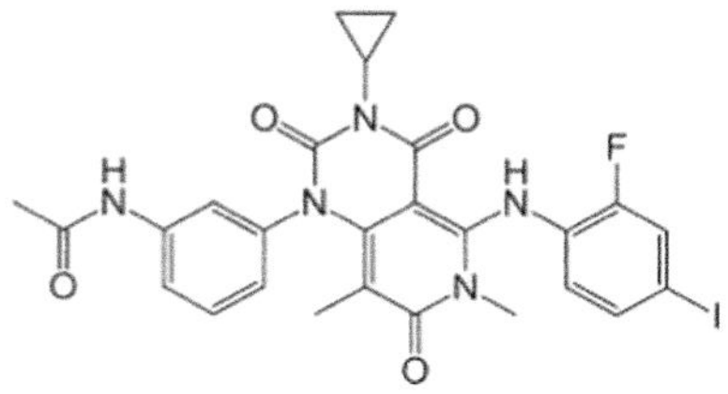	Hesperetin glycoside	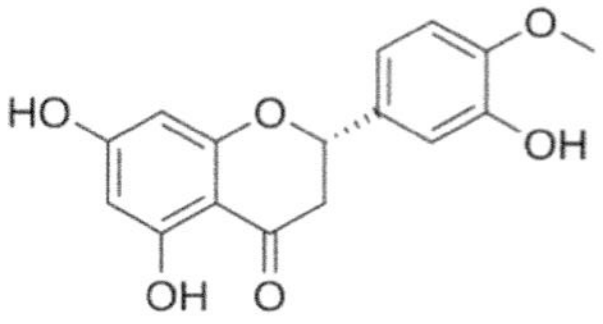

Selumetinib	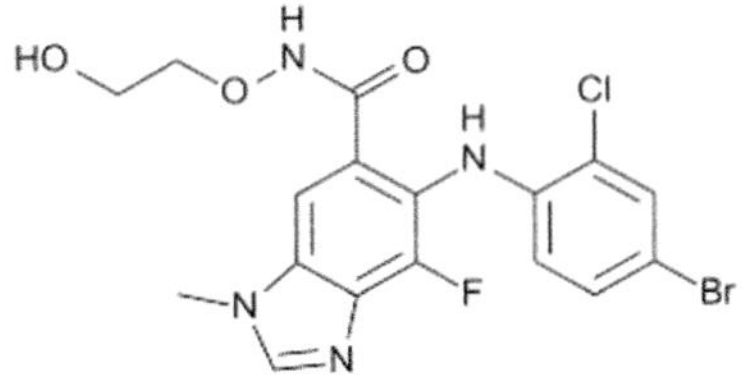	Naringenin	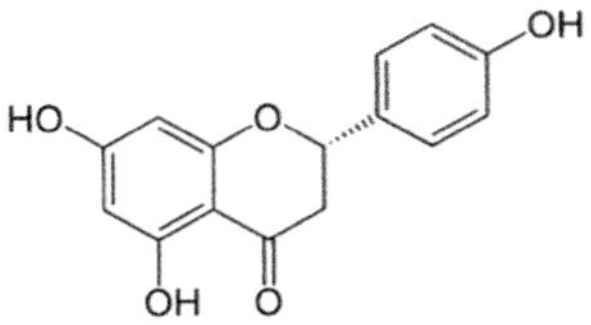
Everolimus	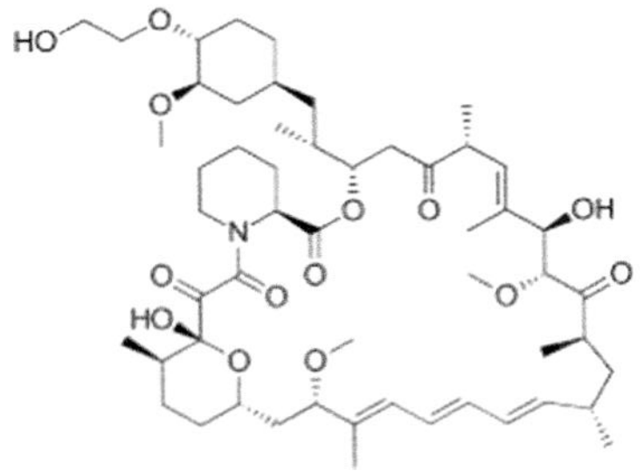	Betulinic acid	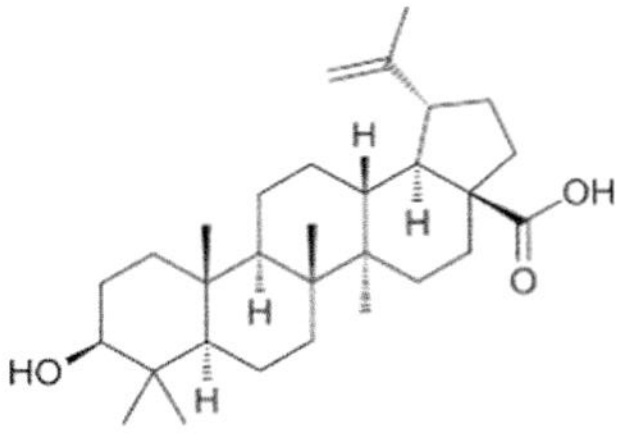
Miltefosine	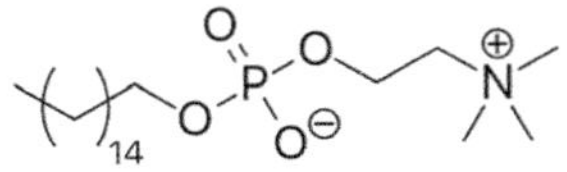	Griffithsin	-
Teriflunamide	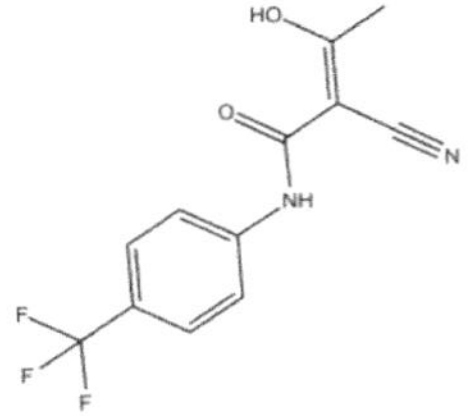	Savinin	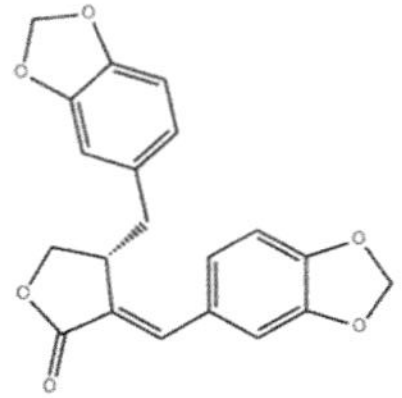
Leflunomide	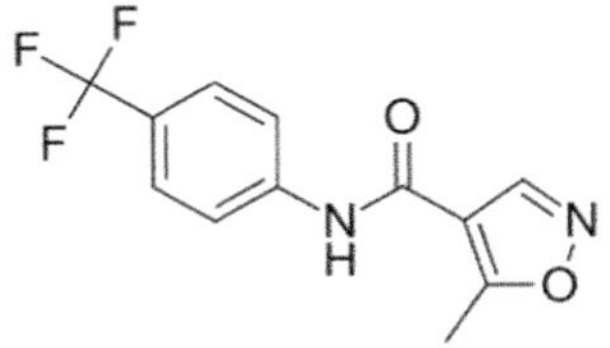	Quercetin	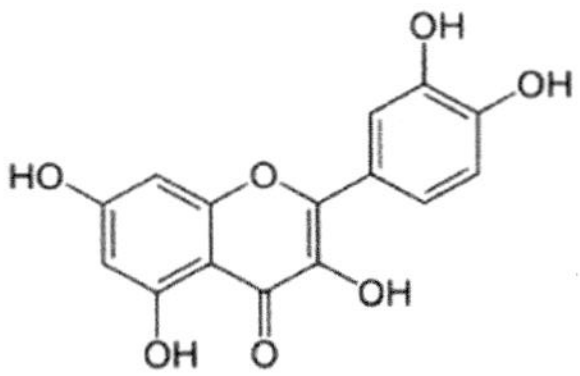

(Continued)

TABLE 12.2 (Continued)
Small-Molecule Inhibitors with Their Structures

Small Molecule	Structure	Small Molecule	Structure
Dasatinib	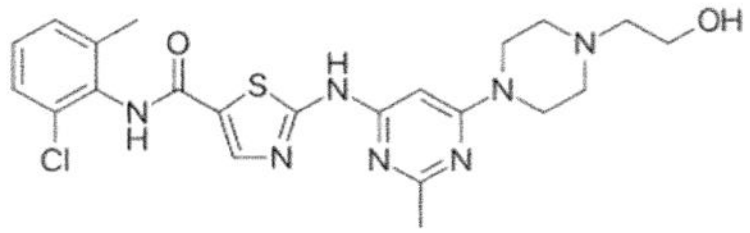	Kaempferol	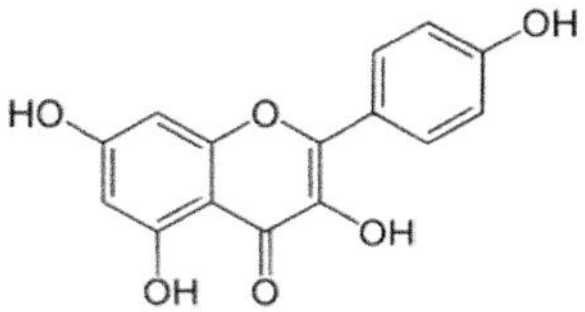
Imatinib	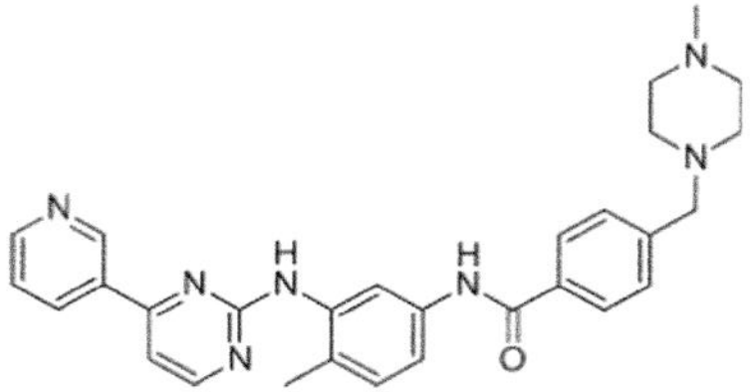	Allicin	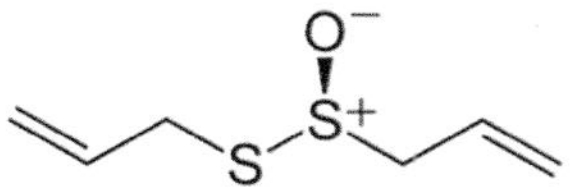
Nilotinib	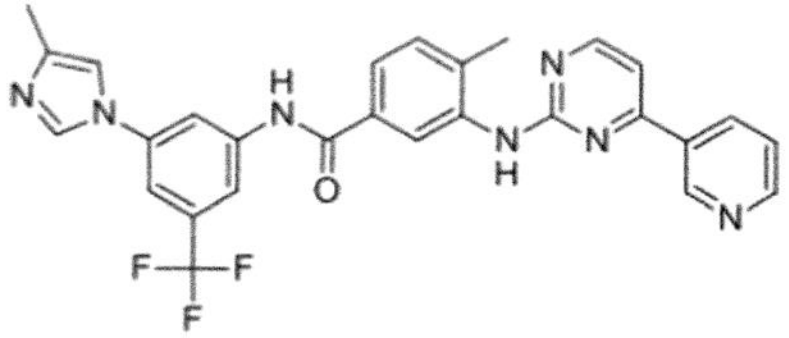	Gingerol	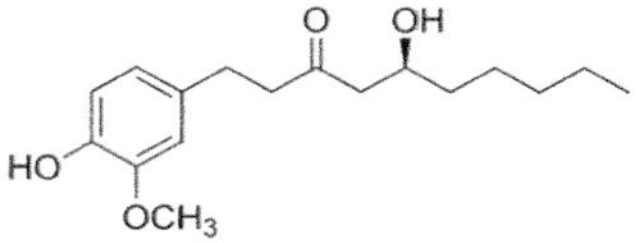
Luteolin-7- glucoside	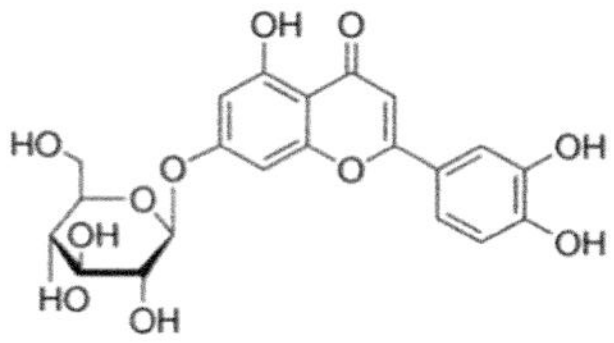	Catechin	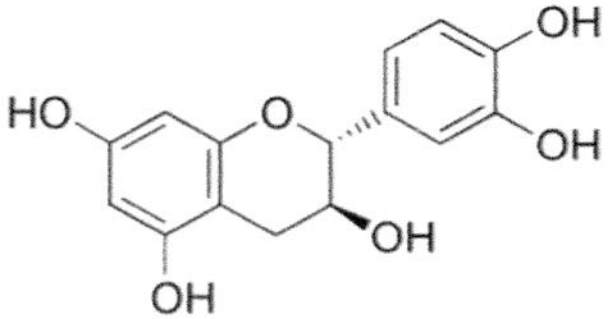
Apigenin-7- glucoside	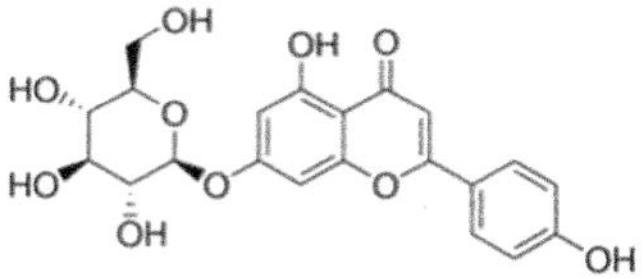	Epicatechingallate	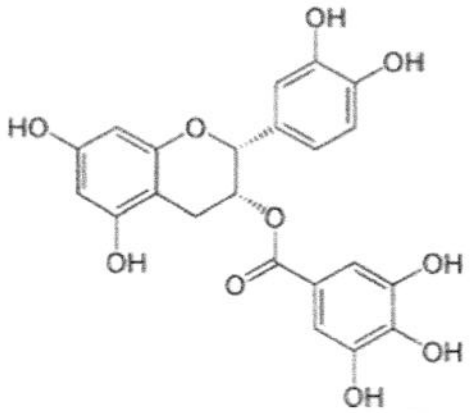

have proven to be highly successful. More importantly, turning peptides into clinically authorized medications is challenging and rarely attempted due to bioavailability, metabolic instability (short half-life), lack of membrane permeability and other difficulties (Bruno et al. 2013; Bulatao et al. 2020; Wang et al. 2021).

12.7 REPURPOSE THERAPEUTIC POTENTIAL FDA-APPROVED DRUGS FOR SARS-CoV-2

The therapeutic potential of different candidate and FDA-approved drugs has been evaluated since the advent of SARS-CoV-2 in 2019. In the absence of a formidable therapeutic regime, repurposing of drugs gained serious attention. These drugs, alone or in combinations, were targeted to reduce severity and mortality among high-risk patients. Some of the most commonly used drugs are remdesivir, hydroxychloroquine (HCQ), chloroquine (CQ), favipiravir, ribavirin, etc. (Guy et al. 2020, Tian et al. 2021, Hosseini and Amanlou 2020).

The positive-sense single-stranded coronavirus genome possesses two open reading frames (ORFs) at 5' terminal, encoded by polyprotein 1a (ORF1a) and 1b (ORF1b). Transcription and replication of these ORFs essentially regulate viral replication, while 16 types of nsps can cleave the polyprotein 1a/1b. 12 of 16 nsps are non-structural RNA-dependent RNA polymerase catalyzes new complementary RNA strand synthesis (Zhu et al. 2020; Flower et al. 2021; Giovanetti et al. 2021). Therefore, efforts have been made to inhibit RNA-dependent RNA polymerase to crumble the viral replication process. Nucleoside and non-nucleoside drugs known to inhibit RNA-dependent RNA polymerase in other viruses were employed as a strategy for SARS-CoV-2 treatment. Studies indicate that non-nucleoside drug is highly susceptible to drug resistance. Showing various degrees of inhibition of RNA-dependent RNA polymerase activity among the lineages of SARS-CoV-2 (Kokic et al. 2021). While nucleoside drugs act as catalytic substrates and bind at the highly conserved pocket of the RNA-dependent RNA polymerase, therefore, inhibit viral replications (Zhu et al. 2020). Remdesivir (GS-5734) is a nucleoside pro-drug. When metabolized, remdesivir is converted into a corresponding active triphosphate structure; that can bind to inhibit replication by the termination of new RNA synthesis in SARS-CoV-2 genome (Gordon et al. 2020). A significant decrease in recovery time seems promising in the absence of vaccination. Remdesivir administrated patient hospital stay was approx. Ten days as compared to the control/placebo group's 15-day average stay (Tchesnokov et al. 2019). The US Food and Drug Administration and European Medicine Agency approved remdesivir for the treatment of SARS-CoV-2 on October 22 (Beigel et al. 2020, Wise 2020). Favipiravir a derivative of pyrazinecarboxamide has been shown to inhibit viral replication. It is effective against influenza, West Nile, Yellow fever and other flaviviruses. Prodrug Favipiravir undergoes phosphoribosylation to generate an intermediate favipiravir-ribofuranosyl-5′-monophosphate, then metabolized into the active form favipiravir-ribofuranosyl-5′-triphosphate, that can bind to the pre-catalytic site of RNA-dependent RNA polymerase (Baranovich et al. 2013). The State Medicine Administration of China approved favipiravir as the first drug for the treatment of SARS-CoV-2. While a multicentre clinical trial result showed increased serum uric levels in patients (Chen et al. 2021). However, Favipiravir remains more effective than other antiviral tested, such as Arbidol-based antipyretic drugs. Ribavirin is a broad-spectrum antiviral medication used to treat respiratory distress caused by a respiratory syncytial virus, hepatitis C, and some hemorrhagic viral infections (Te et al. 2007). Prodrug ribavirin is converted into RBV monophosphate (RMP) by adenosine kinase. Through nucleoside monophosphate and diphosphate kinase, RBV monophosphate is converted into RBV diphosphate and ribavirin triphosphate. Ribavirin mediates SARS-CoV-2 treatment and remains less effective as compared to the previous inhibitor (Peng et al. 2021). However, the magnitude of ribavirin-mediated biological function out way previously mentioned prodrugs. Ribavirin-mediated antiviral effects come from four major pathways; 1. Modulation, and shift from Th2 to

Th1 immune response, therefore stimulating host cytotoxic T cells to effectively clear vial pathogens; 2. RBV monophosphate (RMP) depletes GTP by inhibiting inosine monophosphate dehydrogenase, therefore, depleting resources for viral replication; 3. RBV diphosphate directly binds to the polymerase complex and inhibits viral replication; 4. RBV diphosphate facilitates RNA mutagen to replicate therefore mutating the virus replication system (Tian et al. 2021).

Antimalarial drugs Hydroxychloroquine (HCQ) and chloroquine (CQ) inhibit endosome function and the fusion of autophagosome to the lysosome (Hosseini and Amanlou 2020). Coronavirus effectively uses the endolysosomal pathway to enter inside cells. Within the cells, the virus is uncoated to release RNA particle and hijacking cellular function. Again, in the absence of concrete treatment strategies, the clinician used HCQ and CQ to treat SARS-CoV-2-infected patients (Guy et al. 2020). These drugs showed some promising results but were not consistent among the geographical locations. Small-scale randomized clinical trial results also confirm that HCQ can protect against the severity and hospitalization of SARS-CoV-2 infection. Adding Azithromycin, a broad-spectrum antibiotic known to block autophagosome clearance to the HCQ treatment regime, further improved the severity and hospitalization of SARS-CoV-2 infection compared with HCQ alone. In the United States, HCQ alone or in combination with Azithromycin obtained emergency permission for COVID-19 because to its therapeutic potential and capacity to minimize hospital stay in patients (Hosseini and Amanlou 2020, Tian et al. 2021).

The release of viral genomic RNA from capsid within cells is then used for cap-dependent translation to produce peptides that facilitate the synthesis of new viral particles. Due to protein synthesis's importance in the viral life cycle, targeting the synthesis or assembly of the peptides with an inhibitor seems a viable target against SARS-CoV-2 infections, such as in HIV protease inhibitors for the treatment of HIV (Hosseini and Amanlou 2020). Therefore, efforts were made to assay if the combination of HIV protease inhibitor and ritonavir can be effective against SARS-CoV-2 infection, as the combination was effective against SARS-CoV-1 infection. However, clinical findings indicated that alone, or in combination, the protease inhibitors were ineffective against SARS-CoV-2 (Tian et al. 2021).

The efficiency and efficacy of a drug depend on various factors and are designed for specific diseases; therefore, repurposing those drugs may not meet the required standard. However, in the absence of a therapeutic strategy, these drugs can be repurposed to reduce the severity, including mortality rates. R0 value of SARS-CoV-2 in 2019 exceeded 1, and the death toll reached millions within a short time (Mohanty et al. 2020). Therefore, repurposing drugs was instrumental in appeasing the crisis and provided time for researchers to evaluate various therapeutic options. Among the hundreds of drugs repurposed to treat SARS-CoV-2, HCQ, Camostat, Nafamostat, Azithromycin, Chloroquine (CQ), Remdesivir, Favipiravir and Ribavirin were the most promising ones (Yang et al. 2021).

12.8 CONCLUSION

Due to the lack of selective targeted therapies and vaccination strategies, the life-threatening consequences of the COVID-19 pandemic remain high. The primary reason for this is due to high genomic variability of SARS-CoV-2 as well as its invading mechanism in the host cell. Understanding the structure of SARS-CoV-2 and its domain of spike proteins can inform various strategies for drug designing. Receptor-binding domain is considered to be the most immunogenic region of spike protein. Hence, RBD could be considered a major target for vaccination, drug designing or antibody therapy. Various peptides, peptide inhibitors and small-molecule inhibitors have been designed and tested again SARS-CoV-2 both biologically and computationally. However, many more preclinical and clinical studies are required to discover the therapeutic efficacy of these designed peptides and small molecules.

REFERENCES

Astuti, I.Y., 2020. Severe acute respiratory syndrome coronavirus 2 (SARS-CoV-2): An overview of viral structure and host response. *Diab. Metab. Syndr.* 14, 407–412.

Baranovich, T., Wong, S.S., Armstrong, J., Marjuki, H., Webby, R.J., Webster, R.G., Govorkova, E.A., 2013. T-705 (favipiravir) induces lethal mutagenesis in influenza A H1N1 viruses in vitro. *J. Virol.* 87, 3741–3751.

Bangaru, S., Ozorowski, G., Turner, H.L., Antanasijevic, A., Huang, D., Wang, X., et al., 2020. Structural analysis of full-length SARS-CoV-2 spike protein from an advanced vaccine candidate. *Science* 370, 1089–1094.

Beeraka, N.M., Sadhu, S.P., Madhunapantula, S.V., Rao, P.R., Svistunov, A.A., Nikolenko, V.N., Mikhaleva, L.M., Aliev, G., 2020. Strategies for targeting SARS CoV-2: Small molecule inhibitors-the current status. *Front Immunol.* 11, 1–22.

Beigel, J.H., Tomashek, K.M., Dodd, L.E., Mehta, A.K., Zingman, B.S., Kalil, A.C., Hohmann, E., Chu, H.Y., Luetkemeyer, A., Kline, S. and Lopez de Castilla, D., 2020. Remdesivir for the treatment of Covid-19. *New Eng. J. Med.* 383, 1813–1826.

Belouzard, S., Chu, V.C., Whittaker, G.R., 2009. Activation of the SARS coronavirus spike protein via sequential proteolytic cleavage at two distinct sites. *Proc. Natl. Acad. Sci. USA* 106, 5871–5876.

Belouzard, S., Millet, J.K., Licitra, B.N., Whittaker, G.R., 2012. Mechanisms of coronavirus cell entry mediated by the viral spike protein. *Viruses* 4, 1011–1033.

Bestle, D., Heindl, M.R., Limburg, H., Van, L.T., Pilgram, O., Moulton, H., et al., 2020. TMPRSS2 and furin are both essential for proteolytic activation of SARS-CoV-2 in human airway cells. *Life Sci. Alliance* 3, 1–14.

Boni, M.F., Lemey, P., Jiang, X., Lam, T.T.Y., Perry, B.W., Castoe, T.A., Rambaut, A., Robertson, D.L., 2020. Evolutionary origins of the SARS-CoV-2 sarbecovirus lineage responsible for the COVID-19 pandemic. *Nat. Microbiol.* 5, 1408–1417.

Bojadzic, D., Alcazar, O., Chen, J., Chuang, S.T., Condor, C.J.M., Shehadeh, L.A., Buchwald, P., 2021. Small-molecule inhibitors of the coronavirus spike: ACE2 protein-protein interaction as blockers of viral attachment and entry for SARS-CoV-2. *ACS Infect. Dis.* 7, 1519–1534.

Bosch, B.J., Van Der Zee, R., De, H.C.A., Rottier, P.J., 2003. The coronavirus spike protein is a class I virus fusion protein: Structural and functional characterization of the fusion core complex. *J. Virol.* 16, 8801–8811.

Brian, D.A., Baric, R.S., 2005. Coronavirus genome structure and replication. *CTMI* 287, 1–30.

Bulatao, I., Pinnow, E., Day, B., 2020. Postmarketing safety-related regulatory actions for new therapeutic biologics approved in the United States 2002–2014: Similarities and differences with new molecular entities. *Clin. Pharmacol. Therap.* 108, 1243–1253.

Bruno, B.J., Miller, G.D., Lim, C.S., 2013. Basics and recent advances in peptide and protein drug delivery. *Therap. Deliv.* 4, 1443–1467.

Campos, D.M.O., Fulco, U.L., De Oliveira, C.B.S., Oliveira, J.I.N., 2020. SARS-CoV-2 virus infection: Targets and antiviral pharmacological strategies. *J. Evid. Based Med.* 13, 255–260.

Chen, C., Zhang, Y., Huang, J., Yin, P., Cheng, Z., Wu, J., et al., 2021. Favipiravir versus arbidol for clinical recovery rate in moderate and severe adult COVID-19 patients: a prospective, multicenter, open-label, randomized controlled clinical trial. *Front. Pharmacol.* 12, 683296.

Cueno, M.E., Imai, K., 2021. Structural comparison of the SARS CoV 2 spike protein relative to other human-infecting coronaviruses. *Front. Med.* 7, 1–10.

Daniel, W., Nianshuang, W., Kizzmekia, S.C., Jory, A.G., Ching-Lin, H., Olubukola, A., Barney, S.G., Jason, S.M., 2020. Cryo-EM structure of the 2019-nCoV spike in the prefusion conformation. *Science* 367, 1260–1263.

Duan, L., Zheng, Q., Zhang, H., Niu, Y., Lou, Y., Wang, H., 2020. The SARS-CoV-2 spike glycoprotein biosynthesis, structure, function, and antigenicity: Implications for the design of spike-based vaccine immunogens. *Front Immunol.* 11, 1–12.

El-Sayed, A., Kamel, M., 2020. Climatic changes and their role in emergence and re-emergence of diseases. *Environ. Sci. Poll. Res.* 27, 22336–22352.

Flower, T.G., Buffalo, C.Z., Hooy, R.M., Allaire, M., Ren, X., Hurley, J.H., 2021. Structure of SARS-CoV-2 ORF8, a rapidly evolving immune evasion protein. *Proc. Natl. Acad. Sci. USA* 118, 1–6.

Giovanetti, M., Benedetti, F., Campisi, G., Ciccozzi, A., Fabris, S., Ceccarelli, G., et al., 2021. Evolution patterns of SARS-CoV-2: Snapshot on its genome variants. *Biochem. Biophys. Res. Commun.* 538, 88–91.

Gilg, A., Harms, M., Olari, L.R., 2021. Absence of the CXCR4 antagonist EPI-X4 from pharmaceutical human serum albumin preparations. *J. Transl. Med.* 19, 190–195.

Gordon, C.J., Tchesnokov, E.P., Woolner, E., Perry, J.K., Feng, J.Y., Porter, D.P. Götte, M. 2020. Remdesivir is a direct-acting antiviral that inhibits RNA-dependent RNA polymerase from severe acute respiratory syndrome coronavirus 2 with high potency. *J. Biol. Chem.* 295, 6785–6797.

Guy, R.K., DiPaola, R.S., Romanelli, F., Dutch, R.E., 2020. Rapid repurposing of drugs for COVID-19. *Science* 368, 829–830.

Harrison, A.G., Lin, T., Wang, P., 2020. Mechanisms of SARS-CoV-2 transmission and pathogenesis. *Trends Immunol.* 41, 1100–1115.

Hosseini, F.S., Amanlou, M., 2020. Anti-HCV and anti-malaria agent, potential candidates to repurpose for coronavirus infection: Virtual screening, molecular docking, and molecular dynamics simulation study. *Life Sci.* 258, 118205–118218.

Hu, B., Guo, H., Zhou, P., Shi, Z.L., 2021. Characteristics of SARS-CoV-2 and COVID-19. *Nat. Rev. Microbiol.* 19, 141–154.

Huang, Y., Yang, C., Xu, X.F., Xu, W., Liu, S.W., 2020a. Structural and functional properties of SARS-CoV-2 spike protein: Potential antivirus drug development for COVID-19. *Acta Pharmacol. Sin.* 41, 1141–1149.

Huang, C., Wang, Y., Li, X., Ren, L., Zhao, J., Hu, Y., et al., 2020b. Clinical features of patients infected with 2019 novel coronavirus in Wuhan, China. *Lancet.* 395, 497–506.

Jaimes, J.A., Millet, J.K., Whittaker, G.R., 2020. Proteolytic cleavage of the SARS-CoV-2 spike protein and the role of the novel S1/S2 site. *iScience* 23, 101212–101220.

Kalra, R., Kandimalla, R., 2021. Engaging the spikes: heparan sulfate facilitates SARS-CoV-2 spike protein binding to ACE2 and potentiates viral infection. *Signal Transduct. Target. Ther.* 6, 39–40.

Kahn, J.S., McIntosh, K., 2005. History and recent advances in coronavirus discovery. *Pediatr. Infect. Dis. J.* 24, S223–S227.

Kirtipal, N., Bharadwaj, S., Kang, S.G., 2020. From SARS to SARS-CoV-2, insights on structure, pathogenicity and immunity aspects of pandemic human coronaviruses. *Infect. Genet Evol.* 85, 104502–104518.

Koley, T., Madaan, S., Chowdhury, S.R., Kumar, M., Kaur, P., Singh, T.P., Ethayathulla, A.S., 2021. Structural analysis of COVID-19 spike protein in recognizing the ACE2 receptor of different mammalian species and its susceptibility to viral infection. *3 Biotech* 11, 109–125.

Kokic, G., Hillen, H.S., Tegunov, D., Dienemann, C., Seitz, F., Schmitzova, J., Farnung, L., Siewert, A., Höbartner, C., Cramer, P., 2021. Mechanism of SARS-CoV-2 polymerase stalling by remdesivir. *Nat. Commun.* 12, 1–7.

Letarov, A.V., Babenko, V.V., Kulikov, E.E., 2021. Free SARS-CoV-2 spike protein S1 particles may play a role in the pathogenesis of COVID-19 infection. *Biochemistry.* 86, 257–261.

Li, F., 2016. Structure, function, and evolution of Coronavirus spike proteins. *Annu. Rev. Virol.* 3, 237–261.

Lim, H., Baek, A., Kim, J., 2020. Hot spot profiles of SARS-CoV-2 and human ACE2 receptor protein-protein interaction obtained by density functional tight binding fragment molecular orbital method. *Sci. Rep.* 10, 16862–16870.

Lindahl, J.F., Grace, D., 2015. The consequences of human actions on risks for infectious diseases: a review. *Infect. Ecol. Epidemiol.* 5, 30048–30059.

Liu, D.X., Liang, J.Q., Fung, T.S., 2021. Human Coronavirus-229E, -OC43, -NL63, and -HKU1 (Coronaviridae). In: Bamford, D.H., Zuckerman, M., (Eds.). *Encyclopedia of Virology*. Academic Press, Elsevier. pp. 428–440.

Miao, G., Wenfei, S., Haixia, Z., Jingwei, X., Silian, C., Ye X., Xinquan, W., 2017. Cryo-electron microscopy structures of the SARS-CoV spike glycoprotein reveal a prerequisite conformational state for receptor binding. *Cell Res.* 27, 119–129.

Mohanty, S.K., Satapathy, A., Naidu, M.M., Mukhopadhyay, S., Sharma, S., Barton, L.M., et al., 2020. Severe acute respiratory syndrome coronavirus-2 (SARS-CoV-2) and coronavirus disease 19 (COVID-19)–anatomic pathology perspective on current knowledge. *Diag. Pathol.* 15, 1–17.

Pedro, A., Valiente, H.W., Satra, N., Jinah, L., Hyeon J.K., Jinhee K., et al., 2021. Computational design of potent D-peptide inhibitors of SARS-CoV-2. *J. Med. Chem.* 64, 14955–14967.

Peng, Q., Peng, R., Yuan, B., Wang, M., Zhao, J., Fu, L., et al., 2021. Structural basis of SARS-CoV-2 polymerase inhibition by Favipiravir. *Innovation* 2, 100080–100089.

Piret, J., Boivin, G., 2021. Pandemics throughout history. *Front. Microbiol.* 11, 631736.

Philip, V., Annika, K., Silvio, S., Hanspeter, S., Volker, T., 2021. Coronavirus biology and replication: implications for SARS- CoV-2. *Nat. Rev. Microbiol.* 19, 155–170.

Pruimboom, L. 2021. SARS-CoV 2; Possible alternative virus receptors and pathophysiological determinants. *Med Hypotheses.* 146, 110368–110374.

Sanyal, S., 2020. How SARS-CoV-2 (COVID-19) spreads within infected hosts-what we know so far. *Emerg. Top. Life Sci.* 4, 371–378.

Schütz, D., Ruiz-Blanco, Y.B., Münch, J., Kirchhoff, F., Sanchez-Garcia, E., Müller, J.A., 2020. Peptide and peptide-based inhibitors of SARS-CoV-2 entry. *Adv. Drug Deliv. Rev.* 167, 47–65.

Shang, J., Wan, Y., Luo, C., Ye, G., Geng, Q., Auerbach, A., Li, F., 2020. Cell entry mechanisms of SARS-CoV-2. *Proc. Natl. Acad. Sci. USA* 117, 11727–11734.

Sokkar, P., Harms, M., Stürzel, C. 2021. Computational modeling and experimental validation of the EPI-X4/CXCR4 complex allows rational design of small peptide antagonists. *Commun. Biol.* 4, 1113–1126.

Stone, S., Rothan, H.A., Natekar, J.P., Kumari, P., Sharma, S., Pathak, H., Arora, K., Auroni, T.T., Kumar, M. 2021. SARS-CoV-2 variants of concern infect the respiratory tract and induce inflammatory response in wild-type laboratory mice. *Viruses* 14, 27–34.

Surid, M.C., Shafi, A.T., Akib, M.K., Nadia, A., Ackas, A., Rajib, I., et al., 2020. Antiviral peptides as promising therapeutics against SARS-CoV-2. *J. Phys. Chem. B.* 124, 9785–9792.

Taylor, H., Tom, G., 2012. Ready, set, fuse! the coronavirus spike protein and acquisition of fusion competence. *Viruses* 4, 557–580.

Te, H.S., Randall, G., Jensen, D.M., 2007. Mechanism of action of ribavirin in the treatment of chronic hepatitis C. *Gastroenterol. Hepatol.* 3, 218–326.

Tian, D., Liu, Y., Liang, C., Xin, L., Xie, X., Zhang, D., et al., 2021. An update review of emerging small-molecule therapeutic options for COVID-19. *Biomed. Pharmacother.* 137, 111313–111321.

Tchesnokov, E.P., Feng, J.Y., Porter, D.P., Götte, M., 2019. Mechanism of inhibition of Ebola virus RNA-dependent RNA polymerase by remdesivir. *Viruses* 11, 326–342.

Thomas, P.P., Daniel, H.G., Jie, Z., Laury, B., Rebecca, F., Olivia, C.S., et al. 2021. The furin cleavage site in the SARS-CoV-2 spike protein is required for transmission in ferrets. *Nature Microbiol.* 6, 899–909.

Wang, M.Y., Zhao, R., Gao, L.J., Gao, X.F., Wang, D.P., Cao, J.M., 2020. SARS-CoV-2: Structure, biology, and structure-based therapeutics development. *Front. Cell. Infect. Microbiol.* 10, 1–17.

Wang, X., Ni, D., Liu, Y., Lu, S., 2021. Rational design of peptide-based inhibitors disrupting protein-protein interactions. *Front. Chem.* 9, 682675–682690.

Wise, J., 2020. Covid-19: Remdesivir is recommended for authorisation by European Medicines Agency. *BMJ* 369, m2610.

White, R.J., Razgour, O., 2020. Emerging zoonotic diseases originating in mammals: A systematic review of effects of anthropogenic land-use change. *Mammal Rev.* 50, 336–352.

Xia, S., Lan, Q., Su, S., Wang, X., Xu, W., Liu, Z., et al., 2020. The role of furin cleavage site in SARS-CoV-2 spike protein-mediated membrane fusion in the presence or absence of trypsin. *Signal Transduct. Target. Ther.* 5, 1–3.

Xuhua, X., 2021. Domains and functions of spike protein in SARS-Cov-2 in the context of vaccine design. *Viruses* 13, 109–125.

Yan, C.H., Faraji, F., Prajapati, D.P., Boone, C.E., DeConde, A.S. 2020a. Association of chemosensory dysfunction and Covid-19 in patients presenting with influenza-like symptoms. *Int. Forum Allergy Rhinol.* 10, 806–813.

Yan, R., Zhang, Y., Li, Y., Xia, L., Guo, Y., Zhou, Q., 2020b. Structural basis for the recognition of SARS-CoV-2 by full-length human ACE2. *Science* 367, 1444–1448.

Yang, L., Pei, R.J., Li, H., Ma, X.N., Zhou, Y., Zhu, F.H., et al., 2021. Identification of SARS-CoV-2 entry inhibitors among already approved drugs. *Acta Pharmacol. Sin.* 42, 1347–1353.

Yanxiao, H., Petr, K., 2020. Computational design of ACE2-based peptide inhibitors of SARS-CoV-2. *ACS Nano* 14, 5143–5147.

Ye, Z.W., Yuan, S., Yuen, K.S., Fung, S.Y., Chan, C.P., Jin, D.Y., 2020. Zoonotic origins of human coronaviruses. *Int. J. Biol. Sci.* 16, 1686–1697.

Yu, C., Qianyun, L., Deyin, G., 2020. Emerging coronaviruses: Genome structure, replication, and pathogenesis. *J. Med. Virol.* 92, 418–423.

Zhao, H., To, K.K.W., Lam, H., 2021. Cross-linking peptide and repurposed drugs inhibit both entry pathways of SARS-CoV-2. *Nat. Commun.* 12, 1517–1526.

Zhu, W., Chen, C.Z., Gorshkov, K., Xu, M., Lo, D.C., Zheng, W., 2020. RNA-dependent RNA polymerase as a target for COVID-19 drug discovery. *SLAS Discov.* 25, 1141–1151.

13 Decoding Transcriptomics of Neurodevelopmental Disorders

A Computational Approach

Prekshi Garg and Prachi Srivastava
Amity Institute of Biotechnology, Amity University Uttar Pradesh, Lucknow Campus, Lucknow, India

CONTENTS

13.1 INTRODUCTION

Human brain is considered as one of the most complexed organs of the body with complicated biological system, cell types, neural circuits, functionally distinct regions and approximately ten thousand genes expressed in each region (Wang and Wang 2019). Due to such complexities in the organ, the development of brain has become even more complex. Brain development is dependent on the expression of gene products, RNA and protein. Mutations in these gene products result in altered gene function and structure that consecutively gives rise to neurodevelopmental disorders (NDDs; Tebbenkamp et al. 2014). NDDs are multi-factorial disorders that depict impaired cognition,

DOI: 10.1201/9781003324706-18

communication, behavior and motor skills ultimately leading to abnormal brain development (Mullin et al. 2013). There are various NDDs such as intellectual disability (ID), autism spectrum disorder (ASD), communication disorder, epileptic encephalopathy and attention-deficit hyperactivity disorder (ADHD) present across the globe (Rapoport et al. 2012). These disorders have many common symptoms which makes diagnosis, differentiation and treatment of these diseases challenging (Lugnegård et al. 2013; Owoeye et al. 2013; Domschke and Schizophr 2013; Moreno-de-Luca et al. 2013; Adam 2013). Such overlaps in NDD have given rise to the need of identification of genes that are unique to a certain NDD and should not be related to any other NDD. The prediction of differentially expressed genes (DEGs) was made possible with the advent of whole exome sequencing (WES). Whole exome or transcriptome analysis has helped in increasing our understanding of mechanisms related to NDD, development of precise medicine and targeted treatment (Claussnitzer et al. 2020; Dhindsa and Goldstein 2015; Dhindsa et al. 2021).

Transcriptomes are specific for a certain developmental stage and physiological condition, therefore, at any given time they represent the entire set of RNA transcripts for a given cell at specific conditions (Ozsolak and Milos 2011). For understanding the mechanism of a certain disease and to formulate a proper diagnosis as well as treatment for that disease, it is very important to know the functional elements of the genome which can be analyzed through transcriptome study. Both microarray and RNA-seq techniques have been greatly implied in the field of neuroscience. Advancement in hybridization-based RNA-seq workflow has replaced microarray because of its ability to identify fusion transcripts, spliced events, novel genes and transcripts. There are various computational tools and software designed using different approaches available for transcriptome data analysis. This leads to variations in the workflow of transcriptome data analysis and enables a researcher to design their transcriptome workflow according to their goal (Yang and Kim 2015). A typical RNA-seq workflow is given in Figure 13.1. Various computational packages available for transcriptome analysis have been enlisted in Table 13.1.

13.2 RNA-SEQ WORKFLOW

There are numerous steps involved in the analysis of RNA-Seq data. Each step is described in detail here.

13.2.1 Preprocessing

This step involves quality check and trimming of the raw data. Quality check examines the overall as well as per base quality of each read in the sample. It is an essential step in transcriptome analysis and is usually carried out as the first step of the workflow. Although the step can again be repeated after the trimming of the raw data.

Trimming is an important step carried out to remove duplicated sequences and adaptor sequences from the raw data file. Trimming step is carried out due to the introduction of erroneous sequences during library preparation, sequencing and imaging. FASTQC can again be applied to the trimmed data in order to compare the quality of the raw data and trimmed data.

13.2.2 Alignment

Alignment refers to the process of aligning our query sequence with the reference genome or transcript. Transcript is generally used as a reference in cases where our aim is to identify known exons and junctions as it does not identify splicing events involving novel exons. BWA and Bowtie are commonly used tools for this approach. Genome is used as a reference in places where we want to identify novel transcripts that are generated through alternate splicing. TopHat and STAR are usually used for this approach of alignment.

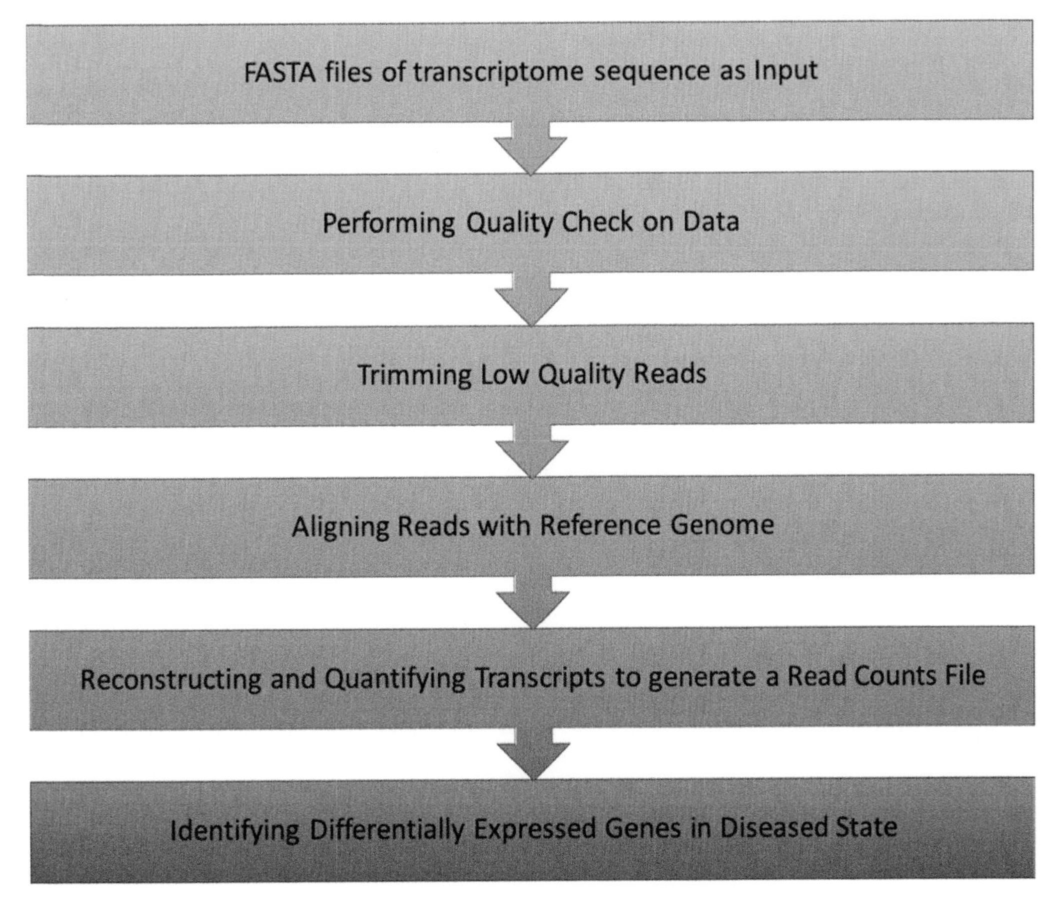

FIGURE 13.1 A typical RNA-seq workflow.

TABLE 13.1
Computational Packages Available for Transcriptome Analysis

Workflow	Package	References
Quality check	FASTQC	Leggett et al. (2013)
	HTQC	Yang et al. (2013)
	FASTX-Toolkit	Liu et al. (2019)
Read alignment	BWA	–
	Bowtie	–
	TopHat	–
	STAR	Dobin et al. (2013)
Transcriptome reconstruction and quantification	Cufflinks	–
	StringTie	Pertea et al. (2015)
	Trinity	Grabherr et al. (2011)
	RSEM	Li and Dewey (2011)
	Sailfish	Patro et al. (2014)
Differential Expression analysis	edgeR	–
	DESeq2	–
	Ballgown	Frazee et al. (2015)
	NOIseq	Tarazona et al. (2015)

13.2.3 Transcriptome Reconstruction

This step focuses on identification of all transcripts expressed in a specimen. It can be carried out by either using reference-guided approach or reference independent approach. In reference, guided approach raw reads are aligned to the reference genome as in alignment and then the overlapping transcripts are assembled. Tools used for this approach are Cufflinks and StringTie. In reference independent approach, the consensus transcripts are directly built from short reads. This approach is commonly used when the reference genome or transcriptome for the species is not available. This approach is also known as de novo assembly algorithm.

13.2.4 Expression Quantification

This step is carried out in order to enhance the expression of reads of the transcript in the raw data. This facilitates better analysis of differential gene expression (DGE) for the sample. There are various gene and isoform level methods available for quantification like RSEM that quantifies transcript level abundance from RNA-Seq data. Cufflinks, StringTie and sailfish are other methods commonly used for quantification.

13.2.5 Differential Expression

Differential expression analysis is carried out to identify genes and miRNA that are significantly expressed in the patient sample when compared to the control sample. Significant expression can be either upregulation or downregulation of genes. The expression of genes helps researchers decode pathways and biological processes that are altered in the diseased condition. DEGs also enable researchers to develop personalized medicine and integrate gene therapy methods for better treatment of the specific disease.

13.3 RECENT TRANSCRIPTOMIC FINDINGS IN NEURODEVELOPMENTAL DISORDERS

The field of transcriptomics research has gained a lot of attention in the past few years. The field of research is now moving toward transcriptomic study and analysis to find solutions to the current problems. The technology is being rapidly applied in the field of healthcare and medicine to find promising diagnosis and treatment protocols for various fatal diseases. The field is now also gaining importance in the neurology sector as well. The neurological diseases are now being studied and analyzed from transcriptomic perspective as well to get a better insight into the disease. The results from such studies are applied in designing gene therapy, personalized medicines and identification of biomarkers for that disease. Whole exome data analysis and chromosomal microarray analysis (CMA) have emerged as important tools for detection of copy number variants including micro-deletions and microduplications (Mullen et al. 2013, Szczałuba and Demkow 2017).

13.3.1 Intellectual Disability

ID is a NDD that has a heterogenous origin and is usually associated with abnormalities of dendrites and dendritic spines. Due to its heterogeneity, ID is generally referred to as a common hallmark for a collection of neurological disorders (Ilyas et al. 2020) that makes its genetic and clinical diagnosis more challenging (Grozeva et al. 2015; Heslop et al. 2014; Vissers et al. 2016; Mefford et al. 2012). ID can be caused due to both genetic as well as environmental factors (Karam et al. 2016). In a recent study by Khouri et al., de novo nonsense variations in CSDE1 gene (Cold Shock Domain containing E1) were known to play a role in ID-ASD patients. The author identified that Wnt/ β-catenin signaling and cellular adhesion pathways were deregulated in patients with ID and ASD (El Khouri et al.

2021). WES and analysis performed by Wang et al. revealed that the DNA variants and phenotypes associated with DDX3X gene account for 1–3% of unexplained ID (Wang et al. 2018).

13.3.2 Autism Spectrum Disorder

ASDs are a group of neurodevelopmental disabilities defined by significant social, communication and behavioral impairments. A tissue-specific transcriptomic analysis conducted by He et al. (2019) revealed that the genes involved in inflammation and immune response are upregulated in brain tissues and downregulated in blood tissues of ASD patients. Another group of researchers DeRosa et al. studied induced pluripotent stem cells (iPSCs) and revealed that the genes involved in neuronal differentiation, axon guidance, cell migration, DNA and RNA metabolism and neural region patterning are concerned with occurrence of ASD. The data was further supported by functional analysis that revealed defects in neuronal migration and electrophysiological activity (DeRosa et al. 2018).

13.3.3 Attention-Deficit/Hyperactivity Disorder (ADHD)

Attention-deficit/hyperactivity disorder (ADHD) is a highly heritable condition that represents the most common NDD in childhood, persisting into adulthood in around 40–65% of the cases. ADHD is characterized by age-inappropriate symptoms of inattention, impulsivity and hyperactivity (Pujol-Gualdo et al. 2021). A study by Lorenzo G et al. conducted using RNA-Seq revealed an abnormal expression of genes associated with Huntington's disease or axonal guidance in patients of ADHD (Lorenzo et al. 2018). Another study by Nuzziello et al. concluded downregulation of TLE1, ANK3, TRIO genes. These genes are associated with neuronal development and ID (Nuzziello et al. 2019). Sánchez-Mora et al. observed an abnormal expression of genes involved in D-myoinositol (1,4,5)-trisphosphate metabolism while working on 93 ADHD patients and 119 controls (Sánchez-Mora et al. 2019).

Therefore, the transcriptomic studies have opened new doors to neurodevelopment research that can help in better understanding, diagnosis and treatment of NDDs.

13.4 APPLICATIONS OF TRANSCRIPTOMICS

The field of transcriptomics data analysis can be used in various different ways in order to explore new research domains and bring out concrete results that can help in early diagnosis and treatment of neuro-related diseases and other diseases as well (Esteve-Codina 2018). Figure 13.2 represents the applications of transcriptomics in the field of research and discovery.

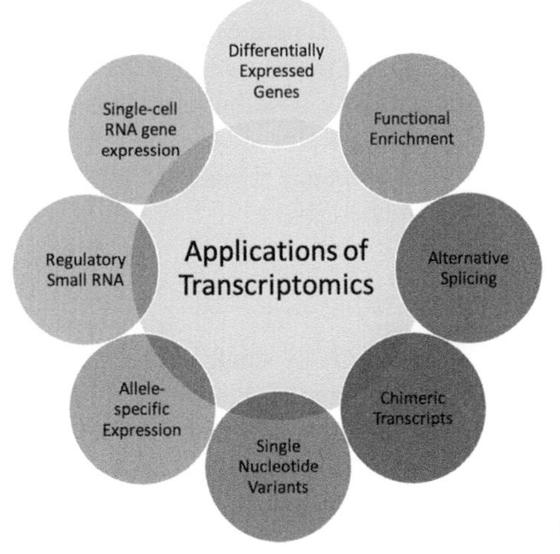

FIGURE 13.2 Applications of transcriptomics.

13.4.1 Identifying Differentially Expressed Genes

Next-generation sequencing techniques enable researchers to access far more massive amounts of data than previously available (Goodwin et al. 2016; Miller et al. 2014; Van et al. 2014). The concept of identification of DEGs opens new doors for the diagnosis of the specific disease. DEGs are genes that have expression levels determined to be significantly differentially expressed across two or more conditions (Pertea et al. 2016; McDermaid et al. 2019). DGE tools perform statistical tests based on quantifications of expressed genes derived from computational analyses of raw RNA-seq reads and assembly to determine which genes have a statistically significant difference along with their expression level (Kim et al. 2015; Goodwin et al. 2015). There are various tools used for DEG analysis like edgeR, DESeq2, Cuffdiff, limma and more.

13.4.2 Functional Enrichment

The genes that are predicted to be differentially expressed in diseased stage can be further explored by identification of biological processes and pathways related to such genes. Functional enrichment further helps in clustering genes into sub-groups that regulate one or more common processes. Targeting processes regulated by these DEGs can help in effective treatment of the specific disease. There are various computational tools available for functional enrichment of DEGs like GO ontology, PANTHER (Mi et al. 2021), KEGG (Kanehisa et al. 2017) and GSEA (Subramanian et al. 2005).

13.4.3 Detection of Alternative Splicing

Alternative splicing is a posttranscriptional processing step applied for increasing protein diversity by including or excluding exons that code for multiexonic genes. The pattern of alternative splicing can be different on the basis of factors such as tissue, gender and developmental stage. A unique pattern of alternative splicing has been found to be associated with cancer. Therefore, alternative splicing patterns can give a good insight into health diseases as well. There are majorly five types of alternative splicing patterns observed, namely, skipped exon, 5' splice site, alternative 3' splice site, mutually exclusive exon and retained intron. rMATS (Shen et al. 2014) is a well-known tool that uses hierarchical framework to model exon inclusion levels, which helps the tool to detect novel splicing events.

13.4.4 Detection of Chimeric Transcripts

Chimeric transcripts are transcriptome complexities that occur in both normal as well as pathologic tissues. These transcripts play various kinds of roles like signal transduction, gene expression regulation, apoptosis and tumorigenesis. These chimeric transcripts are fusion transcript containing segment for two genes formed either due to DNA translocation or by posttranscriptional trans-splicing events. Therefore, these can be formed at both DNA and RNA levels. The chimeric transcripts can be identified by using the extended version of STAR pipeline (Haas et al. 2017) and can be further validated through RT-PCR, FISH, western and southern blot.

13.4.5 Detection of Single Nucleotide Variants

Generally, whole-genome sequencing (WGS) is preferred for the study of single nucleotide variants over RNA-Seq. This is because data from RNA-Seq will only read variants with sufficient read depth, therefore, variants occurring in enhancers, introns, promoters and other lowly expressed regions will not be detected. Still, variant detection is possible through RNA-Seq data STAR 2-pass alignment tool.

13.4.6 Detection of Allele-Specific Expression

Allele-specific expression (ASE) refers to the occurrence of an imbalanced expression in diploid or polyploid genomes, where two or more alleles of a gene are present (Liu et al. 2018, Tung et al. 2011). It is a common phenomenon occurring in humans and majorly contributes to the occurrence of multiple phenotypes and complex traits. Allelic imbalance provides a evidence of cis-regulation of gene expression caused due to epigenetic modification, nucleotide variation at the promoter site or posttranscriptional regulation. MBASED (Mayba et al. 2014) and GeneiASE (Edsgärd et al. 2016) are the two majorly used tools for the detection of ASE.

13.4.7 Small RNA-Seq

Regulatory small RNAs, such as miRNA, siRNA and piRNA, can also be detected through transcriptome sequence analysis. miRNAs and endogenous siRNAs have been shown to regulate gene expression by silencing specific genes, whereas piRNAs have been implicated mainly in genome protection and/or maintenance in germ cells via silencing of transposable elements (Wei et al. 2012). There are various softwares available for detection and analysis of regulatory small RNA-Seq (Riffo-Campos et al. 2016) such as MiRanda, Mireap software and web-based tool like psRNATarget program.

13.4.8 Single-Cell RNA-Seq

Single-cell RNA sequencing (scRNA-seq) technologies allow the dissection of gene expression at single-cell resolution, which greatly revolutionizes transcriptomic studies (Chen et al. 2019). In recent years, scRNA-seq has been applied to various species, especially to diverse human tissues, and these studies revealed meaningful cell-to-cell gene expression variability (Jaitin et al. 2014; Grün et al. 2015; Chen et al. 2016; Cao et al. 2017; Rosenberg et al. 2018). Cell identification and characterization are generally carried out using unsupervised clustering approaches. Computational tools usually used for QC, identification of gene markers and clustering are Seurat (Macosko et al. 2015; Satija et al. 2015), Monocle (Qiu et al. 2017; Trapnell et al. 2014) and Pagoda (Fan et al. 2016).

REFERENCES

Adam, D., 2013. Mental health: On the spectrum. *Nature* 496, 416–418.

Cao, J., Packer, J.S., Ramani, V., Cusanovich, D.A., Huynh, C., Daza, R., et al., 2017. Comprehensive single-cell transcriptional profiling of a multicellular organism. *Science* 357, 661–667.

Chen, G., Ning, B., Shi, T., 2019. Single-cell RNA-Seq technologies and related computational data analysis. *Front. Genet.* 10, 317.

Chen, G., Schell, J.P., Benitez, J.A., Petropoulos, S., Yilmaz, M., Reinius, B., et al., 2016. Single-cell analyses of X Chromosome inactivation dynamics and pluripotency during differentiation. *Gen. Res.* 26, 1342–1354.

Claussnitzer, M., Cho, J.H., Collins, R., Cox, N.J., Dermitzakis, E.T., Hurles, M.E., et al. 2020. A brief history of human disease genetics. *Nature*, 577, 179–189.

DeRosa, B.A., El Hokayem, J., Artimovich, E., Garcia-Serje, C., Phillips, A.W., Van Booven, D., et al., 2018. Convergent pathways in idiopathic autism revealed by time course transcriptomic analysis of patient-derived neurons. *Scientific Rep.* 8, 8423.

Dhindsa, R.S., Goldstein, D.B., 2015. Genetic discoveries drive molecular analyses and targeted therapeutic options in the epilepsies. *Curr. Neurol. Neurosci. Rep.* 15, 70.

Dhindsa, R.S., Zoghbi, A.W., Krizay, D.K., Vasavda, C., Goldstein, D.B., 2021. A transcriptome-based drug discovery paradigm for neurodevelopmental disorders. *Ann. Neurol* 89, 199–211.

Dobin, A., Davis, C.A., Schlesinger, F., Drenkow, J., Zaleski, C., Jha, S., et al., 2013. STAR: ultrafast universal RNA-seq aligner. *Bioinformatics* 29, 15–21.

Domschke, K., 2013. Clinical and molecular genetics of psychotic depression. *Schizophr. Bull.* 39, 766–775.

Edsgärd, D., Iglesias, M.J., Reilly, S.J., Hamsten, A., Tornvall, P., Odeberg, J., Emanuelsson, O., 2016. GeneiASE: Detection of condition-dependent and static allele-specific expression from RNA-seq data without haplotype information. *Sci. Rep.* 6, 21134.

El Khouri, E., Ghoumid, J., Haye, D., Guiliano, F., Drevillon, L., Briand-Suleau A., et al., 2021. Wnt/β-catenin pathway and cell adhesion deregulation in CSDE1-related intellectual disability and autism spectrum disorders. *Mol Psychiat.* 26, 3572–3585.

Esteve-Codina, A., 2018. RNA-seq data analysis, applications and challenges. In: Jaumot, J., Bedia, C., Tauler, R., (Eds.), *Comprehensive Analytical Chemistry*, Elsevier, pp. 71–106.

Fan, J., Salathia, N., Liu, R., Kaeser, G.E., Yung, Y.C., Herman, J.L., et al., 2016. Characterizing transcriptional heterogeneity through pathway and gene set overdispersion analysis. *Nat. Methods* 13, 241–244.

Frazee, A.C., Pertea, G., Jaffe, A.E., Langmead, B., Salzberg, S.L., Leek, J.T., 2015. Ballgown bridges the gap between transcriptome assembly and expression analysis. *Nat. Biotechnol.* 33, 243–246.

Goodwin, S., Gurtowski, J., Ethe-Sayers, S., Deshpande, P., Schatz, M.C., McCombie, W.R., 2015. Oxford Nanopore sequencing, hybrid error correction, and de novo assembly of a eukaryotic genome. *Genome Res.* 25, 1750–1756.

Goodwin, S., McPherson, J.D., McCombie, W.R., 2016. Coming of age: Ten years of next-generation sequencing technologies. *Nature Rev. Genet.* 17, 333–351.

Grabherr, M.G., Haas, B.J., Yassour, M., Levin, J.Z., Thompson, D.A., Amit, I., et al., 2011. Full-length transcriptome assembly from RNA-Seq data without a reference genome. *Nat. Biotechnol.* 29, 644–652.

Grozeva, D., Carss, K., Spasic-Boskovic, O., Tejada, M.I., Gecz, J., Shaw, M., et al., 2015 Targeted next-generation sequencing analysis of 1,000 individuals with intellectual disability. *Hum. Mutat.* 36, 1197–1204.

Grün, D., Lyubimova, A., Kester, L., Wiebrands, K., Basak, O., Sasaki, N., Clevers, H., van Oudenaarden, A. (2015). Single-cell messenger RNA sequencing reveals rare intestinal cell types. *Nature*, 525(7568), 251–255.

Haas, B., Dobin, A., Stransky, N., Li, B., Yang, X., Tickle, T., et al., 2017. STAR-fusion: Fast and accurate fusion transcript detection from RNA-Seq. *BioRxiv*, 120295.

He, Y., Zhou, Y., Ma, W., Wang J., 2019. An integrated transcriptomic analysis of autism spectrum disorder. *Sci. Rep.* 9, 11818.

Heslop, P., Blair, P.S., Fleming, P., Hoghton, M., Marriott, A., Russ, L., 2014. The confidential inquiry into premature deaths of people with intellectual disabilities in the UK: a population-based study. *Lancet.* 383, 889–895.

Ilyas, M., Mir, A., Efthymiou, S., Houlden, H.F., 2020. The genetics of intellectual disability: Advancing technology and gene editing. *F1000 Res. 9, F1000 Faculty Rev-22.*

Jaitin, D.A., Kenigsberg, E., Keren-Shaul, H., Elefant, N., Paul, F., Zaretsky, I., et al., 2014. Massively parallel single-cell RNA-seq for marker-free decomposition of tissues into cell types. *Sci.* 343, 776–779.

Kanehisa, M., Furumichi, M., Tanabe, M., Sato, Y., Morishima, K., 2017. KEGG: new perspectives on genomes, pathways, diseases and drugs. *Nucleic Acids Res.* 45(D1), D353–D361.

Karam, S.M., Barros, A.J., Matijasevich, A., Dos Santos, I.S., Anselmi, L., Barros, F., et al., 2016. Intellectual disability in a birth cohort: Prevalence, etiology, and determinants at the age of 4 years. *Public Health Genom.* 19, 290–297.

Kim, D., Langmead, B., Salzberg, S.L., 2015. HISAT: a fast spliced aligner with low memory requirements. *Nat. Methods.* 12, 357–360.

Leggett, R.M., Ramirez-Gonzalez, R.H., Clavijo, B.J., Waite, D., Davey, R.P., 2013. Sequencing quality assessment tools to enable data-driven informatics for high throughput genomics. *Front Genet.* 4, 288.

Li, B., Dewey, C.N., 2011. RSEM: Accurate transcript quantification from RNA-Seq data with or without a reference genome. *BMC Bioinform.* 12, 323.

Liu, X., Yan, Z., Wu, C., Yang, Y., Li, X., Zhang, G., 2019. FastProNGS: Fast preprocessing of next-generation sequencing reads. *BMC Bioinform.* 20, 345.

Liu, Z., Dong, X., Li, Y. 2018. A genome-wide study of allele-specific expression in colorectal cancer. *Front. Genet.* 9, 570.

Lorenzo, G., Braun, J., Muñoz, G., Casarejos, M.J., Bazán, E., Jimenez-Escrig, A., 2018. RNA-Seq blood transcriptome profiling in familial attention deficit and hyperactivity disorder (ADHD). *Psych. Res.* 270, 544–546.

Lugnegård, T., Unenge Hallerbäck, M., Hjärthag, F., Gillberg, C., 2013. Social cognition impairments in Asperger syndrome and schizophrenia. *Schizophr. Res.* 143, 277–284.
Macosko, E.Z., Basu, A., Satija, R., Nemesh, J., Shekhar, K., Goldman, M., et al., 2015. Highly parallel genome-wide expression profiling of individual cells using nanoliter droplets. *Cell* 161, 1202–1214.
Mayba, O., Gilbert, H.N., Liu, J., Haverty, P.M., Jhunjhunwala, S., Jiang, Z., 2014. MBASED: Allele-specific expression detection in cancer tissues and cell lines. *Genome Biol.* 15, 405.
McDermaid, A., Monier, B., Zhao, J., Liu, B., Ma, Q., 2019. Interpretation of differential gene expression results of RNA-seq data: review and integration. *Brief Bioinform.* 20, 2044–2054.
Mefford, H.C., Batshaw, M.L., Hoffman, E.P., 2012. Genomics, intellectual disability, and autism. *N. Engl. J. Med.* 366, 733–743.
Mi, H., Ebert, D., Muruganujan, A., Mills, C., Albou, L.P., Mushayamaha, T., Thomas, P.D., 2021. PANTHER version 16: A revised family classification, tree-based classification tool, enhancer regions and extensive API. *Nucleic Acids Res.* 49(D1), D394–D403.
Miller, J.A., Menon, V., Goldy, J., Kaykas, A., Lee, C.K., Smith, K.A., 2014. Improving reliability and absolute quantification of human brain microarray data by filtering and scaling probes using RNA-Seq. *BMC Genomics* 15, 154.
Moreno-De-Luca, A., Myers, S.M., Challman, T.D., Moreno-De-Luca, D., Evans, D.W., Ledbetter, D.H., 2013. Developmental brain dysfunction: Revival and expansion of old concepts based on new genetic evidence. *Lancet Neuro.* 12(4), 406–414.
Mullen, S.A., Carvill, G.L., Bellows, S., Bayly, M.A., Trucks, H., Lal, D., et al., 2013. Copy number variants are frequent in genetic generalized epilepsy with intellectual disability. *Neurology* 81, 1507–1514.
Mullin, A.P., Gokhale, A., Moreno-De-Luca, A., Sanyal, S., Waddington, J.L., Faundez, V., 2013. Neurodevelopmental disorders: Mechanisms and boundary definitions from genomes, interactomes and proteomes. *Transl. Psych.* 3, e329.
Nuzziello, N., Craig, F., Simone, M., Consiglio, A., Licciulli, F., Margari, L., et al., 2019. Integrated analysis of microRNA and mRNA expression profiles: An attempt to disentangle the complex interaction network in attention deficit hyperactivity disorder. *Brain Sci.* 9, 288.
Owoeye, O., Kingston, T., Scully, P.J., Baldwin, P., Browne, D., Kinsella, A., et al., 2013. Epidemiological and clinical characterization following a first psychotic episode in major depressive disorder: Comparisons with schizophrenia and bipolar I disorder in the Cavan-Monaghan First Episode Psychosis Study (CAMFEPS). *Schizophr. Bull.* 39, 756–765.
Ozsolak, F., Milos, P.M., 2011. RNA sequencing: advances, challenges and opportunities. *Nat. Rev. Genet.* 12, 87–98.
Patro, R., Mount, S.M., Kingsford, C., 2014. Sailfish enables alignment-free isoform quantification from RNA-seq reads using lightweight algorithms. *Nat. Biotechnol.* 32, 462–464.
Pertea, M., Pertea, G.M., Antonescu, C.M., Chang, T.C., Mendell, J.T., Salzberg, S.L., 2015. StringTie enables improved reconstruction of a transcriptome from RNA-seq reads. *Nat. Biotechnol.* 33, 290–295.
Pertea, M., Kim, D., Pertea, G.M., Leek, J.T., Salzberg, S.L., 2016. Transcript-level expression analysis of RNA-seq experiments with HISAT, StringTie and Ballgown. *Nat. Protoc.* 11, 1650–1667.
Pujol-Gualdo, N., Sánchez-Mora, C., Ramos-Quiroga, J.A., Ribasés, M., Artigas M.S., 2021. Integrating genomics and transcriptomics: Towards deciphering ADHD. *Eur. Neuropsychopharmacol.* 44, 1–13.
Qiu, X., Mao, Q., Tang, Y., Wang, L., Chawla, R., Pliner, H.A., Trapnell, C., 2017. Reversed graph embedding resolves complex single-cell trajectories. *Nat. Methods* 14, 979–982.
Rapoport, J.L., Giedd, J.N., Gogtay, N., 2012. Neurodevelopmental model of schizophrenia: update. *Mol Psyc.* 17, 1228–1238.
Riffo-Campos, Á.L., Riquelme, I., Brebi-Mieville, P., 2016. Tools for sequence-based miRNA target prediction: what to choose? *Int. J. Mol. Sci.* 17, 1987.
Rosenberg, A.B., Roco, C.M., Muscat, R.A., Kuchina, A., Sample, P., Yao, Z., et al., 2018. Single-cell profiling of the developing mouse brain and spinal cord with split-pool barcoding. *Science* 360, 176–182.
Sánchez-Mora, C., Soler Artigas, M., Garcia-Martínez, I., Pagerols, M., Rovira, P., Richarte, V., et al., 2019. Epigenetic signature for attention-deficit/hyperactivity disorder: Identification of miR-26b-5p, miR-185-5p, and miR-191-5p as potential biomarkers in peripheral blood mononuclear cells. *Neuropsychopharmacol.* 44, 890–897.
Satija, R., Farrell, J.A., Gennert, D., Schier, A.F., Regev, A., 2015. Spatial reconstruction of single-cell gene expression data. *Nat. Biotechnol.* 33, 495–502.

Shen, S., Park, J.W., Lu, Z., Lin, L., Henry, M.D., Wu, Y.N., Zhou, Q., Xing, Y., 2014. rMATS: robust and flexible detection of differential alternative splicing from replicate RNA-Seq data. *Proc. Natl. Acad. Sci. USA* 111, E5593–E5601.

Subramanian, A., Tamayo, P., Mootha, V.K., Mukherjee, S., Ebert, B.L., Gillette, M.A., et al., 2005. Gene set enrichment analysis: A knowledge-based approach for interpreting genome-wide expression profiles. *Proc. Natl. Acad. Sci. USA* 102, 15545–15550.

Szczałuba, K., Demkow, U., 2017. Array comparative genomic hybridization and genomic sequencing in the diagnostics of the causes of congenital anomalies. *J. Appl. Genet.* 58, 185–198.

Tarazona, S., Furio-Tari, P., Turra, D., Pietro, A.D., Nueda, M.J., Ferrer, A., Conesa, A., 2015. Data quality aware analysis of differential expression in RNA-seq with NOISeq R/Bioc package. *Nucleic Acids Res.* 43, e140.

Tebbenkamp, A.T., Willsey, A.J., State, M.W., Sestan, N., 2014. The developmental transcriptome of the human brain: implications for neurodevelopmental disorders. *Curr. Opin. Neurol.* 27, 149–156.

Trapnell, C., Cacchiarelli, D., Grimsby, J., Pokharel, P., Li, S., Morse, M., et al., 2014. The dynamics and regulators of cell fate decisions are revealed by pseudotemporal ordering of single cells. *Nat. Biotechnol.* 32, 381–386.

Tung, J., Akinyi, M.Y., Mutura, S., Altmann, J., Wray, G.A., Alberts, S.C., 2011. Allele-specific gene expression in a wild nonhuman primate population. *Mol. Ecol.* 20, 725–739.

van Dijk, E.L., Auger, H., Jaszczyszyn, Y., Thermes, C., 2014. Ten years of next-generation sequencing technology. *Trends Genet.* 30, 418–426.

Vissers, L.E., Gilissen, C., Veltman, J.A., 2016. Genetic studies in intellectual disability and related disorders. *Nat. Rev. Genet.* 17, 9–18.

Wang, W., Wang, G.Z., 2019. Understanding molecular mechanisms of the brain through transcriptomics. *Front. Physiol.* 10, 214.

Wang, X., Posey, J.E., Rosenfeld, J.A., Bacino, C.A., Scaglia, F., Immken, L., et al., 2018. Phenotypic expansion in DDX3X – a common cause of intellectual disability in females. *Ann. Clin. Transl. Neurol.* 5, 1277–1285.

Wei, C., Salichos, L., Wittgrove, C.M., Rokas, A., Patton, J.G., 2012. Transcriptome-wide analysis of small RNA expression in early zebrafish development. *RNA* 18, 915–929.

Yang, I.S., Kim, S., 2015. Analysis of whole transcriptome sequencing data: Workflow and software. *Genom. Inform.* 13, 119–125.

Yang, X., Liu, D., Liu, F., Wu, J., Zou, J., Xiao, X., et al., 2013. HTQC: A fast quality control toolkit for Illumina sequencing data. *BMC Bioinform.* 14, 33.

14 Next-Generation Sequencing Technologies for the Development of Disease-Resistant Plants

Aryan Shukla and Ruchi Yadav
Amity Institute of Biotechnology, Amity University Uttar Pradesh, Lucknow Campus, Lucknow, India

CONTENTS

14.1 INTRODUCTION

Opposite to the contrary belief, the crops production currently done by farmers across the earth is more than enough for the present population and can even feed up to 9.7 billion humans which is the estimated population of humans on earth by 2050 (Crist et al. 2017). Still every single day approximately 25,000 humans, most of which are children, die due to starvation and malnutrition (Shapiro 2021). According to recent estimates, the number of undernourished people rose to around 768 million last year- equivalent to 10% of the world's population and an increase of around 118 million versus 2019. So where is all this extra-produced food going? Most of the food grown by farmers becomes waste due to plant or crop diseases (Sinha and Swain 2022). Surely, we can change our habits, but it would not cause significant compensation enough to make a change.

DOI: 10.1201/9781003324706-19

According to most liberal estimates up to 40% of global crop production is lost to plant pests and diseases (Parlinska and Pagare 2018). Each year, plant diseases cost the global economy more than 18 lakh crore rupees, and invasive insects cost at least 5.5 lakh crore rupees (Alok 2020). With the advent of current genomics and transcriptomics technologies, it is possible to overcome this problem. This chapter discusses about the next-generation sequencing (NGS) technologies and their impact on plant and food production.

14.2 HISTORY OF NGS

NGS, also termed as massively parallel or deep sequencing, is a flexible set of techniques and approaches used for de novo sequencing of eukaryotic genomes. It enables parallel multiplexed analysis of DNA sequences on a massive scale and has reduced both time and cost incurred during large genome sequencing (Sharma et al. 2018). NGS utilizes bioinformatics for converting signals from the machine to meaningful information and allows simultaneous sequencing of enormous amounts of DNA (Behjati and Tarpey 2013; Slatko et al. 2018; Hari and Parthasarathy 2019). In 2002, the whole-genome sequence (WGS) of *Arabidopsis thaliana* was worked out. As the first crop, a new rice variety was engineered with the help of WGS (Kitamura et al. 2022), which led to a rise in the development of new genetically engineered plants as shown in Figure 14.1.

The sequencing technologies so far have made a bank of genomic information on various organisms which helped create a great study tool for further innovation (Li et al. 2019). Scientists across the world getting access to these sequenced genomes of various plants have led to the development for desired crops having resistance to diseases and pests by adding genomic target sequences from resistant plants resistant to the susceptible ones (Jo and Kim 2019). This led to improvement of yield and associated agronomic traits in crops. Diamondback moth (*Plutella xylostella*), an insect pest of cabbage (*Brassica oleracea* var. *capitata*), has been reported to induce losses to the tune of 52% in cabbage (Figure 14.2) (Ayalew 2006; Alvarez et al. 2022). Gene editing carried out to attain desirable variations improves crop production, their resistance against extreme environmental stressors, and a range of biotic stresses (Le Nguyen et al. 2019). This results in the development of novel and resilient plant genotypes. Therefore, genomics is essential for development of new crops to fulfill the agricultural demand of the earth (Zhong et al. 2018). Many strategies regarding genome sequencing have been improvised for crop modifications in recent years.

Nowadays, WGS data on polyploidy crops like hexaploid bread wheat has been generated utilizing NGS strategies like long-read single-molecule sequencing (Dong et al. 2018). This has been possible by the advancement in sequence technologies that provide a firm platform to attain information by genome-based interpretation of epigenomic data (Rimoldi et al. 2018). Epigenomics, the study involving a set of chemical changes which are not encoded within DNA, illustrates when and how genes are expressed and consists of the three-dimensional validation of the proteome, transcriptome, nuclear genome and the huge metabolome (Jia et al. 2018; Chondronasiou et al. 2022). Epigenetics is being increasingly explored to redesign crops with increased yield, stability, stress adaptability and nutritional value (Ali et al. 2022; Yang et al. 2022). Currently in the buzz is CRISPR (clustered regularly interspaced short palindromic repeats)/Cas (CRISPR-associated) protein utilizing technology as it is the best editing tool known to mankind (Aksoy et al. 2022). It can cause modifications in targeted genes and manipulate new genomes with unbelievable precision and ease of operation that helps in exploration of unknown biological processes (Camargo et al. 2019). The de novo (ab initio) method can be used to synthesize entirely new proteins in important crops to produce attributes like high-yielding and short breeding cycle (Sentmanat et al. 2018; Abdallah et al. 2021). Further progress in the genomic techniques is likely to open the gates for crop improvement in the coming years which warrants learning from newer fields and making sense out of inconsistent data to have big practical applications in agriculture (Le Nguyen et al. 2019). These advancements can also help in the development of plant biology and new medicines for humans.

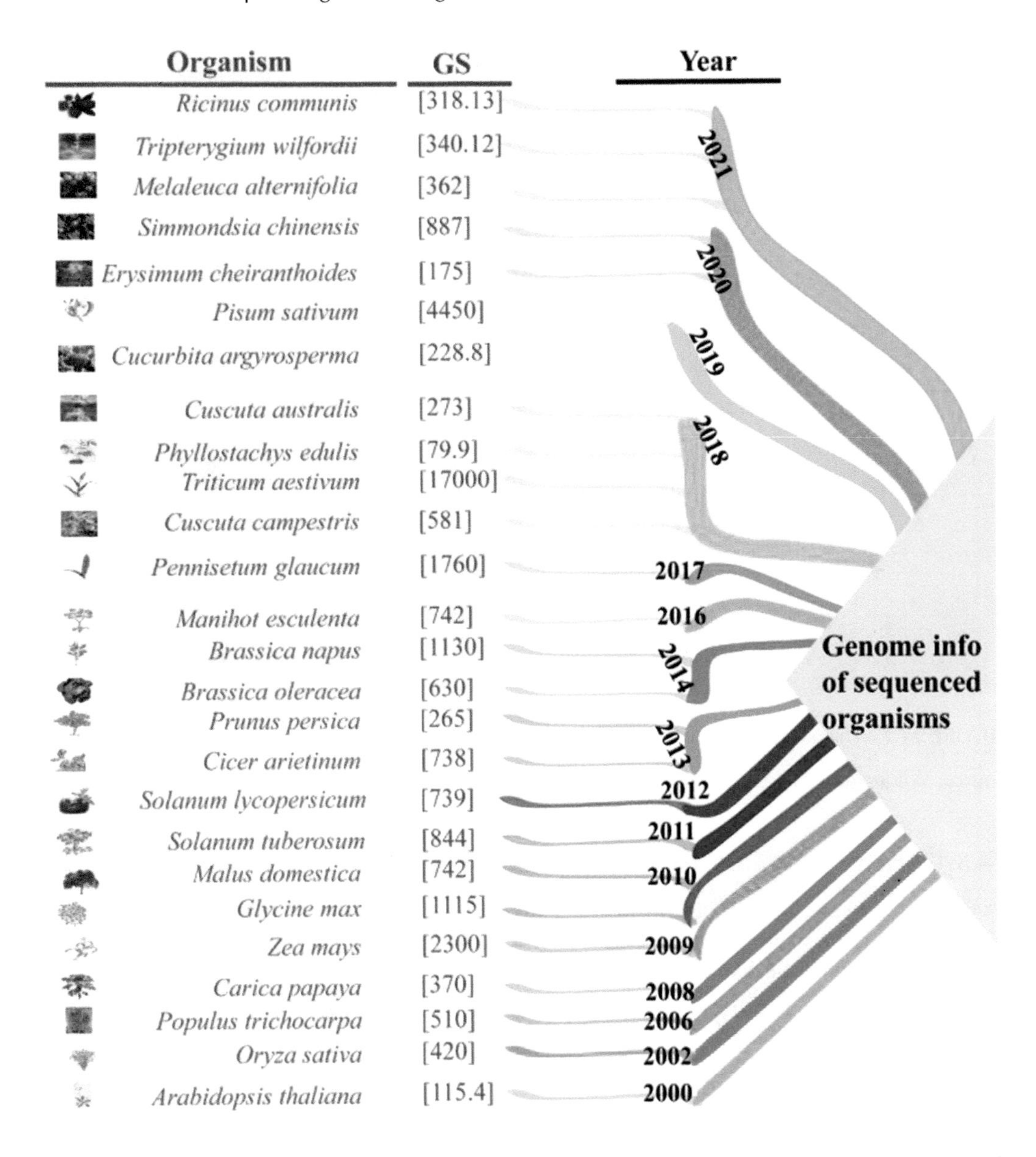

FIGURE 14.1 The image represents the entire timeline of the fully sequenced crop plants along with genome sequences (GS).

14.3 TYPES OF NGS

The small nucleus of microscopic cells contains DNA sequences that store all genetic information of an organism. Therefore, we require techniques that can decode this genetic information which will be of great utility in multiple scientific disciplines. In 1977, Frederick Sanger developed the DNA-seq (DNA sequencing technology) also called Sanger's sequencing or the chain termination method (Sanger and Coulson 1975; Seroussi 2021). Allan Maxam and Walter Gilbert (1977) introduced more improvements in DNA sequencing using chemical advancements of DNA and made cleavage at a particular nucleotide base(s) (Slatko et al. 2018).

FIGURE 14.2 Diamondback moth (DBM).

Source: Wikipedia.

14.3.1 First-Generation Sequencing

Sanger's sequencing was denoted as first-generation sequencing (FGS) which was used on commercial scale in many fields of science (Paul et al. 2018). Sanger's sequencing generated sets of NGS data of organisms like bacteria (small sets) and humans (large sets). But these methods of sequencing were challenging because of radioactive reaction reagents involved in it and issues of scalability. Its biggest disadvantages included high time consumption and exorbitant cost (10 million US dollars were utilized to make an additional human genome). These limitations were dealt with the development of new sequencing techniques (Singh 2020).

In 1987, Applied Biosystems developed the first automatic machine named AB370 containing capillary electrophoresis (CE) which generated a read-length up to 600 bases. This increased the speed and accuracy of sequencing (Su et al. 2018). Researchers also tried to develop more powerful sequencing machines to save cost and time that resulted in the upgraded model named AB373xl which could generate 900 base read lengths by detecting 2.88 megabases per day (Müller et al. 2018). FGS was successfully used to locate and identify the expressed-sequence tags (ESTs), genomic regions of the single-nucleotide-polymorphism (SNP) and simple-sequence repeat (SSR) markers. On different plant species, molecular markers of the agronomic traits had been used to study the effects (Dumschott et al. 2020).

14.3.2 Next-Generation Sequencing

NGS is a term given to all the sequencing technologies post-Sanger sequencing. DNA sequencing had a big breakthrough in understanding the genomic composition of an organism due to the invention of new sequencing strategies in 2007 (Taheri et al. 2018). The new strategies using the high-throughput sequencing methods are referred to as NGS. With this researcher was able to obtain crores of sequenced DNA nucleotides simultaneously due to lakhs of self-directed chemical reactions. These reactions can decode specific target sequences with high quality and higher detailed coverage of short or long sequences of plant species (Sahu et al. 2020). NGS is timesaving and generates huge amounts of sequencing data at incredibly low cost. NGS requires one or two machines to obtain sequence data. Its advantages include flexibility since there is no need for the precloned DNA region (Díaz-Cruz et al. 2019). It is highly competitive for conducting genomic data interpretation in contrast to the microarray strategy that depends on tailored arrays of a subject. NGS forums create genome sequences utilizing data libraries (constructed by fragmented and adapter-attached RNA/DNA/amplicon) (Saidi and Hajibarat 2020).

Sequencing strategies usually have a particular workflow regardless of the sequencing technique. It starts with library construction with the help of nucleic acid. The sequencing machine is run, and aggregation of sequenced data happens. At the end, data interpretation with bioinformatics or related software takes place. Library construction in NGS is an important step to mark it as unique sequencing technology based on the ligation chemistry, chemical composition of synthesis reaction system and single-molecule long read. The short-read sequencing and long-read sequencing are also known as second-generation sequencing (SGS) and third-generation sequencing (TGS), respectively (Singh et al. 2020).

14.3.2.1 Second-Generation Sequencing (SGS)

SGS-short-read forums work on the nucleic acid libraries (construction type) generated by a ligation reaction which integrates DNA strings using linkers or adaptors (Wong et al. 2022). Some common SGS-short-read-associated NGS models used for plant sequencing have been provided below (Yang et al. 2018):

1. Ion-torrent sequencing
2. Oligonucleotide-ligation and detection (SOLiD) sequencing
3. Roche 454 (pyrosequencing)
4. BGI Retrovolocity strategy for DNA-nanoball sequencing
5. Illumina (Solexa) such as HiSeq and MiSeq methods of sequencing

Each technique has its own advantages and disadvantages. The genetic information associated with several plant species such as *Arabidopsis thaliana*, rice (*Oryza sativa*), maize (*Zea mays*) and papaya (*Carica papaya*) have been obtained by sequencing (Liu et al. 2020; Ashraf et al. 2022).

The advantages of SGS over clonal sequencing and FGS are greater number of reactions and high turnout. While SGS can significantly reduce the cost by sample multiplexing, it can also solve the problem of haploid fragment sequencing (Yan et al. 2022). This was a major issue in Sanger's sequencing. The SGS-short-read is versatile and now monopolizes the sequencing business. Many bioinformatic tools are programmed using SGS-short-read data analysis as it is regarded as more accurate than third-generation-long-read sequencing (Madronero et al. 2019). Disadvantages of SGS-short-read sequencing include long running times and structural variations, difficulty in generating de novo assembly and the identification of true isoform of a transcript and inability in sequencing of long fragments of DNA and haplotype phasing (Giraldo et al. 2021).

14.3.2.2 Third-Generation Sequencing (TGS)

A persistent problem was encountered by scientists while studying polyploid genome in crop plants. They found that an assembly of a long chromosome was created by a huge genome size because of the extensive DNA sequence repetition from short-DNA regions (Athanasopoulou et al. 2021). The sequenced DNA data generated from this could not be mapped according to their chromosome and genomic positions. These reasons created a need for the invention of TGS technologies. After its development, it was a common practice to make the sequence of a single-molecule utilizing TGS-long-read technology. Five to 30 kilobases read lengths could be generated using TGS-long-read forums (Wong et al. 2022). Hence, to produce a considerable overlapping read length for sequence assembly, TGS-long-read technology was used to sequence a single molecule. Its biggest advantage was that it removed the amplification bias and could create long-DNA sequences (reads) and scaffolds to encompass the whole genome or chromosome of an organism (Li et al. 2020).

TGS-long-read technologies-based methods are optical mapping, DNA dilution constructed and chromosomal-conformation arrest technologies. TGS-long-read research forums have been provided below (Fan et al. 2020):

1. Electron microscopy helps in generating TGS-long-read data
2. Helico-sequenced data by a genetic analysis system (GAS)
3. SMART (Single-Molecule Real-Time) sequenced data designed by Pacific-Bioscience
4. Nanopore sequenced data created by Oxford-Nanopore platform (Ex-MinION & PromethION)

Advantages of TGS-long-read technologies are very low cost of consumption and more efficiency of the sequenced data compared to any other DNA sequencing technologies.

14.4 APPLICATIONS OF NGS IN CROP SCIENCE

As evident in Figure 14.3, the significance of NGS in developing genetically modified plants includes finding novel genes, disease-resistant genes and better understanding plant–host interaction, metabolome study and effect of environmental conditions on plants (Le Nguyen et al. 2019). It also helps in easing and speeding up pathogen and metagenome identification.

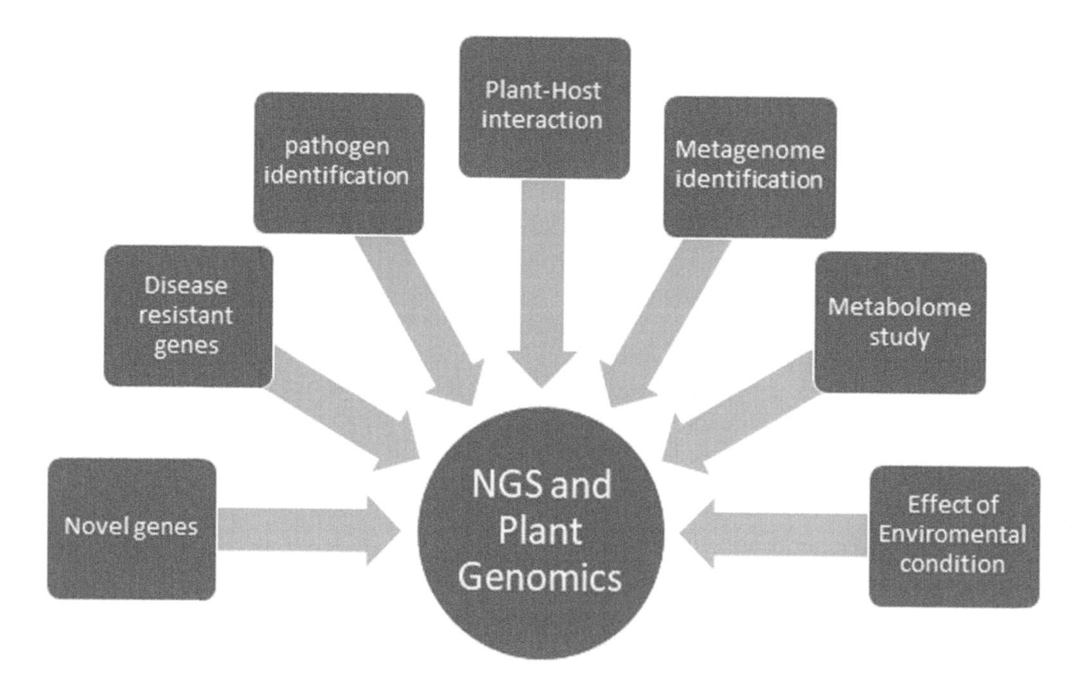

FIGURE 14.3 Various applications of next-generation sequencing (NGS) in plant genomics.

14.4.1 Using Genetic Resources

Globally, genetic resources or gene banks are important assets with broad nutritional prospects that safeguard food security by preserving vital information that can be used to combat a disease outbreak in plants and/or identifying the key genetic factors which allow us to develop crops resistant to harsh environmental conditions (Madronero et al. 2019). These genetic resources constitute new, extinct, and vanishing plant types, primitive landraces, wild species of crops and weeds germplasm/lines under different breeding programs with negligible articulated evidence about germplasm provision for research purposes by gene stock and other national or international organizations around the world (Taheri et al. 2018). As a result of delayed development in underdeveloped nations and slow pace of research, the world's vast population is denied access to the latest genomic discoveries.

The future of mankind depends on crop production within limited land area to feed an ever-increasing population. Uniformity between genetic materials is required to cultivate large-scale areas with the same type of crop or species, which can be planted to combat the increasing events of crop damage by both biotic and abiotic stresses (Wang et al. 2021a, 2021b). The maintained genetic variants can help us locate the source of a disease outbreak by examining genomic data from wild or parent plants (Singh et al. 2020). Many vulnerable and endangered plant species around the world can provide new methods to enhance agriculture's future by using recent developments in DNA sequencing technologies, particularly NGS. As a result, NGS can help scientists reclassify the majority of mislabeled or duplicated accessions by enhancing identification methods, allowing breeders to improve gene bank administration and overcome the frequent obstacles of distinguishing mislabeled genotypes (Abebe 2019). The precise assessment of genetic diversity via NGS can lead researchers to mine the desirable genetic pool, defining the grouping of genetic material. Also designating the main or minor target germplasm for research, which could offer novel cultivars with high resistance to environmental change and assist in increasing crop yield (De Filippis 2018).

The production of sequencing data from accessions (gene banks) using NGS technology and advances in bioinformatic tools is critical to determining or selecting suitable or climate-resilient plants to counteract the terrible catastrophes predicted by various researchers (Jang and Joung 2019). This sequencing data improves comprehension in developing genetic markers and improves precision in molding allelic variants or precise genotyping–phenotyping, which can improve trait-specific breeding strategies. However, gene pools are the treasure that NGS throughout the world must use to establish a comprehensive understanding of inter- or intra-species evolution and disclose the relationship between wild types and contemporary cultivars that can extend germplasm knowledge to achieve sustainable production (Ko et al. 2020). An example for this inter-species evolution is shown in Figure 14.4 in which we can observe red cotton stainer (*Dysdercus cingulatus*) on cotton plant. It has co-evolved in such a way that it cannot prey on any other plant. It drinks sap of the plant and leaves red color stain on the cotton seeds with its excreta (Lo and Shaw 2019).

14.4.2 Mining Novel Genes to Generate Transcriptome

Rapid progress in genomics has established the groundwork for mining desired candidate genes that can aid in the development of climate-resilient plants, and the use of NGS technology can lead researchers to tweak or modify critical agronomic features in crops (Taheri et al. 2018). The most recent technological advancements have also been used to acquire sequencing data at a certain stage or time of a crop. By evaluating the transcriptome obtained by NGS, it is now relatively simple to identify the up-/down-regulated genes or pathways under stress. Several genes (or metabolites) such as grain filling, plant reproduction and assimilate storage get influenced under abiotic stress (Tulsani et al. 2020). Performing transcriptome analysis such as O-acyltransferase, phosphatidylinositol, or/

FIGURE 14.4 *Dysdercus cingulatus* (red cotton stainer) has co-evolved with cotton plant leaf.

and phosphatidylcholine transfer protein (SFH13), glycerol-3-phosphate acyltransferase-2 (GPAT2), phosphatidylcholine-diacylglycerol-choline-phosphotransferase (PDCP), plasmodesmata-callose-binding-protein 3 (PDCB), multiple proteins or genes have been recognized. These genes were activated in plants under stress and a few of them even crypted pyruvate-di-phosphate that participates in shikimate pathways (producing secondary metabolites) (Yadav et al. 2018).

The transcriptome generated by NGS has been used to uncover the role of genetic material (susceptible and resistant) in disease resistance by proving candidate genes such as calcium-transporting ATPase (CT-ATPase), glutamate-receptor 3.2, NPR1 and NPR2, as well as many genes related to phytohormones (jasmonic acid, abscisic acid and gibberellic acid) that play a role in enhancing plant immunity against microbial attack. As a result, such research can be applied to the genetic editing of desirable crops or plants in order to generate climate-resilient plants that can endure poor growing conditions in order to fulfill future food demands (Taheri et al. 2019).

14.4.3 Role of Metabolomics

The study of plant-secondary metabolites that regulate numerous cellular functions in living systems is known as metabolomics. This field denotes the diverse assortment of plant metabolites produced by the plant's metabolic processes (Raza 2020). Metabolome research has been heavily cited in the genomic area, allowing us to exploit different stages of the continuous biomolecular and physiological changes brought on by climate change or genetic instability. Metabolic adjustments are explicitly proposed as postgenomic changes in plants, making phenome or phenotypic determination easier (Kumar et al. 2021). As a result, metabolites such as branched-chain amino acids (BCAAs), respiratory amino acids (glycine and serine) and intermediate products of the tricarboxylic acid cycle observed in various plants are emerging as one of the most dependable tools for understanding stress tolerance and resistance mechanisms in plants (Matich et al. 2019). Metabolomic studies have been carried out in several plants such as Asian rice (*Oryza sativa*), barley (*Hordeum vulgare*) and Thale cress (*Arabidopsis thaliana)* under stress. Changes in the amounts or quantity of proteins or metabolites such as organic acids, proline, tryptophan, glycine-butane, phenolic, phytohormones, as well as sulfur-containing metabolites like methionine, cysteine and glutathione, have also been observed in stressed plants (Villate et al. 2021).

Knowledge of specific metabolites associated with important stages or various time points to investigate the stress response can allow for precise genetic modification by targeting transcripts in a plant. NGS technologies, on the other hand, have emerged as viable breeding tools for describing regulatory processes and/or cellular responses to environmental stimuli such as biotic and abiotic stresses. However, the combination of NGS with metabolomics has improved researchers' predicting skills to uncover preliminary metabolic co-networks via an organism's sequencing data (Hasanpour et al. 2020). In this approach, the data created by NGS technologies and the quantification of the targeted metabolites aid in the development of the most appropriate strategy for increasing agricultural output with greater precision. These technologies have the potential to accelerate genetic manipulation projects aimed at developing climate-resilient smart crops that can provide healthy food while also ensuring global food security (Abdelhafez et al. 2020). There is no doubt that phenotypic or phenome-directed genetic manipulation has been seen to increase the performance of a crop by undertaking metabolic engineering in the same way that genomic fields have demonstrated a significant contribution to furthering genomic research gains (Tuyiringire et al. 2018).

14.4.4 Diagnosing Disease-Causing Pathogens

Multiple pathogens, including bacteria, protozoa, mollicutes, fungi, viroids and viruses, are common disease agents. Researchers have used a variety of molecular approaches to identify and track the infections mentioned above to develop management strategies and boost agricultural yields. Previous methods for determining the minute remnants of an organism, particularly viruses and emerging diseases were ineffective (Díaz-Cruz et al. 2019). Then, with the advent of genomic science, scientists were able to discover the genetic makeup of crops and disease-causing pathogens, as well as provide key clues to monitor plant–pathogen interaction for crop improvement. NGS and metabolomics have been utilized to determine the host genetic factors that the disease-causing agent engineers during attack or infection, as well as the ability to resistance via reprogramming genetic makeup (Hariharan and Prasannath 2021). Sugarcane mosaic virus (SCMV) is a severe threat to China's maize-cultivating farming community (Xie et al. 2016; Xu et al. 2021). Transcriptomic and proteomic studies have helped us to understand the basics of pathogenicity and that SCMV actively participates in downregulating photosynthesis-related genes, resulting in the phenotype of chlorotic lesions (Dong et al. 2017; Akbar et al. 2021; Nehra et al. 2022).

The most current NGS-based deep-sequencing methods produce trustworthy sequenced genomes for genomic analysis using established bioinformatic tools for improved disease control by properly diagnosing plant pathogens. These agricultural technologies (ATech) can be used to generate metagenomics and estimate a crop's infecting or growing microbial population. Furthermore, using the most recent bioinformatics approaches, small RNA (sRNA) families such as small interfering RNAs (siRNAs) could be used to accomplish detection and reconstruction of different virus genomes (DNA or/and RNA) and microvariants (Makeshkumar and Sankar 2022).

NGS technologies are also applicable to find harboring pathogens by insect vectors, crop certification and quarantine programs using diagnostic techniques. Nowadays, it is possible to detect plant metabolites, e.g., flavonoids, cyanogenic glycosides, benzoxazinoids, saponins, terpenes and terpenoids, which are produced under a pathogen attack in various crops such as rice, maize, rye, barley, oat, millet, sorghum, etc. As a result, a combination strategy based on genomic and metabolomics studies is utilized to stabilize the world's diminishing plant production (Nabi et al. 2021).

14.4.5 Integration of Transcriptome and Metabolome

By comparing transcriptome and metabolome data, integrated networks have been created, paving the way for more precise metabolite engineering in plants using genomics (Zhang et al. 2022). This element enlightens us by revealing the role of metabolites and transcripts in plants for stress tolerance, as well as providing further evidence for developing a complete strategy to boost agricultural output (Zhu et al. 2021). Both the metabolome and the transcriptome could reveal critical metabolites and/or cellular processes that can influence plant design and biomass production, as well as contribute to plant adaptations through controlling an organism's physiological condition. Furthermore, by studying the genetics of transgene lines or mutants, recent technological improvements have revealed novel metabolic networks and discovered important regulatory genes (Chen et al. 2020).

14.5 CURRENT RESEARCH IN NGS AND PLANT GENOMICS

14.5.1 PacBio

The third generation, long-read HTS (high-throughput sequencing) technology PacBio, developed by Pacific BioSciences is based on observing fluorescently labeled dNTPs (deoxyribonucleoside triphosphates) when complementary strand synthesis takes place. PacBio has been recently utilized for generating *de novo* genome assembly of animals, plants, viruses, bacteria and fungi, while PacBio sequencing aids in the analysis of metagenomes and transcriptomes (Rhoads and Au 2015). Advantages of PacBio transcriptome sequences are that they are distinguishable between single-nucleotide mutations and alternatively spliced isoforms based on accurate and long enough transcriptome sequences (Cui et al. 2020).

14.5.2 Nanopore Sequencing Methods

Oxford Nanopore Technology (ONT) (Sahlin and Medvedev 2021) developed the best commercial used nanopore-based sequencing method based on observing nucleotide oligomers' signal disruptions where DNA molecules are translocated via nanopores (Dumschott et al. 2020). Its sophistication allows for the detection of nucleotide base modifications while sequencing of RNA and DNA (Liefting et al. 2021). Later MinION came into the scene which was further improvised by the benchtop GridION and PromethION models, all of which increased the throughput. Flongle flow cells (miniature) were introduced in 2019 which reduced single-sample analyses costs. The main advantages of Flongle and MinION are low power requirements, small size and rapid analysis process that enabled analysis in field, camps and other nonlaboratory conditions (Lee et al. 2019).

A prime example of usage of nanopore methods in plant disease detection is in the case of grapevine plants and their respective disease-causing viruses. The real-time detection of the known viruses is possible due to adaptive sequencing by nanopore sequencing devices. Methods like nanopore ligation and rapid sequencing, nanopore direct cDNA and RNA sequencing are used for detection of Grapevine DNA viruses (Hoang et al. 2020). Even identification of long non-coding RNAs (lncRNAs) and circular RNAs (circRNAs) of viruses is possible with nanopore technologies. Detection of viral RNA modifications helps fight newer strands of viruses by sequencing the full length of viroids. Most importantly, it helps us understand their functionality and pathogenicity (Wang et al. 2021a, 2021b).

14.5.3 Synthetic Long-Read (SLR) Sequencing

SLRs represent longer DNA molecules that get generated and assembled by joining multiple short reads. They are joined on the basis of Unique Molecular Identifier (UMI) barcodes located within the reads that link them to the DNA molecule from where they originated (Van Dijk et al. 2018). Compared to PacBio CCS reads, the LoopSeq amplicons (an amplicon library tool kit) of the full-length 16S rRNA gene or 1.5 kb has ten-fold higher accuracy. While in bacterial mock communities, illumina short reads gave approximately perfect (99.9999%) accuracy. In the generation of SLRs, UMI tagging is utilized to have better identification and removal of chimeric molecules produced via. PCR (Callahan et al. 2021).

14.6 ADVANTAGES OF TGS

There are several advantages of TGS. It enables rapid molecular diagnostics which has low cost of apparatus, high portability and less power requirements (Chalupowicz et al. 2019). Other advantages are better identification, better biodiversity assessment of specimens and cultures of pathogens, feature of metabarcoding (a procedure for determining composition and richness of microbial communities), technically nondemanding library preparation steps, easier affordability and real-time access to sequenced data of DNA or RNA molecules (Aragona et al. 2022).

14.7 CONCLUSION

If society continues to function in a similar state as they have in the past, the world population will require about 119% increase in edible crops by the year 2050. By using diagnosis approaches, NGS technology can be used to discover viruses dwelling in insect vectors, crop certification and quarantine programs. Plant metabolites, such as benzoxazinoids, saponins, terpenes, glycosides, flavonoids, cyanogenic and terpenoids, that are formed in response to pathogen infection in diverse crops, such as millet, sorghum, rice, barley, rye, oat and maize, may now be detected. As a result, a combination strategy based on genomic and metabolomics studies can be utilized to stabilize the world's diminishing plant production. Long reads are advantageous over short reads in intra-specific population genetics analyses because of their capacity to distinguish among genotypes, phase haplotypes and resolve repeat-rich regions. Metagenomics is broadly used for analyses of taxonomic and functional potential of microbial communities and recovery of individual viral and prokaryote genomes. The unison of Sanger's sequencing method, automatic-sequencing machine and linked data-analyzing software has led to improvements in sequencing strategies. For further improvements robots, artificial intelligence and a strong integration of up-to-date chemicals, database setup, instruments along with bioinformatics would be of key importance. One thing is for sure that with the pace of current developments, we can hope for a future where everyone can get their daily bread.

REFERENCES

Abdallah, N.A., Hamwieh, A., Radwan, K., Fouad, N., Prakash, C., 2021. Genome editing techniques in plants: A comprehensive review and future prospects toward zero hunger. *GM Crops and Food* 12, 601–615.

Abdelhafez, O.H., Othman, E.M., Fahim, J.R., Desoukey, S.Y., Pimentel-Elardo, S.M., Nodwell, J.R., et al., 2020. Metabolomics analysis and biological investigation of three Malvaceae plants. *Phytochem. Anal.* 31, 204–214.

Abebe, K.S., 2019. Genotype by sequencing method and its application for crop improvement (A Review). *Adv. Biosci. Bioeng.* 7, 1–7.

Akbar, S., Yao, W., Qin, L., Yuan, Y., Powell, C.A., Chen, B., Zhang, M., 2021. Comparative analysis of sugar metabolites and their transporters in sugarcane following sugarcane mosaic virus (SCMV) infection. *Int. J. Mol. Sci.* 22, 13574.

Aksoy, E., Yildirim, K., Kavas, M., Kayihan, C., Yerlikaya, B.A., Çalik, I., Sevgen, İ., Demirel, U., 2022. General guidelines for CRISPR/Cas-based genome editing in plants. *Mol. Biol. Rep.* 49, 12151–12164.

Ali, S., Khan, N., Tang, Y., 2022. Epigenetic marks for mitigating abiotic stresses in plants. *J. Plant Physiol.* 275, 153740.

Alok, A., 2020. Problem of poverty in India. *Int. J. Res. Rev.* 7, 174–181.

Alvarez, M.V.N., Alonso, D.P., Kadri, S.M., Rufalco-Moutinho, P., Bernardes, I.A.F., de Mello, A.C.F., et al., 2022. *Nyssorhynchus darlingi* genome-wide studies related to microgeographic dispersion and blood-seeking behavior. *Parasites Vect.* 15, 1–11.

Aragona, M., Haegi, A., Valente, M.T., Riccioni, L., Orzali, L., Vitale, S., Luongo, L., Infantino, A., 2022. New-generation sequencing technology in diagnosis of fungal plant pathogens: A dream comes true? *J. Fungi* 8, 737.

Ashraf, M.F., Hou, D., Hussain, Q., Imran, M., Pei, J., Ali, M., et al., 2022. Entailing the next-generation sequencing and metabolome for sustainable agriculture by improving plant tolerance. *Int. J. Mol. Sci.* 23, 651.

Athanasopoulou, K., Boti, M.A., Adamopoulos, P.G., Skourou, P.C., Scorilas, A., 2021. Third-generation sequencing: The spearhead towards the radical transformation of modern genomics. *Life* 12, 30.

Ayalew, G., 2006. Comparison of yield loss on cabbage from diamondback moth, *Plutella xylostella* L. (Lepidoptera: Plutellidae) using two insecticides. *Crop Prot.* 25, 915–919.

Behjati, S., Tarpey, P.S., 2013. What is next generation sequencing? Archives of disease in childhood. *Edu. Prac. Ed.* 98, 236–238.

Callahan, B.J., Grinevich, D., Thakur, S., Balamotis, M.A., Yehezkel, T.B., 2021. Ultra-accurate microbial amplicon sequencing with synthetic long reads. *Microbiome* 9, 1–13.

Camargo, J.F., Ahmed, A., Morris, M.I., Anjan, S., Prado, C.E., Martinez, O.V., et al., 2019. Next generation sequencing of microbial cell-free DNA for rapid noninvasive diagnosis of infectious diseases in immunocompromised hosts. *Biol. Blood Marrow Transplant.* 25, S356–S357.

Chalupowicz, L., Dombrovsky, A., Gaba, V., Luria, N., Reuven, M., Beerman, A., et al., 2019. Diagnosis of plant diseases using the nanopore sequencing platform. *Plant Pathol.* 68, 229–238.

Chen, J., Wang, J., Wang, R., Xian, B., Ren, C., Liu, Q., et al., 2020. Integrated metabolomics and transcriptome analysis on flavonoid biosynthesis in safflower (*Carthamus tinctorius* L.) under MeJA treatment. *BMC Plant Biol.* 20, 1–12.

Chondronasiou, D., Gill, D., Mosteiro, L., Urdinguio, R.G., Berenguer-Llergo, A., Aguilera, M., et al., 2022. Multi-omic rejuvenation of naturally aged tissues by a single cycle of transient reprogramming. *Aging Cell* 21, e13578.

Crist, E., Mora, C., Engelman, R. 2017. The interaction of human population, food production, and biodiversity protection. *Science* 356, 260–264.

Cui, J., Lu, Z., Xu, G., Wang, Y., Jin, B., 2020. Analysis and comprehensive comparison of PacBio and nanopore-based RNA sequencing of the Arabidopsis transcriptome. *Plant Methods* 16, 1–13.

De Filippis, L.F., 2018. Underutilized and neglected crops: Next generation sequencing approaches for crop improvement and better food security. In: Ozturk, M., Hakeem, K., Ashraf, M., Ahmad, M. (Eds.) *Global Perspectives on Underutilized Crops*. Springer, Cham. pp. 287–380.

Díaz-Cruz, G.A., Smith, C.M., Wiebe, K.F., Villanueva, S.M., Klonowski, A.R., Cassone, B.J., 2019. Applications of next-generation sequencing for large-scale pathogen diagnoses in Soybean. *Plant Disease* 103, 1075–1083.

Dong, M., Cheng, G., Peng, L., Xu, Q., Yang, Y., Xu, J., 2017. Transcriptome analysis of sugarcane response to the infection by sugarcane steak mosaic virus (SCSMV). *Trop. Plant Biol.* 10, 45–55.

Dong, W., Wu, D., Li, G., Wu, D., Wang, Z., 2018. Next-generation sequencing from bulked segregant analysis identifies a dwarfism gene in watermelon. *Scientific Rep.* 8, 1–7.

Dumschott, K., Schmidt, M.H., Chawla, H.S., Snowdon, R., Usadel, B., 2020. Oxford nanopore sequencing: new opportunities for plant genomics? *J. Exp. Bot.* 71, 5313–5322.

Fan, X., Tang, D., Liao, Y., Li, P., Zhang, Y., Wang, M., et al., 2020. Single-cell RNA-seq analysis of mouse preimplantation embryos by third-generation sequencing. *PLoS Biol.* 18, 3001017.

Giraldo, P.A., Shinozuka, H., Spangenberg, G.C., Smith, K.F., Cogan, N.O., 2021. Rapid and detailed characterization of transgene insertion sites in genetically modified plants via nanopore sequencing. *Front. Plant. Sci.* 11, 602313.

Hari, R., Parthasarathy, S., 2019. Next generation sequencing data analysis. *Encycl. Bioinform. Comput. Biol.* 3, 157–163.

Hariharan, G., Prasannath, K., 2021. Recent advances in molecular diagnostics of fungal plant pathogens: a mini review. *Front. Cell. Inf. Microbiol.* 10, 600234.

Hasanpour, M., Iranshahy, M., Iranshahi, M., 2020. The application of metabolomics in investigating anti-diabetic activity of medicinal plants. *Biomed. Pharmacother.* 128, 110263.

Hoang, P.T., Fiebig, A., Novák, P., Macas, J., Cao, H.X., Stepanenko, A., et al., 2020. Chromosome-scale genome assembly for the duckweed *Spirodela intermedia*, integrating cytogenetic maps, PacBio and Oxford Nanopore libraries. *Scientific Rep.* 10, 1–14.

Jang, G., Joung, Y.H., 2019. CRISPR/Cas-mediated genome editing for crop improvement: current applications and future prospects. *Plant Biotech. Rep.* 13, 1–10.

Jia, M., Guan, J., Zhai, Z., Geng, S., Zhang, X., Mao, L., Li, A., 2018. Wheat functional genomics in the era of next generation sequencing: An update. *Crop J.* 6, 7–14.

Jo, Y.D., Kim, J.B., 2019. Frequency and spectrum of radiation-induced mutations revealed by whole-genome sequencing analyses of plants. *Quantum Beam Sci.* 3, 7.

Kitamura, S., Satoh, K., Oono, Y., 2022. Detection and characterization of genome-wide mutations in M1 vegetative cells of gamma-irradiated *Arabidopsis*. *PLoS Genet.* 18, e1009979.

Ko, H.Y., Salem, G.M., Chang, G.J.J., Chao, D.Y., 2020. Application of next-generation sequencing to reveal how evolutionary dynamics of viral population shape dengue epidemiology. *Front. Microbiol.* 11, 1371.

Kumar, M., Kumar Patel, M., Kumar, N., Bajpai, A.B., Siddique, K.H., 2021. Metabolomics and molecular approaches reveal drought stress tolerance in plants. *Int. J. Mol. Sci.* 22, 9108.

Le Nguyen, K., Grondin, A., Courtois, B., Gantet, P., 2019. Next-generation sequencing accelerates crop gene discovery. *Trends Plant Sci.* 24, 263–274.

Lee, Y.G., Choi, S.C., Kang, Y., Kim, K.M., Kang, C.S., Kim, C., 2019. Constructing a reference genome in a single lab: The possibility to use Oxford Nanopore Technology. *Plants* 8, 270.

Li, J., Manghwar, H., Sun, L., Wang, P., Wang, G., Sheng, H., et al., 2019. Whole genome sequencing reveals rare off-target mutations and considerable inherent genetic or/and somaclonal variations in CRISPR/Cas9-edited cotton plants. *Plant Biotech. J.* 17, 858–868.

Li, Y., He, X.Z., Li, M.H., Li, B., Yang, M.J., Xie, Y., et al., 2020. Comparison of third-generation sequencing approaches to identify viral pathogens under public health emergency conditions. *Virus Genes* 56, 288–297.

Liefting, L.W., Waite, D.W., Thompson, J.R., 2021. Application of Oxford nanopore technology to plant virus detection. *Viruses* 13, 1424.

Liu, Z., Zhang, G., Jingyuan, Z., Yu, L., Sheng, J., Zhang, N., Yuan, H., 2020. Second-generation sequencing with deep reinforcement learning for lung infection detection. *J. Healthc. Eng.* 2020, 3264801.

Lo, Y.T., Shaw, P.C., 2019. Application of next-generation sequencing for the identification of herbal products. *Biotechnol. Adv.* 37, 107450.

Madronero, L.J., Corredor-Rozo, Z.L., Escobar-Perez, J., Velandia-Romero, M.L., 2019. Next generation sequencing and proteomics in plant virology: how is Colombia doing? *Acta Biol. Colomb.* 24, 423–438.

Makeshkumar, T., Sankar, M.S.A., 2022. Recent developments in the diagnosis of geminiviruses. In: Gaur, R.K., Sharma, P., Czosnek, P. (Eds.) *Geminivirus: Detection, Diagnosis and Management*. Elsevier. pp. 33–42.

Matich, E.K., Soria, N.G.C., Aga, D.S., Atilla-Gokcumen, G.E., 2019. Applications of metabolomics in assessing ecological effects of emerging contaminants and pollutants on plants. *J. Hazard. Mat.* 373, 527–535.

Maxam, A.M., Gilbert, W., 1977. A new method for sequencing DNA. *Proc. Natl. Acad. Sci. USA* 74, 560–564.

Müller, P., Alonso, A., Barrio, P.A., Berger, B., Bodner, M., Martin, P., Parson, W., 2018. Systematic evaluation of the early access applied biosystems precision ID Globalfiler mixture ID and Globalfiler NGS STR panels for the ion S5 system. *Forensic Sci. Int. Genet.* 36, 95–103.

Nabi, S.U., Yousuf, N., Yadav, M.K., Choudhary, D.K., Ahmad, D., Kirmani, S.N., Sheikh, P., Ahmed, I., 2021. Recent insights in detection and diagnosis of plant viruses using next-generation sequencing technologies. In: Singh, R.K., Gopala (Eds.) *Innovative Approaches in Diagnosis and Management of Crop Diseases*. Apple Academic Press. pp. 85–100.

Nehra, C., Verma, R.K., Petrov, N.M., Stoyanova, M.I., Sharma, P., Gaur, R.K., 2022. Computational analysis for plant virus analysis using next-generation sequencing. In: Sharma, P., Yadav, D., Gaur, R.K. (Eds.) *Bioinformatics in Agriculture*. Academic Press. pp. 383–398.

Parlinska, M., Pagare, A. 2018. Food losses and food waste versus circular economy. Food losses and food waste versus circular economy. *Probl World Agric/problemy Rolnictwa Światowego* 18, 228–237.

Paul, F., Otte, J., Schmitt, I., Dal Grande, F., 2018. Comparing Sanger sequencing and high-throughput metabarcoding for inferring photobiont diversity in lichens. *Sci. Rep.* 8, 1–7.

Raza, A., 2020. Metabolomics: A systems biology approach for enhancing heat stress tolerance in plants. *Plant Cell Rep.* 41, 741–763.

Rhoads, A., Au, K.F., 2015. PacBio sequencing and its applications. *Genom. Proteom. Bioinformatics* 13, 278–289.

Rimoldi, S., Terova, G., Ascione, C., Giannico, R., Brambilla, F., 2018. Next generation sequencing for gut microbiome characterization in rainbow trout (*Oncorhynchus mykiss*) fed animal by-product meals as an alternative to fishmeal protein sources. *PLoS One* 13, e0193652.

Sahlin, K., Medvedev, P., 2021. Error correction enables use of Oxford Nanopore technology for reference-free transcriptome analysis. *Nat. Commun.* 12, 1–13.

Sahu, P.K., Sao, R., Mondal, S., Vishwakarma, G., Gupta, S.K., Kumar, V., Singh, S., Sharma, D., Das, B.K., 2020. Next generation sequencing based forward genetic approaches for identification and mapping of causal mutations in crop plants: A comprehensive review. *Plants* 9, 1355.

Saidi, A., Hajibarat, Z., 2020. Application of Next Generation Sequencing, GWAS, RNA seq, WGRS, for genetic improvement of potato (*Solanum tuberosum* L.) under drought stress. *Biocatalysis Agric. Biotechnol.* 29, 101801.

Sanger, F., Coulson, A.R., 1975. A rapid method for determining sequences in DNA by primed synthesis with DNA polymerase. *J. Mol. Biol.* 94, 441–448.

Sentmanat, M.F., Peters, S.T., Florian, C.P., Connelly, J.P., Pruett-Miller, S.M., 2018. A survey of validation strategies for CRISPR-Cas9 editing. *Sci. Rep.* 8, 1–8.

Seroussi, E., 2021. Estimating copy-number proportions: The comeback of Sanger sequencing. *Genes* 12, 283.

Shapiro, P.A., 2021. Food supply, starvation, and food as a weapon in the camps and ghettos of Romanian-occupied Bessarabia and Transnistria, 1941–44. *East/West* 8, 43–80.

Sharma, T.R., Devanna, B.N., Kiran, K., Singh, P.K., Arora, K., Jain, P., et al., 2018. Status and prospects of next generation sequencing technologies in crop plants. *Curr. Issues Mol. Biol.* 27, 1–36.

Singh, N., Bhatt, V., Rana, N. and Shivaraj, S.M., 2020. Advances of next-generation sequencing (ngs) technologies to enhance the biofortifications in crops. In: Sharma, T.R., Deshmukh, R., Sonah, H. (Eds.) *Advances in Agri-Food Biotechnology*, Springer, Singapore. pp. 427–450.

Singh, R.R., 2020. Next-generation sequencing in high-sensitive detection of mutations in tumors: challenges, advances, and applications. *J. Mol. Diagn.* 22, 994–1007.

Sinha, S., Swain, M., 2022. Response and resilience of agricultural value chain to COVID-19 pandemic in India and Thailand. In: Shaw, R., Pal, I., (Eds.) *Pandemic Risk, Response, and Resilience*, Elsevier. pp. 363–381.

Slatko, B.E., Gardner, A.F., Ausubel, F.M., 2018. Overview of next-generation sequencing technologies. *Curr. Protoc. Mol. Biol.* 122, e59.

Su, J.P., Tan, L.Y., Garland, S.M., Tabrizi, S.N., Mokany, E., Walker, S., et al., 2018. Evaluation of the SpeeDx ResistancePlus MG diagnostic test for *Mycoplasma genitalium* on the Applied Biosystems 7500 fast quantitative PCR platform. *J. Clin. Microbiol.* 56, e01245– 17.

Taheri, S., Lee Abdullah, T., Yusop, M.R., Hanafi, M.M., Sahebi, M., Azizi, P., Shamshiri, R.R., 2018. Mining and development of novel SSR markers using next generation sequencing (NGS) data in plants. *Molecules* 23, 399.

Taheri, S., Abdullah, T.L., Rafii, M.Y., Harikrishna, J.A., Werbrouck, S.P., Teo, C.H., Sahebi, M., Azizi, P., 2019. De novo assembly of transcriptomes, mining, and development of novel EST-SSR markers in Curcuma alismatifolia (Zingiberaceae family) through Illumina sequencing. *Sci. Rep.* 9, 1–14.

Tulsani, N.J., Hamid, R., Jacob, F., Umretiya, N.G., Nandha, A.K., Tomar, R.S., Golakiya, B.A., 2020. Transcriptome landscaping for gene mining and SSR marker development in Coriander (*Coriandrum sativum* L.). *Genomics* 112, 1545–1553.

Tuyiringire, N., Tusubira, D., Munyampundu, J.P., Tolo, C.U., Muvunyi, C.M., Ogwang, P.E., 2018. Application of metabolomics to drug discovery and understanding the mechanisms of action of medicinal plants with anti-tuberculosis activity. *Clin. Translational Med.* 7, 1–12.

Van Dijk, E.L., Jaszczyszyn, Y., Naquin, D., Thermes, C., 2018. The third revolution in sequencing technology. *Trends Genet.* 34, 666–681.

Villate, A., San Nicolas, M., Gallastegi, M., Aulas, P.A., Olivares, M., Usobiaga, A., et al., 2021. Metabolomics as a prediction tool for plants performance under environmental stress. *Plant Sci.* 303, 110789.

Wang, C., Liu, R., Liu, Y., Hou, W., Wang, X., Miao, Y., et al., 2021a. Development and application of the Faba bean-130K targeted next-generation sequencing SNP genotyping platform based on transcriptome sequencing. *Theor. Appl. Genet.* 134, 3195–3207.

Wang, F., Chen, Z., Pei, H., Guo, Z., Wen, D., Liu, R., Song, B., 2021b. Transcriptome profiling analysis of tea plant (*Camellia sinensis*) using Oxford Nanopore long-read RNA-Seq technology. *Gene* 769, 145247.

Wong, L.L., Razali, S.A., Deris, Z.M., Danish-Daniel, M., Tan, M.P., Nor, S.A.M., et al., 2022. Application of second-generation sequencing (SGS) and third generation sequencing (TGS) in aquaculture breeding program. *Aquaculture* 548, 737633.

Xie, X., Chen, W., Fu, Q., Zhang, P., An, T., Cui, A., An, D., 2016. Molecular variability and distribution of sugarcane mosaic virus in Shanxi, China. *PLoS One* 11, e0151549.

Xu, X.J., Zhu, Q., Jiang, S.Y., Yan, Z.Y., Geng, C., Tian, Y.P., Li, X.D., 2021. Development and evaluation of stable sugarcane mosaic virus mild mutants for cross-protection against infection by severe strain. *Front. Plant Sci.* 12, 788963.

Yadav, R., Lone, S.A., Gaikwad, K., Singh, N.K., Padaria, J.C., 2018. Transcriptome sequence analysis and mining of SSRs in Jhar Ber (*Ziziphus nummularia* (Burm. f.) Wight & Arn) under drought stress. *Sci. Rep.* 8, 1–10.

Yan, C., Zhang, N., Wang, Q., Fu, Y., Zhao, H., Wang, J., et al., 2022. Full-length transcriptome sequencing reveals the molecular mechanism of potato seedlings responding to low-temperature. *BMC Plant Biol.* 22, 1–20.

Yang, L., Jin, Y., Huang, W., Sun, Q., Liu, F., Huang, X., 2018. Full-length transcriptome sequences of ephemeral plant Arabidopsis pumila provides insight into gene expression dynamics during continuous salt stress. *BMC Genom.* 19, 1–14.

Yang, L., Zhang, P., Wang, Y., Hu, G., Guo, W., Gu, X., Pu, L., 2022. Plant synthetic epigenomic engineering for crop improvement. *Sci. China Life Sci.* 65, 2191–2204.

Zhang, Q., Li, T., Gao, M., Ye, M., Lin, M., Wu, D., et al. 2022. Transcriptome and metabolome profiling reveal the resistance mechanisms of rice against brown planthopper. *Int. J. Mol. Sci.* 23, 4083.

Zhong, C., Sun, S., Li, Y., Duan, C., Zhu, Z., 2018. Next-generation sequencing to identify candidate genes and develop diagnostic markers for a novel *Phytophthora* resistance gene, RpsHC18, in soybean. *Theor. Appl. Genet.* 131, 525–538.

Zhu, Q., Chen, L., Chen, T., Xu, Q., He, T., Wang, Y., et al., 2021. Integrated transcriptome and metabolome analyses of biochar-induced pathways in response to *Fusarium* wilt infestation in pepper. *Genomics* 113, 2085–2095.

Section VI

Biochemistry

15 Alcohol Dehydrogenase

Structural and Functional Diversity

Upagya Gyaneshwari, Kumari Swati, Akhilesh Kumar Singh, and Anand Prakash
Departments of Biotechnology, School of Life Sciences, Mahatma Gandhi Central University, Motihari, Bihar, India

Priti Pal
Shri Ramswaroop Memorial College of Engineering and Management, Tiwariganj, Lucknow, Uttar Pradesh, India

Babli Kumari and Brijesh Pandey
Departments of Biotechnology, School of Life Sciences, Mahatma Gandhi Central University, Motihari, Bihar, India

CONTENTS

15.1 INTRODUCTION

An oxidation (loss of electrons) and reduction (gain of electrons) or redox reaction is the transfer of electrons between two chemical entities. The species that gains electrons are known as an oxidizing agent, and the species that provides electrons is known as a reducing agent. In other words, the reducing agent is oxidized and the oxidizing agent is reduced in a redox reaction. A redox pair

DOI: 10.1201/9781003324706-21

refers to the oxidizing and reducing agent pair in a redox process, whereas a redox couple refers to the matching reduced or oxidized species (Franco and Vargas 2018). Photosynthesis, respiration, combustion and corrosion or rusting are all examples of redox reactions that are common and essential to life. Redox processes are necessary for life, but when they are disrupted, they can accelerate disease progression (Franco et al. 2018). Cellular respiration allows organisms to release the energy stored in glucose's molecular bonds. The development of oxidoreductase (OR)-based techniques for synthesis of polymers and functionalized organic substrates together with the bioreactors construction for biodegradation of pollutants and biomass processing has made substantial progress (Kareem 2020). ORs are a group of enzymes, which catalyze oxidation–reduction or redox reactions in living organisms, i.e., plants, animals and microorganisms. These reactions may be conducted *in vivo* or *in vitro* for academic, research as well as industrial purposes including monitoring, assessing and diagnosing the situations. The ORs are involved in life-sustaining activities such as glycolysis, the tricarboxylic acid (TCA cycle)/ citric acid cycle, oxidative phosphorylation, amino acid metabolism and so on (Legesse Habte and Assefa Beyene 2021).

B = oxidant (electron acceptor)

$$A^{-} + B = A + B^{-}$$

A = reductant (electron donor)

According to the international classification of enzymes, ORs are enzymes that catalyze the transfer of electrons, such as hydride ions or hydrogen atoms. Enzyme commission (EC) has characterized/classified the ORs by the way of EC1, including oxidase, oxygenase, peroxidase and dehydrogenase. The other types of redox reactions involve the addition of oxygen, proton extraction, hydride ion transfer and other essential steps (Legesse Habte and Assefa Beyene 2021; Kareem 2020). The importance of ORs can be realized by the fact that investigators target these enzymes for both fundamental as well as applied research. ORs are a distinctive pathological hallmark of several diseases (Berry and Hare 2004). For instance, the ORs such as cytochrome oxidase, succinate dehydrogenase, monoamine oxidase and catalase are dramatically down-regulated in the host when compared to healthy controls (Ngoka 2008). The most significant free radical scavenger systems, such as catalase, Superoxide dismutase–SOD and glutathione peroxidase, are all ORs. ORs are employed in the bleaching, dyeing and waste management of cotton fibers in the textile industry (Kareem 2020). Their role in wood formation and secondary metabolism has been also studied well (Pandey et al. 2011, 2014; Sibout et al. 2005). ORs have been successfully immobilized on the surface of simple and modified magnetic nanoparticles, as well as entrapped within the network of polymeric magnetic nano matrices, in a variety of ways. As a biosensor, the immobilized ORs demonstrated remarkable operating stability and reusability. This feature has been observed during the synthesis and transformation of valuable chemicals in both batch and continuous bioreactors (Husain 2017). Apart from these, many enzymes from natural or recombinant sources have been isolated and characterized for various kinetic parameters. Enzyme catalysis has been employed commercially in a variety of industries, including pharmaceuticals, foods, biofuel manufacturing, natural gas conversion, and others (Chapman et al. 2018). Different kinds of ORs are nowadays exploited in the agricultural industry for the manufacture of fertilizer, food packaging, dairy processing and other food processing, and a number of studies have proved their cost-effectiveness and product quality (Legesse Habte and Assefa Beyene 2021). One such ORs of immense industrial and academic importance is alcohol dehydrogenase (ADH). Its structural, evolutionary, kinetic and functional diversity makes it an enzyme of choice in the field of biotechnology. The ADH, an alcohol: NAD OR (EC1.1.1.1), is a metalloprotein class of conjugated protein that contains zinc as a prosthetic group. It catalyzes the oxidation of hydroxyl group to carbonyl group (Wesolowski and Lyerla 1983). Primary alcohols are converted into

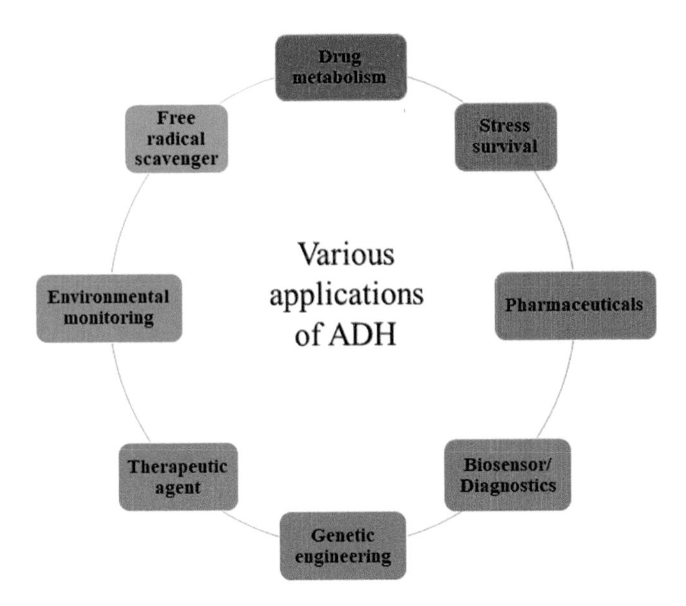

FIGURE 15.1 Applications of ADHs in diverse areas.

aldehydes, while secondary alcohols are converted to ketones in a reversible/irreversible reaction. ADHs are part of the OR superfamily and they catalyze the interconversion of alcohols to aldehydes or ketones with good stereoselectivity. ADHs are commonly utilized as biocatalysts for racemic substrate kinetic resolution. Several enantiomerically pure chemicals have also been made with the help of ADHs (Zheng et al. 2017). Figure 15.1 shows the applications of ADH in several fields. It is worth mentioning that these preparations by enzymatic method dominate over chemical methods for efficacy, purity and safety.

Since the enzymes catalyze the loss of two electrons and two hydrogen ions (that is, two hydrogen atoms) from the substrates (process is called dehydrogenations) they are called dehydrogenases. Major role of dehydrogenases in metabolism is recycling of coenzymes and the intermediate products linked to such reactions are useful secondary metabolites of pharmaceutical uses such as mevinic acid (Legesse Habte and Assefa Beyene 2021). Pellagra is a condition in which niacin shortage affects all NAD(P)-dependent dehydrogenases, resulting in rough skin. Dehydrogenase-related illnesses are not new to modern scientific literature. The historical discovery could be traced back to the investigation of Batelli and Stern (1910), where they showed that various organs of animals could oxidize alcohol. For instance, they found that the liver has a great potential to metabolize alcohol. Hamill (1910) demonstrated that a rabbit heart perfused with an alcohol-containing ringer solution eliminated a portion of the alcohol during the experiment. In studies with cat hearts, Fisher (1916) made a similar observation, as did Klewitz (1923) with rabbit hearts (Lutwak-Mann 1938; Forsander 1971). Looking into the importance of alcohol in metabolic disorders and the importance of enzymes involved in industrial synthesis of alcohol, many ADH enzymes were targeted for purification and characterization. Yeast ADH was first purified in 1937, while the first mammalian form was purified in 1948 (Jörnvall 2008). One of the earliest dehydrogenases whose structure was established was ADH from horse liver. Crystallographic studies of the ethanol-active E-isozyme, ADH1E of horse liver ADH, yielded this structure. ADH1E is a dimeric class 1 MDR-ADH, whereas comparable fungal, bacterial and yeast ADHs are tetrameric (Eklund and Ramaswamy 2008). ADHs are cytosolic enzymes found in higher amounts in the liver. Although, they are also present in other tissues including gastrointestinal tract and adipose (Di et al. 2021). Every living entity has an activity that is formed from enzymes of several families and is frequently observed interacting in multiple ways in a complex manner. ADH is also a component of our defense mechanism against a variety of reactive chemicals (Jörnvall 2008).

15.2 CLASSIFICATION OF ADHs

Based on differences in sequence motifs, structural comparisons, length of the protein chain and mechanistic features, dehydrogenases can be classified as (i) Short-chain dehydrogenases/reductases (SDRs), (ii) Medium-chain dehydrogenases/reductases (MDRs) and (iii) Long-chain dehydrogenases/reductases (LDRs) (Jörnvall et al. 1995). The SDRs with 250 residue subunits are a broad family of NAD(P)(H)-dependent ORs with comparable processes and sequence features. The SDR enzymes are involved in the metabolism of carbohydrates, lipids, amino acids, hormones, cofactors and xenobiotics, as well as redox sensor systems (Kavanagh et al. 2008). SDRs contain one residue (Tyr 151) that is conserved and crucial for enzymatic function (Jörnvall et al. 1995). The typical liver ADH, quinone reductase, leukotriene B4 dehydrogenase and many other variants belong to the MDR superfamily, which has 350-residue subunits (Persson et al. 2008). They can be found in all kingdoms of life and play a role in metabolism, regulatory processes and cell damage defense. All MDRs are dimers or tetramers that utilize NAD (H) or NADP(H) as a cofactor (Jörnvall et al. 1999). MDRs are further divided into zinc-containing and zinc-free MDRs. Dimeric ADHs, cinnamyl ADHs (CADs), yeast ADHs/tetrameric ADHs (YADHs), and polyol dehydrogenases (PDHs) are zinc-containing MDRs, whereas quinone ORs (QORs), leukotriene B4 dehydrogenases (LTD) (ACRs), mitochondrial response proteins (MRFs) are non-zinc-containing MDRs (Knoll and Pleiss 2008). However, the vertebrate ADH is a cytosolic, zinc-containing, NAD-dependent dimeric enzyme with a 40-kDa subunit. In vertebrates, eight different classes have been established based on sequence alignment, phylogenetic analysis, catalytic characteristics and gene expression patterns.

Table 15.1 summarizes the different type of ADH that occurs in diverse animal tissues. These classes exhibit around 60% sequence identity at the amino acid level. Within class, the multiple ADH isoenzymes exhibit above 90% identity. In humans, all ADH gene loci are found in a single gene cluster, however, the classes are functionally separate and have significant variances (Jörnvall 2008). To be more specific among vertebrates, the mammalian ADHs belong to the medium-chain dehydrogenases/reductases (MDR) superfamily of protein. ADH is a group of typical dimeric metalloproteins having zinc with subunit mass of approximately 40 kDa (Edenberg 2000). Human ADH appears to have evolved from an ancestral formaldehyde dehydrogenase (ALDH) line by serial gene duplication during vertebrate evolution, with the first one dating back to 500 MYA (million years ago) (Persson et al. 2008). ADHs are common enzymes that can work with a variety of substrates. They are divided into three groups based on their cofactor preferences. The zinc-dependent ADH, which requires NAD(P)+ as a cofactor, is the first class. The second class relies on the cofactors pyrroloquinoline quinone (PQQ), heme, or cofactor F420, whereas the third class depends on flavin adenine dinucleotide (FAD) (Guo et al. 2019). ADH is further classified based on enzymic characteristics and sequence similarity. The members of the ADH family include seven enzymes, ADH1–7 (ADH1A (earlier called α), ADH1B (earlier called β) and ADH1C (earlier called γ), ADH4, ADH5, ADH6 and ADH7). These seven ADH enzyme-encoding genes are grouped in a short region of chromosome 4 (4q21–24) in a head-to-tail array of around 370 kb in length in humans (Crabb et al. 2004; Eklund and Ramaswamy 2008; Jelski et al. 2008; Edenberg and McClintick 2018).

15.3 SUBSTRATE SPECIFICITY AND DIVERSITY IN ADHs

ADHs help detoxification pathways by catalyzing the reversible oxidation of alcohols to aldehydes on a variety of substrates ranging from methanol to long-chain alcohols and sterols (Edenberg 2000). The tiny and cylindrical active regions of ADHs may explain the narrow substrate

TABLE 15.1
ADH Types, Their Occurrence, Substrate and Tissue Distribution

ADH Type	Occurrence	Substrate	Tissue Distribution	References
ADHIA	Human	Ethanol, secondary alcohols	Liver (mostly) gastrointestinal tract, kidneys and lungs	Jelski et al. (2008); Staab et al. (2008); Edenberg and McClintick (2018)
ADHIB	Human	Ethanol	Liver	Jelski et al. (2008); Edenberg and McClintick (2018)
ADHIC (Glutathione dependent)	Vertebrate, invertebrate, plant fungi and prokaryotes	Formaldehyde, steroid substrates,	Almost all tissue type	Jelski et al. (2008); Edenberg and McClintick (2018)
ADH3	From bony fish to humans	Long-chain alcohols, i.e., omega hydroxy fatty acid, S-hydroxy-methyl-glutathione, Formaldehyde	Liver and other tissues	Niederhut et al. (2001); Pares et al. (2008); Eklund and Ramaswamy (2008)
ADHII/4	Human	Ethanol at higher concentration	Upper part of digestive tract	Edenberg (2007); Jelski et al. (2008); Edenberg and McClintick (2018)
ADHIII/5	Human	S-nitrosoglutathione, Formaldehyde	Liver, GI tract and brain	Edenberg (2007); Edenberg and McClintick (2018); Zhao et al. (2020)
ADHIV/6	Human, *S. cerevisiae*	Broad substrate specificity	Liver	Edenberg and McClintick (2018); Meng et al. (2020)
ADHV/7	Human	Ethanol, Retinol, alcohols with aliphatic chain, hydroxysteroids, lipid peroxidation products	Esophagus, stomach	Edenberg (2007); Edenberg and McClintick (2018); Meng et al. (2020)
ADH8	Amphibians	Retinaldehyde	Gastric tissue (stomach)	Peralba et al. (1999); Parés et al. (2008)

specificity. This helps to explain why secondary alcohols are often poor substrates for ADHs (Di et al. 2021). The ADH1A, ADH1B and ADH1C subunits of human class I enzymes are the most comparable to the horse E-subunit. The three class I ADH isoenzymes have around 93-sequence identity with each other and with the horse enzyme, but they differ in substrate specificity, developmental expression and consequently substrate specificity (Niederhut et al. 2001; Eklund and Ramaswamy 2008). When Phe-93 is replaced with Ala in the substrate pocket, the active site at the catalytic zinc is enlarged, resulting in selectivity for branching substrates and inhibitors, as well as greater activity with secondary alcohols, in the ADH1A isozyme (Niederhut et al. 2001). ADH activity is influenced by a number of variables, including the enzyme's overall activity and product inhibition by NADH and acetaldehyde. In both intact animals and cultured hepatocytes, growth hormone boosted ADH activity. It was reduced by androgens and thyroid hormones. Fasting and protein restriction both reduce liver ADH activity significantly (Crabb et al. 2004). The class I isoenzymes are the most important contributors to beverage ethanol metabolism and are thought to have a role in the detoxification of a variety of biogenic and dietary alcohols and aldehydes (Niederhut et al. 2001). ADH1E from horse liver is the first dehydrogenase whose structure was determined. This is a dimeric class 1 MDR-ADH, whereas comparable fungal, bacterial and yeast ADHs are tetrameric. Aromatic alcohols/aldehydes, as well as substrate-like molecules like trifluoroethanol, were investigated (Eklund and Ramaswamy 2008). ADH3 (ADH3) is found in both prokaryotic and eukaryotic organisms in large amounts. ADH3 is the ancestral form, according to species variability determined by extensive genome screens and sequence comparisons among members of the ADH family. Because of its universal occurrence and structural conservation, ADH3 is thought to play an important role in living creatures. As a result, ADH3 is linked to important signaling and metabolic networks. ADH3 is a crucial enzyme in the detoxification of endogenous and exogenous formaldehyde by oxidizing S-hydroxymethyl glutathione (HMGSH), the spontaneous glutathione (GSH) adduct of formaldehyde (Staab et al. 2008). In one of the studies, Tsigos et al. (1998) described the purification and characteristics of an ADH *Moraxella* sp. TAE12*3* (isolated from the Antarctic psychrophile). Apart from its principal catalytic activity, which is the oxidation of primary and secondary alcohols to their carbonyl compounds, this ADH was also active against unusual, non-natural substrates. When ethanol was utilized as a substrate, the fastest reaction rate was recorded. By lengthening and branching the carbon chain of the alcohol, a progressive decrease in rate was noticed (Tsigos et al. 1998). Several phenol and terpene alcohols were shown to be substrates for the polyextremophilic ADH/A1a. Cinnamyl alcohol had the fastest reaction rates, followed by 1,5-pentanediol, and 1-heptanol. Surprisingly, the secondary alcohol 3-buten-2-ol oxidized slowly as well. The MDH (methanol dehydrogenase) *Geobacillus stearothermophilus* DSM 2334 (GsADH) demonstrated the maximum catalytic efficiency of 60.27 per Mol per sec toward isopropanol, followed by ethanol, 2-butanol, 1-butanol and tert-butanol. Furthermore, GsADH was found to have action toward 1,3-propanediol. With glycerol, no action was seen (Guo et al. 2019).

15.4 REACTION MECHANISM OF ADHs

The mode of action of ADH involves the oxidation of alcohol into aldehyde concomitant to addition of released hydrogen atoms to NAD/ NADP (Svensson et al. 1999; Pandey et al. 2014). The zinc attaches to the oxygen in the alcohol, whereas NAD binds to the isoleucine. The reaction catalyzed by ADH is a two-step process. The first step involves hydrogen removal from the OH on the alcohol. While in the second step, a carbon-oxygen double bond formation occurs, hydrogen is then removed from the carbon and added to NAD/NADP to form $NADH^+$/$NADPH^+$. ADH dimers are found in all mammalian class I–IV enzymes. Each subunit of dimeric ADH has two zinc atoms, one catalytic and the other structural. The active site contains the catalytic zinc, which is coupled with two Cys and one His (Östberg et al. 2016; Pandey et al. 2011). Zinc remains surrounded by Sulfur of Cys and Nitrogen of His. Actually, the active site of ADH includes zinc, serine, histidine (His), isoleucine and cysteine (Cys). Enzyme activity has been reported to be dependent on Zinc (Pandey et al. 2011; Akal et al. 2019).

15.5 PHYSIOLOGICAL ROLE IN MICROBES AND PLANTS

ADH activity is normally modest under normal oxygen tension, but it spikes when there is a lack of oxygen, which leads to ethanolic fermentation. It is essential for the plant's response to anaerobiosis. In aerobic conditions, ADH also plays a crucial part in the correct development of organs. In germinating seeds, pollen and tree stems, ADH activity has been identified and reported. Aromatic ADHs have their defined role in lignin monomer biosynthesis, which governs the quality and quantity of lignin. These traits are of industrial significance as they influence the quality of pulp and paper and determine their durability (Pandey et al. 2014). Changes in ADH activity or expression during the normal fruit ripening process have also been reported (Tesniere and Abbal

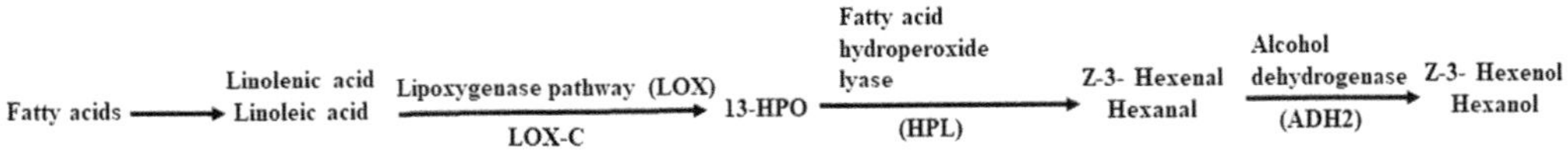

Fatty acids → Linolenic acid / Linoleic acid → Lipoxygenase pathway (LOX) / LOX-C → 13-HPO → Fatty acid hydroperoxide lyase (HPL) → Z-3- Hexenal / Hexanal → Alcohol dehydrogenase (ADH2) → Z-3- Hexenol / Hexanol

FIGURE 15.2 Biochemical origin of volatile compounds in fruits that contribute to flavor/aroma production (Redrawn and modified from Li et al. 2021: open access article distributed under the terms and conditions of the Creative Commons Attribution license (http://creativecommons.org/licenses/by/3.0/).

Retinol →(ADH) Retinal →(ALDH) Retinoic acid

FIGURE 15.3 Conversion of retinol to retinal and finally to retinoic acid in skin and other epithelial tissues with the help of alcohol dehydrogenase.

2009). Figure 15.2 shows the formation of volatile compounds during fruit ripening and the role of ADH in this process.

15.5.1 ADH in the Development of Skin and Other Epithelial Tissues in Animals

Retinoids are necessary for cell growth and development, along with the maintenance of proper skin and epithelial tissue function. ADH and ALDH are enzymes that govern the oxidation of retinol to retinal and finally to retinoic acid. By acting as a ligand of the nuclear receptor family, Retinoic acid, which is produced from retinol oxidation, regulates embryonic development, spermatogenesis and epithelial differentiation (Crabb et al. 2004). The cytosolic enzyme Retinol dehydrogenase (ADH) is involved in the first step of retinol metabolism by catalyzing the oxidation of retinol to retinaldehyde (Figure 15.3).

Two of the ADH isoenzymes, ADH I and IV, have been found to take part in this reaction. According to further research, it has been observed that class IV ADH is the most effective isozyme for retinol oxidation. The conversion of retinaldehyde to retinoic acid by ALDH is the next stage in retinol metabolism (Haselbeck et al. 1997; Orywal et al. 2008). ADHIV/4 may have two significant physiological roles: it may be a key contributor to first-pass ethanol metabolism in the stomach. It also acts as a rate-limiting step in the synthesis of local retinoic acid in the upper digestive tract epithelium, skin and neural tissue of developing embryos (Yin et al. 2003). According to rigorous kinetic investigations of all human ADH family members, ethanol can effectively prevent the oxidation of retinol to retinoic acid at quantities reported in heavy drinkers. In chronic alcoholics, ethanol's inhibition of retinol oxidation may lead to testicular shrinkage, oligospermia, psoriasis, and an increased incidence of oral, esophageal and colorectal malignancies (Crabb et al. 2004). ADH1 and ADH3 are enzymes that can oxidize retinol in all vertebrate taxa, from bony fish to humans. In addition, ADHs that are retinoid-active have been found in lower vertebrates. An ADH (ADH-F) active with retinol and steroids is found in fetal and adult chicken tissues. The principal enzymes involved in retinol detoxification in the liver are ADH1 and ADH2, while ADH3 (less active) and ADH4 (most active) are involved in retinoic acid production in tissues (Parés et al. 2008).

15.5.2 Metabolism of Drugs

ADHs are enzymes that help ethanol be oxidized and eliminated. ADHs are also important in the metabolism of medicines and metabolites having alcohol as functional groups, such as hydroxyzine (antihistamine), abacavir (HIV/AIDS) and ethambutol (antituberculosis) (Di et al. 2021). After absorption and passage through the liver, ethanol is largely metabolized to acetaldehyde with the help of ADH in the cytosol and cytochrome P450IIE1 in microsomes. ALDH converts acetaldehyde to acetate, which is released from the liver and digested by the heart and muscle (Crabb et al. 2004).

15.5.3 Nitric Oxide Metabolism

ADH3 is involved in important signaling and metabolic processes. S-nitrosothiol/S-nitrosoglutathione reduction by ADH3 has been identified (GSNO). Because it acts as a nitric oxide (NO) reservoir (with a short biological half-life), GSNO maintains equilibrium in airway smooth muscle tone and inflammation. GSNO is the main repository of cellular S-nitrosothiol (SNO) species that modulate total and local NO bioavailability *in vivo*. S-nitrosothiol (SNOs) and GSNO act as functional NO stores. NO and GSNO affect on bronchial smooth muscle tone and responsiveness, adrenergic receptor activity and anti-inflammatory. All these activities help in maintaining normal lung physiology and function (Blonder et al. 2014). As a GSNO reductase, ADH3 appears to play a role in protein S-Nitrosation control by establishing a dynamic equilibrium between S-Nitrosothiols and GSNO. ADH3 has been identified as having a housekeeping role in living creatures due to its widespread expression pattern in mammalian tissues (Calabrese et al. 2007). Catabolism of GSNO is mediated by a class III ADH, S-Nitrosoglutathione reductase (GSNOR). As a result, GSNOR plays a critical function in the regulation of intracellular SNOs. As shown in respiratory and other disorders, dysregulation of this enzyme might have negative consequences. Studies have observed a drop in levels of GSNO, SNOs and some activities related to them after increased GSNOR activity. This suggests that ADH could be used as a therapeutic target for the treatment of respiratory disorders such as asthma (Blonder et al. 2014).

15.5.4 Defense Against Reactive Compounds

ADHs are a component of our defense mechanism against a variety of reactive chemicals. As a result, they act similarly to cytochrome P450s. ADHs have the extra benefit of not producing reactive oxygen species, which can harm tissues. Because many of the aldehydes generated are reactive and hazardous, an efficient battery of ALDHs, in combination with many of the medium, and short-chain dehydrogenases/reductases, forms integrated metabolic pathways (Jörnvall 2008).

15.5.5 Stress Survival

ADH is a zinc-dimeric enzyme that uses $NAD^+/NADH^+$ as a cofactor to catalyze the reversible conversion of aldehyde (acetaldehyde) to alcohol (ethanol). In anaerobic settings, this is the final phase of fermentative metabolism, when pyruvate, created following the glycolyic break down of glucose, leads to the synthesis of alcohol. The development of this enzyme's activity and gene expression in response to anaerobiosis (living without oxygen) has been extensively researched in the plant kingdom, notably in organs such as roots, tubers, seeds and fruit (Wesolowski and Lyerla 1983; Tesniere and Abbal 2009). The role of ADHs in plant responses to different stressors, elicitors, and abscisic acid has been studied all over the world (Tesniere and Abbal 2009). ADH converts acetaldehyde to ethanol and recycles NADH to NAD in hypoxic environments. With the support of ADH, anaerobic glycolysis can be employed as an energy-generating pathway. It also plays a role in the maintenance of equilibrium in cytoplasm, as ADH activity is greatly influenced by pH (Roberts et al. 1984).

15.6 INDUSTRIAL APPLICATIONS

Several industries use ADHs for the formation of economically, therapeutically and other important chemicals or intermediates.

15.6.1 Nootkatone

Nootkatone is a natural ingredient and a high-value sesquiterpenoid responsible for the characteristic smell of grapefruit (Milhim et al. 2019). It is commonly used to manufacture perfumes and colognes in the fragrance industry. Ticks, mosquitoes and a range of other biting pests are all repelled and killed by it. It can be found in trace amounts in Alaska yellow cedar trees and grapefruit peel. Nootkatone has a wide range of biological activities and qualities that make it a good candidate for aromatics, medicines and biofuels (CDC 2020; Meng et al. 2020). It has limited yields when isolated from natural plant sources, and chemical syntheses use carcinogenic or toxic chemicals. A two-enzyme biocatalytic method for the production of (+)-nootkatone is described here. In the first step, sesquiterpene, i.e., (+)-valencene, undergoes selective allylic hydroxylation leading to the production of intermediate alcohol nootkatol, where a cytochrome P450 monooxygenase catalyzes this step. In the second step, nootkatol is further oxidized to (+)-nootkatone by ADH (Schulz et al. 2015; Milhim et al. 2019). The difficult activity of developing a proper cofactor regeneration system was accomplished by carefully selecting an acceptable cosubstrate for the ADH, which has two functions (Schulz et al. 2015). A whole-cell biocatalyst system based on a new short-chain dehydrogenase from *Bacillus megaterium* has been created *in vivo*. This biotransformation transforms the industrially useful (+)-nootkatone from the less valuable intermediate nootkatol. The newly discovered dehydrogenase selectively and efficiently transformed nootkatol (100% conversion) into the final product, providing a novel tool to overcome the limitations of the two-step enzymatic biotechnological process for the manufacture of (+)-nootkatone (Milhim et al. 2019).

15.6.2 Butane-2,3-diol (2,3-BD)

2, 3-Butanediol is a volatile constituent of sweet corn, fermented soybean curds, whole and ground grains and rotten mussels. The 2,3-Butanediol is found in the fruit of sweet pepper plants and produced in the fermentation of many fruits and grains (Duke 2021). It is used in food, medicine and chemical industries more specifically as a flavoring agent, solvent for dyes, in resins, as an intermediate, and as a blending agent in perfume, fumigants and pharmaceutical carrier (Zhang et al. 2019). Wang et al. (2014) created the *Rhodococcus erythropolis* WZ010 (*R. erythropolis*) strain with the potential to transform diacetyl into optically pure 2,3-butanediol. In this conversion of diacetyl to 2,3-butanediol via (S)-acetoin, purified ADH showed 100% stereospecificity (Figure 15.4) (Wang et al. 2014). In another study, Yu et al. (2015) developed and characterized a novel strain *R. erythropolis* WZ010, which has the capacity to produce both R- and S-stereospecific chiral alcohols such as 2,3-butanediol (S) and 2,3-butanediol (R). This reveals the

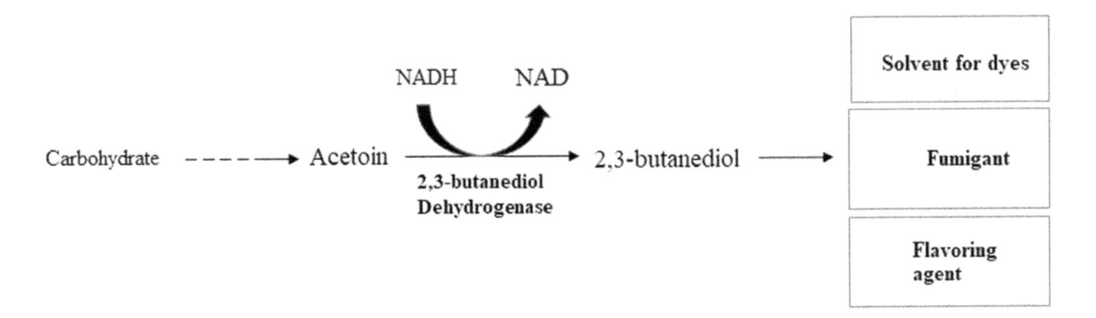

FIGURE 15.4 Pathway showing formation of 2,3-butanediol. With carbohydrate as the starting molecule, the action of enzymes of Embeden–Meyerhof and pentose phosphate pathway and several others enzymes leads to formation of acetoin forms which get converted to 2,3-butanediol with the help of 2,3-butanediol dehydrogenase, an ADH.

presence of the gene encoding 2,3-butanediol dehydrogenase (an ADH) in its genome. Using meso-2,3-Butanediol as a substrate, a novel *Escherichia coli* (pET-rrbdh-nox-vgb) biocatalyst was developed for efficient production of (3)-Acetoin and 2,3-Butanediol (He et al. 2018). By using a one-step bioconversion approach, synthetically altered genes of *Corynebacterium crenatum* (*C. crenatum*) were produced to preferentially manufacture acetoin or 2,3-butanediol. In one-step biocatalysis, these *C. crenatum* strains produced 88.83 g per L of 2,3-BD with a high yield (Zhang et al. 2019). The human pathogen *Neisseria gonorrhoeae* FA1090's 2,3-butanediol dehydrogenase (ADH) has been generated and described in *E. coli*. The findings are helpful in understanding the 2,3-BD metabolism of *N. gonorrhoeae*. It provides the way for further research into the impact of 2,3-BD metabolism on pathogenicity (Tang et al. 2021).

15.6.3 Γ-VALEROLACTONE (GVL)

Gamma-valerolactone (GVL) is a value-added intermediate generated from lignocellulosic biomass that can be used to make biofuels, fuel additives and polymers (Kim et al. 2018). ADHs can be used to make GVL from methyl levulinate. It can also be made by employing methanol, ethanol or 2-propanol as the H-donor/solvent in the catalytic conversion of methyl and ethyl levulinate by catalytic transfer hydrogenation (CTH) (Tabanelli et al. 2020). Using various ADHs from various sources, attempts have been undertaken to make a number of enantiopure hydroxyl esters and lactones. Using *Lactobacillus brevis* ADH (LBADH) and ADH-A as biocatalysts, enantioenriched GVL was produced with >99% conversion and >99% enantiomeric excess. During this method, isopropanol was used to recycle the cofactor (Díaz-Rodríguez et al. 2014).

15.6.4 PHENYLETHANOL (PE)

Phenylethanol (PE) is a key intermediary in the production of enantiomerically pure medicines, fine chemicals for food applications and a possible alcohol for next-generation biofuels (Zhang et al. 2014). *Thermus thermophilus* HB27 ADH (Tt27-ADH2) was used for the asymmetric reduction of different substituted acetophenones to obtain the chiral alcohols. Tt27-ADH2 is the first anti-Prelog selective OR (ADH) from a thermophilic source. This selectivity is for aryl ketones, which create R-profen derivatives preferentially. Pharmaceutical chemistry is also interested in R-profens derivatives because they have been shown to have medicinal characteristics (Rocha-Martín et al. 2012). As model molecules for pharmaceutically relevant building blocks, Adebar N et al concentrated ADHs on preferred solvents and ketone substrates. As such, the reduction of acetophenone to PE in the presence of an ADH from *Lactobacillus brevis* (LB-ADH) is used as a model reaction for an ADH-catalyzed reduction (Figure 15.5). Because the needed cofactor, NADPH is a costly chemical, a cofactor regeneration mechanism based on glucose dehydrogenase (GDH) was used. The reduction (in situ) of $NADP^+$ to gluconolactone and subsequent ring opening (irreversible) to gluconic acid is the basis of this cofactor regeneration system (Adebar et al. 2019).

15.7 APPLICATIONS IN HEALTHCARE

The (S)-Duloxetine is an antidepressant that works by inhibiting the reuptake of both serotonin and norepinephrine. It's also being looked at as a possible treatment for stress urine incontinence. Initially, Eli Lilly Company defined the stereoselective reduction of prochiral ketones to (S)-Chiral alcohols as the chemical synthesis of (S)-Duloxetine. The chiral metal-ligand catalysts utilized in this stage are expensive, and the method left trace metal contamination in the finished goods. The ADH-mediated reduction of ketones to (S)-Duloxetine was developed to address the limitations of chemical techniques (Chen et al. 2016). ADH from *Lactobacillus kefir* was utilized as a template, and Codexis (Redwood City, CA, USA) used protein engineering to create a series of ADH variations. This ADH variation could convert a ketone molecule to an alcohol compound (Zheng

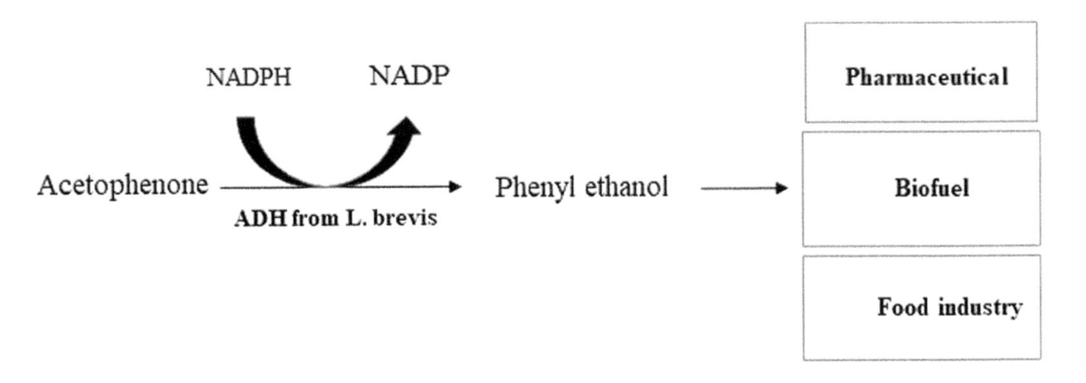

FIGURE 15.5 Conversion of acetophenone to phenylethanol using ADH.

et al. 2017). The montelukast was identified as a cysteinyl leukotriene-1 receptor antagonist by Merck. It was developed as a treatment for asthma. Merck's earlier process, which followed chemical production of montelukast, had significant drawbacks. Enzyme-catalyzed techniques for this reduction were investigated to solve these drawbacks. The first successful reaction produced (S)-hydroxy ester with >95 % enantiomeric excess (ee) using complete *Microbacterium* sp. MB 5614 cells harboring a ketoreductase. *L. kefir* ADH was created by Codexis, which was capable of reducing keto ester asymmetrically to the corresponding (S)-hydroxy ester through directed enzyme evolution (Zheng et al. 2017). Pharmaceuticals like talsaclidine and revatropate, which are utilized as bronchodilators, utilize 3-Quinuclidinol as a valuable chiral building block. The resolution of racemic 3-quinuclidinol and asymmetric reduction of 3-quinuclidinone have both been reported as methods for obtaining 3-quinuclidinol (Tsutsumi et al. 2009). The direct conversion of 3-quinuclidinone to 3-quinuclidinol using ADH received more attention. The maximal theoretical yield of this enzymatic process was 100%, compared to the racemic resolution approach's bottleneck of 50% theoretical yield (Isotani et al. 2012). Taxol (Paclitaxel) is an FDA-approved anticancer medication for the treatment of ovarian and breast cancers. This medication works by preventing microtubule disintegration. N-Benzoyl-3-phenyl isoserine ethyl ester was produced by stereoselective reduction of 2-keto-3-(N-benzoylamino)-3-phenylpropionic acid ethyl ester, a chiral intermediate of the C-13 side chain of taxol (Zheng et al. 2017). The reduction of 3-phenyl-2-chloro-3-oxo propionic acid ethyl ester was carried out using a novel ADH cloned from *Clostridium acetobutylicum*. This asymmetric reduction yielded a building block 2-chloro-3-hydroxy 3-phenylpropionic acid ethyl ester (Applegate et al. 2011).

15.8 APPLICATIONS IN DIAGNOSTICS AND ENVIRONMENTAL MONITORING

When compared to healthy controls, a substantial down-regulation of ORs such as cytochrome oxidase, succinate dehydrogenase, monoamine oxidase and catalase in the host was found (Ngoka. 2008). Therefore, ORs are a distinctive pathological hallmark of several diseases (Berry and Hare 2004). Jelski and Szmitkowski (2008) investigated variations in the activity of specific ADH isoenzymes in the sera of cancer patients. They discovered that cancer cells secrete ADH isoenzymes, which could be valuable for cancer diagnosis. Cancer tissues from liver, stomach, esophagus and colorectum have considerably higher overall ADH activity than healthy organs. Furthermore, ADH activity is substantially higher than ALDH activity (Jelski and Szmitkowski 2008). According to one of the studies, patients with liver cancer had increased activity of classes III and IV of ADH isoenzymes in their serum; however, the differences were insignificant. There were nearly three times increased total serum ADH activity in patients having secondary malignancies than in controls. The ALDH activity of primary and secondary tumor patients, as well as the control group, did not differ significantly. According to the findings, the activity of total ADH and class I ADH isoenzymes vary significantly between primary and metastatic cancers. In another study, Jelski et al. (2008) reported significantly elevated class I ADH activity in the sera of heavy drinkers with cancer (3.43 mU/L) as compared to moderate drinkers with cancer (2.73 mU/L) (Jelski et al. 2008). Jelski et al. (2011) have already demonstrated that the level of activity of class IV ADH (the predominant class in the stomach) in the serum changes in response to gastric cancer. ADH I and ADHIV may be useful in detecting colorectal and stomach cancers, respectively. In the Chinese population, it was discovered that TNBC patients had higher ALDH1 expression than non-TNBC patients, meaning that Triple-negative breast cancer (TNBC) patients have a higher cancer stem cell potential, which could be clinically significant for a better understanding of TNBC patients' poor prognosis. ALDH1 can thus be used to tell the difference between cancer stem cells and breast cancers (Li et al. 2013; Jelski et al. 2018). The possible predictive significance of ADH genes in patients with hepatocellular carcinoma (HCC) was investigated by one group of researchers. When compared to normal liver samples, expression of ADH1A, ADH1B, ADH1C, ADH4 and ADH6 was considerably down-regulated in HCC samples, but raised in alcohol intake HCC patients. The level of expression of ADH1A–ADH4, and ADH6 were considerably reduced in HCC tissues compared to normal liver tissues, according to the findings. ADH1A–ADH4 and ADH6 may act as tumor suppressors in HCC, according to the researchers. In HCC tissues, however, ADH5 expression was substantially higher than in normal liver tissues. As a result, ADHs lacking ADH5 may act as a tumor suppressor in HCC by blocking oncogenic signaling pathways (Liu et al. 2020).

The oral cavity, throat, larynx, esophagus, stomach, liver, pancreas, colon, rectum and breast have all been linked to alcohol consumption. When ADH breaks down ethanol, it produces acetaldehyde, which is very toxic and carcinogenic. Acetaldehyde is further metabolized to acetate in normal/healthy cells (Rodriguez and Coveñas 2021). After alcohol ingestion, the amount of acetaldehyde and its exposure to cells or tissues may be of great importance and it might affect carcinogenesis. Furthermore, acetaldehyde is oxidized and acts as an electrophile that reacts with nucleophiles. This leads to generating covalent crosslinking with lipids, proteins, and nucleic acids (Orywal and Szmitkowski 2017; Rodriguez and Coveñas 2021). By altering retinoic acid production, alcohol shows its effect in the genesis of alcohol-induced oral or esophageal cancer, skin disorders, and fetal alcohol syndrome. These are some of the most typical medical consequences of excessive drinking (Yin et al. 2003). Women who consume more than one or two standard drinks per day may have a higher chance to get breast cancer, according to epidemiological study. Ethanol metabolism in the breast tissue may be confined because ADH is over-expressed in the mammary epithelial cells and it also lacks Low-Km ALDH (Triano et al. 2003). It has been observed that the ADH gene is amongst the most common cold-induced genes in cereals and *Arabidopsis*. Pollen grain seed formation, fruit growth and aerobic metabolism are all affected by ADHs. They also help plants withstand floods, droughts, cold and salt stress. The plant ADH enzyme (ADH-P) was found to be active during hypoxia. The ADH gene is active throughout plant growth and under a variety of stresses. Low temperature, osmotic stress, thirst, salt, mechanical trauma and the exogenous hormone abscisic acid are all known to promote ADH gene expression (ABA). Some major factors such as hypoxia, dehydration, low temperature and plant hormone ABA enhanced the ADH expression in *Arabidopsis* (Su et al. 2020). The ADH gene has been recognized as a marker for plant fragrance and hypoxia sensitivity for a long time. Overexpression of Mango ADH1 in tomato plants increased the fermentative pathway in roots and mimicked the hypoxic response by forming adventitious roots from the stem as an adaptive strategy. Hypoxia, low temperatures, osmotic stress and pathogen infection all require ADH to function properly (Singh et al. 2018).

15.9 APPLICATIONS IN GENETIC ENGINEERING

Manipulation of gene codes in plants for various ORs can also influence plant properties, such as productivity and resistance to herbicides and environmental changes. Glyphosate herbicide side effect tolerance can be achieved by increasing the expression of the glyphosate OR (GOX) enzyme. GOX catalyzes the oxidation and cleavage of the CN bond on the carboxyl side of glyphosate, which results in the production of amino methyl phosphonic acid (AMPA) and glyoxylate. Plants use ORs like xanthine dehydrogenase to metabolize pathogen-associated reactive oxygen species and protect them from oxidative stress. Plants with higher levels of xanthine dehydrogenase expression are more productive (Legesse Habte and Assefa Beyene 2021). Cloning, development and characterization of a very powerful ADH for the production of Ethyl 4-Chloro-3-Hydroxybutyrate (CHBE) ADH were all successful. The isolated *Bartonella apis* ADH (BaADH) showed reasonably high enzyme activity and adequate storage stability over a wide pH range. Because BaADH has a high specific activity for ethyl 4-chloroacetoacetate (COBE), it might be used to generate CHBE. The efficiency of creating CHBE was established by combining BaADH with a very active GDH to build a coenzyme regeneration system (Zhu et al. 2019). Shah et al. (2020) identified and described a novel *Paenibacillus pini* glucose-1-dehydrogenase (GDH) that prefers NAD^+ or $NADP^+$ as a cofactor. Pure recombinant *P. pini* GDH had a specific activity of 247.5 U/mg. The enzyme was stable in the pH range of 4–8.5 even in the absence of NaCl or glycerol, and it had good thermal stability up to 50 °C for 24 hours. Organic solvent resistance was also found in *P. pini* GDH, suggesting that it could be exploited to recycle cofactors for biotransformation. In this study, *P. pini* GDH was found to be very successful in recycling NADH in redox biocatalysis (Shah et al 2020). The yeast *Arxula adeninivorans* was utilized to overexpress an ADH gene from *Lactobacillus brevis* coding for (R)-specific ADH (LbADH). This LbADH was used to manufacture enantiomerically pure 1-(R)-PE. It was discovered that *A. adeninivorans* is a feasible host for the production of LbADH, BpG6PDH or BmGDH, as well as a simple method for synthesizing alcohols that are enantiomerically pure (Rauter et al. 2015).

15.10 CONCLUSION

ADHs are metal proteins and significant key enzymes that convert alcohols to aldehydes or ketones in moderate circumstances. Although the sequences of all seven human ADH isoforms (ADH1–7) are similar, their tissue distributions, catalytic activity and substrate specificity differ. Significant research and widespread uses of ADHs in numerous aspects of life, such as industry, healthcare, stress survival, defense against reactive compounds and the development of biosensors, have garnered a lot of consideration recently. With the introduction of screening processes that use a cocktail of omics technologies such as technologies with metagenome analysis and genome mining, novel ADHs could be found and used as prospective candidates for their differential role. More research is needed to determine its utility in environmental monitoring and as biocatalysts. A blend of rational and irrational protein design approaches must be devised. It will enhance the catalytic properties of ADHs for industrial uses in the synthesis of high-value compounds and medicines.

REFERENCES

Adebar, N., Choi, J.E., Schober, L., Miyake, R., Iura, T., Kawabata, H., Gröger, H., 2019. Overcoming work-up limitations of biphasic biocatalytic reaction mixtures through liquid-liquid segmented flow processes. *ChemCatChem* 11, 5788–5793.

Akal, A.L., Karan, R., Hohl, A., Alam, I., Vogler, M., Grötzinger, S.W., et al., 2019. A polyextremophilic alcohol dehydrogenase from the Atlantis II Deep Red Sea brine pool. *FEBS Open Bio* 9, 194–205.

Applegate, G.A., Cheloha, R.W., Nelson, D.L., Berkowitz, D.B., 2011. A new dehydrogenase from *Clostridium acetobutylicum* for asymmetric synthesis: dynamic reductive kinetic resolution entry into the Taxotere side chain. *Chem. Commun.* 47, 2420–2422.

Batelli, F., Stern, L., 1910. Die Alkoholoxydase in Den Tiergeweben. *Biochem.* 28, 145–168.

Berry, C.E., Hare, J.M., 2004. Xanthine oxidoreductase and cardiovascular disease: molecular mechanisms and pathophysiological implications. *J. Physiol.* 555, 589–606.

Blonder, J.P., Mutka, S.C., Sun, X., Qiu, J., Green, L.H., Mehra, N.K., et al., 2014. Pharmacologic inhibition of S-nitrosoglutathione reductase protects against experimental asthma in BALB/c mice through attenuation of both bronchoconstriction and inflammation. *BMC Pulmon. Med.* 14, 1–15.

Calabrese, V., Mancuso, C., Calvani, M., Rizzarelli, E., Butterfield, D., Stella, A., 2007. Nitric oxide in the central nervous system: neuroprotection versus neurotoxicity. *Nat. Rev. Neurosci.* 8, 766–775.

Centers for Disease Control and Prevention (CDC). 2020. Nootkatone Now Registered by EPA (CDC discovers active ingredient for development into new mosquito/tick insecticides and repellents). *CDC Newsroom.* https://www.cdc.gov/media/releases/2020/p0810-nootkatone-registered-epa.html.

Chapman, J., Ismail, A.E., Dinu, C.Z., 2018. Industrial applications of enzymes: Recent advances, techniques, and outlooks. *Catalysts* 8, 238.

Chen, X., Liu, Z.Q., Lin, C.P., Zheng, Y.G., 2016. Chemoenzymatic synthesis of (S)-duloxetine using carbonyl reductase from *Rhodosporidium toruloides*. *Bioorg. Chem.* 65, 82–89.

Crabb, D.W., Matsumoto, M., Chang, D., You, M., 2004. Overview of the role of alcohol dehydrogenase and aldehyde dehydrogenase and their variants in the genesis of alcohol-related pathology. *Proc. Nutr. Soc.* 63, 49–63.

Díaz-Rodríguez, A., Borzecka, W., Lavandera, I., Gotor, V., 2014. Stereodivergent preparation of valuable γ- or δ-hydroxy esters and lactones through one-pot cascade or tandem chemoenzymatic protocols. *ACS Catal.* 4, 386–393.

Di, L, Balesano, A., Jordan, S., Shi, S. M. 2021. The role of alcohol dehydrogenase in drug metabolism: Beyond ethanol oxidation. The AAPS Journal, 23(1), 20. 10.1208/s12248-020-00536-y.

Dr. Duke's Phytochemical and Ethnobotanical Databases. 2021. 2,3-Butanediol. Available from, as of July 31, 2021. https://phytochem.nal.usda.gov/phytochem/search.

Edenberg, H.J., 2000. Regulation of the mammalian alcohol dehydrogenase genes. *Prog. Nucleic Acid Res. Mol. Biol.* 64, 295–341.

Edenberg, H.J., 2007. The genetics of alcohol metabolism: Role of alcohol dehydrogenase and aldehyde dehydrogenase variants. *Alcohol Res. Health* 30, 5–13.

Edenberg, H.J., McClintick, J.N., 2018. Alcohol dehydrogenases, aldehyde dehydrogenases, and alcohol use disorders: A critical review. *Alcohol Clin. Exp. Res.* 42, 2281–2297.

Eklund, H., Ramaswamy, S., 2008. Medium- and short-chain dehydrogenase/reductase gene and protein families. *Cell. Mol. Life Sci.* 65, 3907.

Forsander, O.A., 1971. Extrahepatic oxidation of alcohol and alcohol metabolites. In: Martini, G.A., Bode, C., (Eds.) *Metabolic Changes Induced by Alcohol.* Springer, Berlin, Heidelberg. pp. 14–22.

Fischer, W. 1916.Untersuchung über die Wirkung kleinster Gaben von Äthylalkohol auf das isolierte Herz. *Arch. Exp.Pathol.Pharmakol.* 80, 93–130.

Franco, R., & Vargas, M. R. (2018). Redox biology in neurological function, dysfunction, and aging. *Antioxidants & Redox Signaling*, 28(18), 1583–158610.1089/ars.2018.7509.

Guo, X., Feng, Y., Wang, X., Liu, Y., Liu, W., Li, Q., et al., 2019. Characterization of the substrate scope of an alcohol dehydrogenase commonly used as methanol dehydrogenase. *Bioorg. Med. Chem. Lett.* 29, 1446–1449.

Hamil, P. , 1910. Cardiac metabolism of alcohol.*J. Physiol.* 39, 476–484.

Haselbeck, R.J., Ang, H.L., Duester, G., 1997. Class IV alcohol/retinol dehydrogenase localization in epidermal basal layer: Potential site of retinoic acid synthesis during skin development. *Develop. Dynam.* 208, 447–453.

He, Y., Chen, F., Sun, M., Gao, H., Guo, Z., Lin, H., et al., 2018. Efficient (3 S)-acetoin and (2 S, 3 S)-2, 3-butanediol production from meso-2, 3-butanediol using whole-cell biocatalysis. *Molecules* 23, 691.

Husain, Q., 2017. High yield immobilization and stabilization of oxidoreductases using magnetic nanosupports and their potential applications: An update. *Curr. Catal.* 6, 168–187.

Kareem, H.M., 2020. Oxidoreductases: significance for humans and microorganism. *In Oxidoreductase. IntechOpen.* doi: 10.5772/intechopen.93961.

Isotani, K., Kurokawa, J., Itoh, N., 2012. Production of (R)-3-quinuclidinol by *E. coli* biocatalysts possessing NADH-dependent 3-quinuclidinone reductase (QNR or bacC) from *Microbacterium luteolum* and *Leifsonia* alcohol dehydrogenase (LSADH). *Int. J. Mol. Sci.* 13, 13542–13553.

Jelski, W., Szmitkowski, M., 2008. Alcohol dehydrogenase (ADH) and aldehyde dehydrogenase (ALDH) in the cancer diseases. *Clin. Chim. Acta* 395, 1–5.

Jelski, W., Kutylowska, E., Laniewska-Dunaj, M., Szmitkowski, M., 2011. Alcohol dehydrogenase (ADH) and aldehyde dehydrogenase (ALDH) as candidates for tumor markers in patients with pancreatic cancer. *J. Gastrointestin. Liver Dis.* 20, 255–259.

Jelski, W., Wolszczak-Biedrzycka, B., Zasimowicz-Majewska, E., Orywal, K., Lapinski, T.W., Szmitkowski, M., 2018. Alcohol dehydrogenase isoenzymes and aldehyde dehydrogenase activity in the serum of patients with non-alcoholic fatty liver disease. *Anticancer Res.* 38, 4005–4009.

Jelski, W., Zalewski, B., Szmitkowski, M., 2008. Alcohol dehydrogenase (ADH) isoenzymes and aldehyde dehydrogenase (ALDH) activity in the sera of patients with liver cancer. *J. Clin. Lab. Anal.* 22, 204–209.

Jörnvall, H., 2008. Medium-and short-chain dehydrogenase/reductase gene and protein families: MDR and SDR gene and protein superfamilies. *Cell. Mol. Life Sci.* 65, 3873–3878.

Jörnvall, H., Höög, J.O., Persson, B., 1999. SDR and MDR: completed genome sequences show these protein families to be large, of old origin, and of complex nature. *FEBS Lett.* 445, 261–264.

Jörnvall, H., Persson, B., Krook, M., Atrian, S., Gonzalez-Duarte, R., Jeffery, J., Ghosh, D., 1995. Short-chain dehydrogenases/reductases (SDR). *Biochemistry* 34, 6003–6013.

Kavanagh, K.L., Jörnvall, H., Persson, B., Oppermann, U., 2008. Medium-and short-chain dehydrogenase/reductase gene and protein families: The SDR superfamily: functional and structural diversity within a family of metabolic and regulatory enzymes. *Cell. Mol. Life Sci.* 65, 3895–3906.

Kim, J., Byun, J., Ahn, Y., Han, J., 2018. Catalytic production of gamma-valerolactone from two different feedstocks. In *Computer Aided Chemical Engineering*. Vol. 44. Elsevier. 295–300.

Klewitz, F., 1923.Über Alkoholverbrauch durch das überlebende Warmblüterherz. *Arch. Exp. Pathol. Pharmakol.* 99, 250–252.

Knoll, M., Pleiss, J., 2008. The medium-chain dehydrogenase/reductase engineering database: A systematic analysis of a diverse protein family to understand sequence–structure–function relationship. *Protein Sci.* 17, 1689–1697.

Li, H., Ma, F., Wang, H., Lin, C., Fan, Y., Zhang, X., et al., 2013. Stem cell marker aldehyde dehydrogenase 1 (ALDH1)-expressing cells are enriched in triple-negative breast cancer. *Int. J. Biol. Markers* 28, 357–364.

Li, S., Chen, K., Grierson, D., 2021. Molecular and hormonal mechanisms regulating fleshy fruit ripening. *Cells* 10, 1136.

Liu, X., Li, T., Kong, D., You, H., Kong, F., Tang, R., 2020. Prognostic implications of alcohol dehydrogenases in hepatocellular carcinoma. *BMC Cancer* 20, 1–13.

Lutwak-Mann, C., 1938. Alcohol dehydrogenase of animal tissues. *Biochem. J.* 32, 1364–1374.

Meng, X., Liu, H., Xu, W., Zhang, W., Wang, Z., Liu, W., 2020. Metabolic engineering *Saccharomyces cerevisiae* for de novo production of the sesquiterpenoid (+)-nootkatone. *Microb. Cell Factories* 19, 1–14.

Legesse Habte, M., Assefa Beyene, E., 2021. Biological application and disease of oxidoreductase enzymes. *Oxidoreductase*. doi: 10.5772/intechopen.93328

Milhim, M., Hartz, P., Gerber, A., Bernhardt, R., 2019. A novel short chain dehydrogenase from *Bacillus megaterium* for the conversion of the sesquiterpene nootkatol to (+)-nootkatone. *J. Biotechnol.* 301, 52–55.

Ngoka, L., 2008. Dramatic down-regulation of oxidoreductases in human hepatocellular carcinoma hepG2 cells: Proteomics and gene ontology unveiling new frontiers in cancer enzymology. *Proteome Sci.* 6, 1–21.

Niederhut, M.S., Gibbons, B.J., Perez-Miller, S., Hurley, T.D., 2001. Three-dimensional structures of the three human class I alcohol dehydrogenases. *Protein Sci.* 10, 697–706.

Orywal, K., Szmitkowski, M., 2017. Alcohol dehydrogenase and aldehyde dehydrogenase in malignant neoplasms. *Clin. Exp. Med.* 17, 131–139.

Orywal, K., Jelski, W., Szmitkowski, M., 2008. The participation of dehydrogenases in retinol metabolism. *Polski Merkuriusz Lekarski: Organ Polskiego Towarzystwa Lekarskiego* 25, 276–279.

Östberg, L.J., Persson, B., Höög, J. O. 2016. Computational studies of human class V alcohol dehydrogenase - the odd sibling. *BMC Biochem.*, 17(1), 16.

Pandey, B., Pandey, V.P., Dwivedi, U.N., 2011. Cloning, expression, functional validation and modeling of cinnamyl alcohol dehydrogenase isolated from xylem of *Leucaena leucocephala*. *Protein Exp. Purif.* 79, 197–203.

Pandey, B., Pandey, V.P., Shasany, A.K., Dwivedi, U.N., 2014. Purification and characterization of a zinc-dependent cinnamyl alcohol dehydrogenase from *Leucaena leucocephala*, a tree legume. *Appl. Biochem. Biotechnol.* 172, 3414–3423.

Pares, X., Farres, J., Kedishvili, N., Duester, G., 2008. Medium-and short-chain dehydrogenase/reductase gene and protein families: Medium-chain and short-chain dehydrogenases/reductases in retinoid metabolism. *Cell. Mol. Life Sci.* 65, 3936–3949.

Peralba, J.M., Cederlund, E., Crosas, B., Moreno, A., Julià, P., Martínez, S.E., et al., 1999. Structural and enzymatic properties of a gastric NADP (H)-dependent and retinal-active alcohol dehydrogenase. *J. Biol. Chem.* 274, 26021–26026.

Persson, B., Hedlund, J., Jörnvall, H., 2008. Medium-and short-chain dehydrogenase/reductase gene and protein families. *Cell. Mol. Life Sci.* 65, 3879–3894.

Rauter, M., Prokoph, A., Kasprzak, J., Becker, K., Baronian, K., Bode, R., et al., 2015. Coexpression of *Lactobacillus brevis* ADH with GDH or G6PDH in *Arxula adeninivorans* for the synthesis of 1-(R)-phenylethanol. *Appl. Microbiol. Biotechnol.* 99, 4723–4733.

Roberts, J.K., Callis, J., Jardetzky, O., Walbot, V., Freeling, M., 1984. Cytoplasmic acidosis as a determinant of flooding intolerance in plants. *Proc. Natl. Acad. Sci. USA* 81, 6029–6033.

Rocha-Martín, J., Vega, D., Bolivar, J.M., Hidalgo, A., Berenguer, J., Guisán, J.M., López-Gallego, F., 2012. Characterization and further stabilization of a new anti-prelog specific alcohol dehydrogenase from *Thermus thermophilus* HB27 for asymmetric reduction of carbonyl compounds. *Bioresour. Technol.* 103, 343–350.

Rodriguez, F.D., Coveñas, R., 2021. Biochemical mechanisms associating alcohol use disorders with cancers. *Cancers* 13, 3548.

Schulz, S., Girhard, M., Gaßmeyer, S.K., Jäger, V.D., Schwarze, D., Vogel, A., Urlacher, V.B., 2015. Selective enzymatic synthesis of the grapefruit flavor (+)-nootkatone. *ChemCatChem* 7, 601–604.

Shah, S., Sunder, A.V., Singh, P., Wangikar, P.P., 2020. Characterization and application of a robust glucose dehydrogenase from *Paenibacillus pini* for cofactor regeneration in biocatalysis. *Ind. J. Microbiol.* 60, 87–95.

Sibout, R., Eudes, A., Mouille, G., Pollet, B., Lapierre, C., Jouanin, L., Séguin, A. 2005. CINNAMYL ALCOHOL DEHYDROGENASE-C and -D are the primary genes involved in lignin biosynthesis in the floral stem of Arabidopsis. *The Plant Cell* 17, 2059–2076.

Singh, R.K., Srivastava, S., Chidley, H.G., Nath, P., Sane, V.A., 2018. Overexpression of mango alcohol dehydrogenase (MiADH1) mimics hypoxia in transgenic tomato and alters fruit flavor components. *Agri Gene* 7, 23–33.

Staab, C.A., Hellgren, M., Höög, J.O., 2008. Medium-and short-chain dehydrogenase/reductase gene and protein families: Dual functions of alcohol dehydrogenase 3: implications with focus on formaldehyde dehydrogenase and S-nitrosoglutathione reductase activities. *Cell. Mol. Life Sci.* 65, 3950–3960.

Su, W., Ren, Y., Wang, D., Su, Y., Feng, J., Zhang, C., Tang, H., Xu, L., Muhammad, K., Que, Y., 2020. The alcohol dehydrogenase gene family in sugarcane and its involvement in cold stress regulation. *BMC Genom.* 21, 1–17.

Svensson, S., Some, M., Lundsjo, A., Helander, A., Cronholm, T., Hoog, J.-O. 1999. Activities of human alcohol dehydrogenases in the metabolic pathways of ethanol and serotonin. *Eur. J. Biochem.*, 262(2), 324–329. 10.1046/j.1432-1327.1999.00351.x.

Tabanelli, T., Vásquez, P.B., Paone, E., Pietropaolo, R., Dimitratos, N., Cavani, F., Mauriello, F., 2020. Improved Catalytic Transfer Hydrogenation of levulinate esters with alcohols over ZrO2 catalyst. *Chem. Proc.* 2, 28.

Tang, W., Lian, C., Si, Y., Chang, J., 2021. Purification and characterization of (2 R, 3 R)-2, 3-butanediol dehydrogenase of the human pathogen *Neisseria gonorrhoeae* FA1090 produced in *Escherichia coli*. *Mol. Biotechnol.* 63, 491–501.

Tesniere, C., Abbal, P., 2009. Alcohol dehydrogenase genes and proteins in grapevine. In: Roubelakis-Angelakis, K.A., (Ed.) *Grapevine Molecular Physiology and Biotechnology*, Springer, Dordrecht. pp. 141–160.

Triano, E.A., Slusher, L.B., Atkins, T.A., Beneski, J.T., Gestl, S.A., Zolfaghari, R., et al., 2003. Class I alcohol dehydrogenase is highly expressed in normal human mammary epithelium but not in invasive breast cancer: Implications for breast carcinogenesis. *Cancer Res.* 63, 3092–3100.

Tsigos, I., Velonia, K., Smonou, I., Bouriotis, V., 1998. Purification and characterization of an alcohol dehydrogenase from the Antarctic psychrophile *Moraxella sp.* TAE123. *Eur. J. Biochem.* 254, 356–362.

Tsutsumi, K., Katayama, T., Utsumi, N., Murata, K., Arai, N., Kurono, N., Ohkuma, T., 2009. Practical asymmetric hydrogenation of 3-quinuclidinone catalyzed by the xylSkewphos/PICA– ruthenium (II) complex. *Organic Proc. Res. Dev.* 13, 625–628.

Wang, Z., Song, Q., Yu, M., Wang, Y., Xiong, B., Zhang, Y., Zheng, J., Ying, X., 2014. Characterization of a stereospecific acetoin (diacetyl) reductase from *Rhodococcus erythropolis* WZ010 and its application for the synthesis of (2 S, 3 S)-2, 3-butanediol. *Appl. Microbiol. Biotech.* 98, 641–650.

Wesolowski, M.H., Lyerla, T.A., 1983. Alcohol dehydrogenase isozymes in the clawed frog, *Xenopus laevis*. *Biochem. Genet.* 21, 1003–1017.

Yin, S.J., Chou, C.F., Lai, C.L., Lee, S.L., Han, C.L., 2003. Human class IV alcohol dehydrogenase: Kinetic mechanism, functional roles and medical relevance. *Chemico-Biological Interac.* 143, 219–227.

Yu, M., Huang, M., Song, Q., Shao, J., Ying, X., 2015. Characterization of a (2 R, 3 R)-2, 3-butanediol dehydrogenase from *Rhodococcus erythropolis* WZ010. *Molecules* 20, 7156–7173.

Zhang, H., Cao, M., Jiang, X., Zou, H., Wang, C., Xu, X., Xian, M., 2014. De-novo synthesis of 2-phenylethanol by *Enterobactersp.* CGMCC 5087. *BMC Biotechnol.* 14, 1–7.

Zhang, X., Han, R., Bao, T., Zhao, X., Li, X., Zhu, M., et al. 2019. Synthetic engineering of *Corynebacterium crenatum* to selectively produce acetoin or 2, 3-butanediol by one step bioconversion method. *Microb. Cell Fact.* 18, 1–12.

Zhao, J., Wei, Q., Gu, X.R., Ren, S.W., Liu, X.N., 2020. Alcohol dehydrogenase 5 of *Helicoverpa armigera* interacts with the CYP6B6 promoter in response to 2-tridecanone. *Insect Sci.* 27, 1053–1066.

Zheng, Y.G., Yin, H.H., Yu, D.F., Chen, X., Tang, X.L., Zhang, X.J., et al., 2017. Recent advances in biotechnological applications of alcohol dehydrogenases. *Appl. Microbiol. Biotechnol.* 101, 987–1001.

Zhu, Y.H., Liu, C.Y., Cai, S., Guo, L.B., Kim, I.W., Kalia, V.C., et al., 2019. Cloning, expression and characterization of a highly active alcohol dehydrogenase for production of ethyl (S)-4-chloro-3-hydroxybutyrate. *Ind. J. Microbiol.* 59, 225–233.

Index

V

W

X

Z